State and Environmer

American and Comparative Environmental Policy
Sheldon Kamieniecki and Michael E. Kraft, series editors

State and Environment

The Comparative Study of Environmental Governance

Edited by Andreas Duit

The MIT Press
Cambridge, Massachusetts
London, England

MIT Press books may be purchased at special quantity discounts for business or sales promotional use. For information, please email special_sales@mitpress.mit.edu.

This book was set in Sabon LT Std by Toppan Best-set Premedia Limited. Printed and bound in the United States of America.

Library of Congress Cataloging-in-Publication Data

State and environment : the comparative study of environmental governance / edited by Andreas Duit.
    pages cm. — (American and comparative environmental policy)
    Includes bibliographical references and index.
    ISBN 978-0-262-02712-0 (hardcover : alk. paper) — ISBN 978-0-262-52581-7 (pbk. : alk. paper)   1. Environmental policy.   2. Environmental management.   3. Sustainable development.   I. Duit, Andreas.
    GE170.S7165 2014
    333.7—dc23
                                    2013037678

10   9   8   7   6   5   4   3   2   1

Contents

Series Foreword

The study of comparative environmental politics and policy has come a long way from its beginnings in the 1970s. At the dawn of the modern environmental movement, scholars took up the fascinating question of why nations approached environmental problems in sometimes strikingly different ways. What difference does it make, for example, if political systems are open and encourage citizen participation in policymaking processes or, instead, rely heavily on government administrators and technical experts? Does provision for greater access to the political process by organized groups, both environmental and business groups, affect the kinds of environmental policy choices that are made? Do nations that rely on consensual policymaking processes produce different kinds of policies than those that are more adversarial, and are those policies more likely to achieve their objectives? Does variation in institutional structures and the centralization of policymaking help explain the level of success in environmental policymaking across nations? Does the nature and structure of party systems, including the presence of a green party, influence environmental politics and policymaking? The answers to such questions may shape practical political strategies as well as help to build knowledge of how the characteristics of political systems affect policymaking processes and policy outcomes.

As the world struggles to address monumental environmental challenges of the twenty-first century, such as climate change, these kinds of inquiries should have great value to both academics and practitioners. Potentially, there is much to be learned from countries that are able to reduce pollution and conserve natural resources in a cost-effective manner. At the same time, past failures in pollution control and natural resource conservation by countries can provide guidance as to what policy approaches should be avoided in the future.

Despite the apparent value, however, the comparative nation approach to the study of environmental politics and policy was not very common in the 1970s and 1980s, when comparative studies largely focused on single nations rather than offer a systematic comparison of actions across nations. Quite often such single-nation studies appeared in edited volumes that brought together a diversity of analyses focusing on disparate issues in both developed and developing nations without, however, any systematic comparison of the patterns of, for example, agenda setting, policy adoption, policy implementation, or policy change.

Until recently, what has been missing in this literature is genuine comparison of national actions that is well grounded in theory and rigorous empirical analysis linked to the broader study of comparative politics. In 2012, a book in this series, Paul Steinberg and Stacy VanDeveer's *Comparative Environmental Politics: Theory, Practice, and Prospects*, sought to advance just this kind of comparative analysis of environmental politics, and, as a consequence, it received considerable praise from the academic community. In this volume, Andreas Duit and his colleagues seek to advance further the comparative study of environmental governance.

They begin by acknowledging that global environmental change today may well threaten citizen well-being in ways that few thought possible in earlier decades, and that environmental policy and management are now core responsibilities in most developed nations, even if they all are not yet sufficiently strong to address the problems effectively. Hence, they analyze how nations to date have dealt with some fundamental issues of environmental governance and what kinds of changes we might expect in future years.

The contributors to this volume rely on a variety of theoretical and analytical perspectives drawn from the field of comparative politics, and they make use of a wide selection of methodologies and data sources and types as they explore the role of the state in environmental governance. In particular, the authors examine patterns of environmental performance across nations, governance and citizenship, natural resource management, policy diffusion and change, and the role of the state in managing complex and unpredictable ecosystems as the world's nations chart an uncertain path toward sustainable development. All of the contributors are alert to the limited value of descriptive case studies, and, consequently, they seek to draw from the analytic power of the comparative method to illuminate better what nations have done to respond to environmental problems, why they chose the policies and actions they did, what impact those governmental actions have had to date, and what the implications are for future environmental governance and the role of the state.

One of the major purposes of the volume is to help advance a comparative approach to the study of environmental politics by demonstrating the analytical utility of the comparative method when applied to empirical investigations of environmental issues. The contributors are convinced that careful empirical studies of environmental politics will provide the kinds of reliable evidence that citizens and policymakers need to devise workable solutions to the myriad environmental challenges they face. We agree. Too often environmental policy decisions are affected by political ideology, partisan agendas, and narrow and short-sighted interpretations of both national and global needs. We should do much better, and the kinds of study exemplified in this collection highlight the value of comparative political analysis for clear thinking about the role of citizens and civil society, political institutions, policy analysis, policymaking and implementation processes, and governmental management.

The book illustrates well our purpose in the MIT Press series in American and Comparative Environmental Policy. We encourage work that examines a broad range of environmental policy issues. We are particularly interested in volumes that incorporate interdisciplinary research and focus on the linkages between public policy and environmental problems and issues both within the United States and in cross-national settings. We welcome contributions that analyze the policy dimensions of relationships between humans and the environment from either a theoretical or empirical perspective.

At a time when present environmental policies are increasingly seen as controversial and new, alternative approaches are being implemented widely, we especially encourage studies that assess policy successes and failures, evaluate new institutional arrangements and policy tools, and clarify new directions for environmental politics and policy. The books in this series are written for a wide audience that includes academics, policymakers, environmental scientists and professionals, business and labor leaders, environmental activists, and students concerned with environmental issues. We hope they contribute to public understanding of environmental problems, issues, and policies of concern today and also suggest promising actions for the future.

Sheldon Kamieniecki, University of California, Santa Cruz
Michael Kraft, University of Wisconsin-Green Bay
American and Comparative Environmental Policy Series Editors

Preface

This book grew out of a conviction that the study of environmental politics has much to gain from rediscovering two central features in general political science: the state and the comparative method. In scholarship on environmental issues in recent years, the role of the state in addressing environmental problems has been overshadowed by a focus on small-scale natural resource management, green norms and behavior among citizens, and international environmental treaties and regimes. In fact, one could argue that with few exceptions, students and practitioners of environmental policy and politics have looked everywhere *but* to the state in search of solutions for looming environmental disasters.

The studies collected in this volume all illustrate the continued and pivotal role that the state can play in contemporary environmental dilemmas. As many commentators have pointed out before, solutions that the state offers to environmental problems are often incomplete, insufficient, and biased toward continued economic growth rather than long-term sustainability. Nevertheless, there are two reasons why the state deserves a spot in the analytical limelight. The first is that over the last four decades, most states in industrialized countries have developed extensive administrative and regulatory responses to environmental problems. Policies to protect the environment have been issued, and administrative structures for environmental policy implementation, monitoring, and knowledge generation have been erected. The driving forces behind this regulatory expansion, as well as cross-country variations, remain poorly understood and require more scholarly attention.

The second reason why this development is analytically relevant is that although this process has been slow, gradual, marked by setbacks, and far from sufficient in halting environmental degradation, it is also the most comprehensive response issued to environmental problems by society writ large, dwarfing the environmental efforts of markets, international organizations, and individuals in both scope and impact. As such, this process of regulatory and organizational growth in the environmental area is an important object of study for the purpose of assessing society's ability to address environmental problems.

Analyzing issues and hypotheses in political science by systematically comparing the differences and similarities among countries lies at the heart of a discipline in which true experimental designs can be applied only to a limited extent. The comparative method in political science rests on the idea that cross-national variations can be exploited to investigate the causes and effects of a wide range of political phenomena. Perhaps as a consequence of the simultaneous appearance of environmental problems in most industrialized countries in the 1960s, the comparative method has a long history in the study of environmental governance. By building on this tradition and adding a systematic comparative approach, this volume not only hopes to advance research on environmental matters, but also to enhance our knowledge of more general questions regarding the limits and possibilities of human governance in overcoming large-scale problems.

Acknowledgments

The papers collected in this volume were first presented and discussed at the workshop "Mapping the Politics of Ecology," held in Stockholm on June 28–29, 2010. This workshop was generously funded by the Swedish Foundation for Strategic Environmental Research (Mistra) as part of the Mapping the Politics of Ecology (MAPLE) project hosted by the Department of Political Science at Stockholm University. The editor would also like to thank Sweden's National Science Council for a grant funding the work with the volume subsequent to the workshop.

The editor and the authors also wish to express their gratitude to American and Comparative Environmental Policy series editors Sheldon Kamieniecki and Michael E. Kraft for support and constructive comments. We are also grateful to Clay Morgan at MIT Press for his helpfulness and encouragement throughout the publication process. Finally, the criticism and comments offered by three anonymous reviewers were greatly appreciated and were very helpful in improving the volume.

1

Introduction: The Comparative Study of Environmental Governance

Andreas Duit

Global environmental change is threatening prosperity and well-being in developed and developing countries alike, and environmental management is now considered a core area of state responsibility in most countries. Indeed, many states now devote substantial proportions of their public spending to environmental monitoring, protection, and restoration, and many have developed considerable administrative, institutional, regulatory, and legislative capacities in the environmental area. In spite of this expansion of regulatory capacity, society's efforts are far from sufficient for reversing, or even just slowing, the ongoing processes of environmental degradation. It is generally recognized that mitigating the ecological crisis must entail a reorganization of the social and political world on par with the previous great transformations, such as the emergence of the nation state system, the market economy, liberal democracy, or the welfare state.

Reflecting the ongoing transformative process of the nature-society relationship is the emergence in the 1960s and 1970s of subdisciplines in social science focusing on environmental issues. In contemporary political science, the study of environmental politics, natural resource management, and environmental policy is a well-established part of the discipline, complete with subfields ranging from international to local political scales, and from large-N cross-sectional studies to green political theory. As a result, significant progress has been made in the understanding of environmental governance. In particular, advances have been achieved in the areas receiving most attention in recent years—that is to say, research focused on subnational and supranational scales in which scholars have analyzed patterns of institution building, policymaking, and regime formation in international arenas (Meyer et al. 1997; Young 1999; Bäckstrand 2008; Paterson 2009), or questions of local or regional policymaking and the role of institutions and stakeholder participation in resource management (Ostrom 1990; Agrawal and Gibson 2001; Ostrom 2005).

In the meantime, however, environmental governance on the meso level connecting macro and micro levels—the state—has been largely overlooked by social science. As Barry and Eckersley point out, this neglect of the state is, at least in part, linked to its contested role in green political thinking (Barry and Eckersley 2005). The state has been criticized on several accounts, most of them well known to students of and actors in environmental politics. The first point of critique has to do with the perceived inadequacy of the state as a type of political organization for dealing with problems on a global scale (Biermann and Dingwerth 2004). Most processes of environmental degradation are paradigmatic examples of truly globalized problems that individual states are thought to lack both the ability and the incentive to address, which has spurred both researchers and environmentalists to turn to the global arena in search for solutions to the environmental crisis. A second criticism focuses on the linkages between capitalism and electoral democracies. As the familiar argument goes, representative democracies, due to electoral pressures, will tend to promote economic growth, tax revenues, or employment opportunities whenever these conflict with environmental protection. Thus, we can only expect the state to supply a basic level of environmental regulation compatible with sustained economic growth (Buttel 2004). In a similar vein, the liberal democratic state has also been criticized for not allowing civic society and social movement representatives access to environmental decision- and policy-making processes, thereby restricting the representation of nature, as well as those groups in society which depend the most on their natural environment for their livelihood (Dryzek and Niemeyer 2008). In addition, a heavily "scientized" political discourse within the framework of liberal democracy tends to marginalize views and standpoints not compatible with paradigms of sustained growth or technocratic forms of reasoning (Bäckstrand 2004). Another argument against the state is that liberal democracy has a strong tendency to prefer short-term over long-term gains, especially when long-term gains imply some sort of short-term reduction of well-being (Underdal 2010). Finally, critics have argued that the state's foremost tool—public administration—is, due to its reliance on control-and-command management templates and an hierarchical and expert-dominated organization, inadequate for managing complex and unpredictable ecosystems (Holling and Meffe 1996) as well as for responding to the demands of citizens and stakeholders (Durant et al. 2004).

Bringing the State Back into Environmental Governance

In light of this fairly long list of complaints lodged at the state, why should it be brought back into studies of environmental governance? A first answer is that although most of these critiques of the state are well founded to some extent, it would nevertheless be unwise to rule it out as an important object of study just because it is not a reliable source of much-needed solutions to environmental problems. In fact, achieving a deepened understanding of how and under which circumstances the state fails and succeeds when addressing different types of environmental problems can be considered a key research objective, with potentially far-ranging implications for environmental policymaking in practical settings. A second answer is that for the foreseeable future, the state will continue to play a key role in structuring society's relationship to nature. Although the focus of environmental politics is gradually shifting from the national to the international policy arena, the state remains both a primary site for and an important actor in environmental governance. As Barry and Eckersley (2005) and Steinberg and VanDeveer (2012) argue, states are still deeply involved in managing or mismanaging natural resources and in reducing or increasing emissions of harmful substances into air, water, and soils. A powerful illustration of the state's growing involvement in environmental problem solving can be found in recent studies on regulatory expansion in the environmental policy area (Tews et al. 2003; Holzinger et al. 2008), and it is clearly the case that the role of the state in combating environmental degradation has undergone a steady and rapid expansion during the last four decades (Meadowcroft 2012). In addition, it is states that engage in, and therefore can choose to cooperate with or defect from, international environmental treaties (Weiss and Jacobson 2000). States, moreover, play an important role in stimulating or repressing environmentally beneficial behavior among their citizens (John et al. 2011), as well as influence the mobilization of green social movements (Dryzek et al. 2003; Dalton 2005).

From a variety of theoretical perspectives, and drawing on a wide selection of data types, methodologies, and study formats, all the contributions to this volume explore the role of the state in environmental governance. Among the collected chapters, a broad distinction can be made between two types of approaches, reflecting the state's dual role as actor and arena for environmental governance. State-centric studies consist of investigations of the state as an agent in relation to environmental matters, and they generally employ some sort of measure of state behavior or policy outcomes as a dependent variable. State-centric studies are typically interested in understanding some aspect of public government (e.g., policies, institutions, or bureaucratic organizations) in relation to environmental matters. The literature on environmental performance is a good example of state-centric type of studies in which the aim is to explain differences in the ability of the central public environmental administration to address environmental problems. (See the chapters in this book by Detlef Jahn; Roger Karapin; Christoph Knill, Susumu Shikano, and Jale Tosun; Susan Baker and Katarina Eckerberg; and Thomas Sommerer for examples of state-centric studies.) Non-state-centric studies, on the other hand, seek to understand some facet of environmental governance by introducing the behavior or the characteristics of the state as an independent variable for understanding things such as citizens' environmental behavior (see the chapter by Michele Micheletti, Dietlind Stolle, and Daniel Berlin in this volume) and environmental policymaking and resource management on the local level (see the chapters by Krister Andersson et al., Martin Sjöstedt; and Andreas Duit and Ola Hall in this volume).

Environmental Governance in a Comparative Perspective

In addition to a shared focus on the role of the state in environmental governance, both state-centric and non-state-centric studies collected in this book employ a distinctly comparative approach to the study of environmental governance. In this sense, they are building on a well-established tradition within studies of environmental politics and policy. Early classics in the field (e.g., Enloe 1975; Lundqvist 1980) often had a pronounced comparative outlook when trying to understand how the political system was responding to environmental problems, and the regular appearance of similar studies (e.g., Jänicke and Weidner 1997; Desai 1998; Hanf and Jansen 1998; Desai 2002; Weidner and Jänicke 2002; Jordan et al. 2003; Rootes 2003) during the last two decades indicate that comparative environmental studies has remained a vital genre. However, many of these earlier studies took the form of collections of descriptive case studies of single countries guided by a common framework. As such, they have provided an important first mapping of environmental politics and policy in a comparative perspective, but their design format does not fully harness the analytic power of the comparative method of systematic comparison.

This volume hopes to contribute to a growing body of scholarly work that seeks to analyze patterns of how different societies are responding politically, administratively, and institutionally to the challenge of global environmental change. Moreover, the studies collected here illustrate the large methodological variation compatible with a comparative outlook on environmental governance. Some of the chapters are case studies based on qualitative data from a smaller number of countries, whereas others are statistical analyses of quantitative data from a larger sample of countries and longer time periods. Both types of study designs are comparative in the sense that inferences about theoretical issues are derived from a systematic analysis of empirical data collected from a number of strategically selected countries. Despite the homogeneity in basic study design, a large degree of diversity in terms of theoretical frameworks and methods of data analysis can be found among the contributions to this volume. For example, methods range from process tracing, used in Roger Karapin's analysis of wind energy policy in the United States and Germany, and game theoretical models in Martin Sjöstedt's study of African fisheries, to the innovative and novel statistical techniques used to analyze large-N data on policy diffusion patterns in the chapters by Thomas Sommerer and Christoph Knill, Susumu Shikano, and Jale Tosun. Thus, a comparative perspective is compatible with a wide range of different theoretical frameworks, topics of analysis, and methodological approaches and therefore should not be perceived as a study format that allows only a restricted set of analytical formats and research topics.

The method of cross-national comparison is not without its critics among scholars of environmental governance. Few would take issue with the method of comparison in social science, but the practice of comparing nations is a different matter. In a fairly recent article, Spaargaren and Mol (2008) argue that because the nation-state no longer plays a central role in environmental protection and policymaking, cross-national study designs—or "methodological nationalism" as they term it—to environmental politics should be abandoned:

While (networks of) nation-states have been the primary actor(s) through which most environmental reform took shape in the early 1990s, this is no longer the case today. Nation-states in the global network society have to reinvent their political roles and cannot but share responsibilities for the governance of environmental flows with economic and civil-society actors, both at the local and the global level [. . .]. Consequently, methodological nationalism—the tendency in the social sciences to implicitly take the nation–state as central unit of analysis in both theoretical and empirical works—needs to be abandoned if we are to analyse and understand today's politics and reforms, also with respect to sustainable consumption." (p. 350)

Contrary to this assertion, one of the aims of this volume is to advance a comparative approach to issues in environmental politics, primarily by demonstrating the analytical utility of the comparative method applied in empirical investigations of environmental issues. The studies reported in the individual chapters all bear witness to the fruitfulness of adopting a comparative perspective on a wide range of issues in environmental politics. For instance, Thomas Sommerer is able to debunk the idea of a "race to the bottom" in terms of environmental policy standards; James Meadowcroft reviews the shortcomings of existing measures and suggests an alternative approach to estimating comparative environmental performance; and Christoph Knill, Susumu Shikano, and Jale Tosun go beyond a previous focus in the literature on "explosive" diffusion processes and are able to demonstrate the key role of international institutions in the spread of environmental policies between countries. Further, Krister Andersson et al. analyze data from three Latin American countries and find that decentralization reforms need to include the transfer of control over local resources if they are to lead to sustainable forest management; Michele Micheletti, Dietlind Stolle, and Daniel Berlin investigate the emergence of sustainable citizenship norms in a comparative perspective and demonstrate that there is an ongoing shift in the way citizens perceive their responsibilities toward each other and the natural world; and Roger Karapin reveals how structures, problems, and politics streams, as well as advocacy coalitions, interact to produce policy outcomes in wind energy sectors of Germany and the United States. Susan Baker and Katarina Eckerberg draw on evidence from twenty-seven European countries and conclude that the state still plays a pivotal role in many of the environmental policy processes normally sorted under the "governance" label; Martin Sjöstedt finds evidence in his study of fisheries regulation in three African countries that the trustworthiness and lack of corruption of the state are important but hitherto largely overlooked factors in understanding the management of common pool resources; and Andreas Duit and Ola Hall identify an effect of political rights on levels of stakeholder participation in conservation programs, as well as a link between the participation of local stakeholders and better conservation performance.

All these findings, each of which has high relevance to its respective scholarly debate, would have been difficult to discover through a noncomparative approach. This is not to argue that novel and relevant research cannot be conducted through other types of study designs, or that all types of issues can and should be addressed through a comparative approach. The argument, rather, is that the comparative method seems to be especially rewarding in addressing key issues in the field of environmental politics and policy. In fact, one argument against the critique about "methodological nationalism" is that for the precise reason that environmental problems tend to affect many countries at the same time and in roughly similar ways, cross-national analyses of policy responses, institution building, social movements, party formations and the emergence of environmental norms are crucial for getting a deeper understanding of the differences between how societies influence and shape responses to the environmental crisis.

Thus Spaargarten and Mol´s argument is not empirically substantiated by the studies in this volume, but their argument also can be challenged on a purely theoretical basis. The comparative method is not primarily posited on the idea of comparing nation-states as such. Instead, the rationale is to use the fact that the political world is divided into separate cultural, administrative, and political entities that are different from each other, together with the assumption that cross-national differences in political cultures, institutions, and structures are the result of natural experiments, as a means of analyzing political phenomena.

The term "methodological nationalism," therefore, is based on a misunderstanding of the comparative method: the entities being compared are not nations in themselves, but actor behavior, policy processes, and institutions in different structural and cultural contexts. Neither does a comparative approach presuppose that elements such as the central governmental administration or national policies and institutions are the only valid units of analysis: as demonstrated by the non-state-centric contributions to the volume, individuals, local communities, and stakeholder groups are just as plausible as objects of analysis in a cross-national comparative study. For example, Michele Micheletti et al. are using data from different nations to map the emergence of norms of sustainable citizenship; Andreas Duit and Ola Hall analyze local-level stakeholder involvement in conservation programs in fifty-five different political contexts; and Krister Andersson et al. seek to understand the actions of local mayors in three Latin American countries.

Wherein lies the "added value" of the comparative approach to environmental politics? There are two compelling answers to this question. The first is the distinct advantage that the comparative method offers to almost any object of political inquiry. Treating countries as units of analysis in a natural experiment of political processes and structures, it allows for inquiries into how underlying political and social factors shape the effects of political institutions and determine the outcome of policies. As Lichbach and Zuckerman (1997) write in their introductory chapter to an edited volume on comparative politics:

> Because events of global historical importance affect so many countries in so short a period of time, studies of single countries and abstract theorizing are woefully inadequate to capture epoch-shaping developments. More than three decades ago, when the founders of the contemporary field of comparative politics initiated the most recent effort to merge theory and data in the study of politics they therefore established another of the field's guiding principles: The proper study of politics requires systematic comparisons. (pp. 4–5)

The comparative approach consequently has yielded an impressive bulk of knowledge in key areas of political inquiry. Examples of path-breaking works applying a comparative perspective include Arend Lijphart's studies of the effects of consensual and majoritarian types of democracies (Lijphart 1999), Theda Skocpol's analyses of the causes of revolutions (Skocpol 1979), Barrington Moore's studies of democratization processes (Moore 1966), Gøsta Esping-Andersen's work on welfare state regimes (Esping-Andersen 1990), and Peter Hall and David Soskice's explorations of the varieties of capitalism (Hall and Soskice 2003).

The second reason why the study of environmental politics calls for a comparative approach is that many of the core questions in environmental politics are inherently comparative. Why are some countries so much better at addressing environmental problems than others? Are different environmental policy regimes evolving in different political systems, and how do they compare? Various sociopolitical contexts seem to influence local resource management efforts differently—but how exactly do they do that? As Lichbach and Zuckerman point out, the comparative method has often been employed in studies of the "big questions" in political science, such as why countries differ in prosperity, the causes behind large-scale processes of social change, and patterns of power and representation (Lichbach and Zuckerman 1997, p. 5). Given the momentous challenges presented by ongoing processes of environmental degradation, and the comprehensive reorganization of modern society required to overcome them, it can hardly be an overstatement to say that the issue of how societies are responding to environmental problems is one of the epoch-shaping developments of great concern for contemporary political inquiry.

Current Directions in Research on Comparative Environmental Politics and Policy

Through their explicit focus on the role on the state, the studies collected in this volume build on previous work in the general field of comparative environmental politics. Three main clusters of studies can be distinguished within this field, albeit with a considerable degree of overlap. The first cluster—and also the most prominent cluster in terms of the number of publications—involve efforts to explore and understand cross-national patterns of *environmental governance*. Within this larger category, at least three subdivisions can be discerned. The first contains studies that focus on the evolution of policymaking and policy instruments within the environmental area. A second related but yet distinct subcategory of studies has revolved around processes of *diffusion of policy instruments*. A third group of studies deals with changing patterns of *civic engagement in environmental governance*, as expressed by green social movements and emerging notions of environmental citizenship.

Debates in all three subdivisions have to a large extent been carried out against the backdrop of ecological modernization theory, which highlights the role of changing patterns of consumption and production, non-state actors, and new policy instruments in addressing environmental degradation. Work found in a second cluster seeks to explain cross-national differences in *environmental performance*, typically in the format of large-N studies of countries in the Organisation for Economic Cooperation and Development (OECD). A third discussion in contemporary comparative environmental politics deals with issues of *natural resource management*. Compared to the two previous research topics, there are relatively few studies that approach natural resource management from a comparative perspective, and the chapters presented in this book represents an important contribution to this field.

In the following sections, a brief overview of the most influential works within each category will be presented, along with a description of how the studies collected in this volume add to the research agendas within each category. Before proceeding, two things should be emphasized. First, neither the review nor the works collected in this volume claim to be representative of the entire field of environmental governance. A number of research debates (e.g., environmental parties, green voters, and green social movements), as well as some regions of the world (e.g., East- and Southeast Asia and Oceania), are not represented in the volume. Second, the review of the three clusters of research in comparative environmental politics is, by necessity, based on a selection of influential works in the comparative tradition, rather than an exhaustive overview of published studies in the field.

Environmental Governance in a Comparative Perspective

A dominant theme within comparative studies on the environmental policymaking process is the issue of how economic growth and environmental sustainability can, or cannot, coexist: Is the market economy inevitably linked to environmental harms, or is it possible to discern certain forms of economies that ultimately are not based on accelerating resource depletion? The debate around ecological modernization theory (EMT) is framed in just such an opposition between economy and ecology. Proponents of the ecological modernization school tend to adopt a more optimistic outlook on the prospects of combining economy and ecology and tend to be particularly interested in new policy instruments and solutions that generate win-win solutions and synergies between markets and environmental regulations and policies (Mol and Spaargaren 2000; Mol et al. 2009).

One of the most ambitious undertakings within the EMT paradigm thus far is the two-volume collection of case studies of "environmental capacity building" in 30 countries commissioned and edited by Martin Jänicke and Helmut Weidner (Jänicke and Weidner 1997; Weidner and Jänicke 2002). Jänicke and Weidner were interested in taking stock of how far the process toward environmental capacity building—defined as society's capacity to identify and solve environmental problems (Weidner and Jänicke 2002, p. 1)—has progressed in both developing and developed countries. Jänicke and Wiedner's studies are prime examples of works that seek to understand how different stages in the evolution of the market economy are associated with different levels and types of environmental impacts, as well as different forms of policy responses. What they found was a relatively high level of capacity for dealing with the end-of-pipe type of pollution (primarily through technical solutions), an increasing pace of policy diffusion between different countries, a move toward governance-type policy instruments, and increasing environmental capacity in many developing countries. They also found an overall low level of capacity for addressing problems such as sustainable land use practices, soil degradation, and climate protection (Weidner and Jänicke 2002). Another compilation of case studies is Kenneth Hanf and Alf-Inge Jansen's edited volume *Governance and Environment in Western Europe* (1998), in which the interplay between institutional contexts and the strategic behavior of various actors engaged in environmental politics is linked to outcomes of environmental policies in a historical perspective. The collaborators place particular emphasis on understanding the emergence of policies associated with EMT.

A further set of case studies of environmental policymaking in Western countries can be found in an edited volume by Udai Desai, in which contributions center on cataloguing how the effectiveness of state interventions into environmental matters are affected by three central institutions (business, government, and international organizations) in seven different countries. The overall conclusion here is that economic interests have a significant degree of influence over environmental protection in all the surveyed countries (Desai 2002). Using a similar focus on the level of economic development as the overarching framework, Desai has also edited a volume of case studies of environmental politics in nine developing countries (Desai 1998).

The EMT comes close to filling the role of a theoretical framework for comparative environmental governance. It identifies some general mechanisms that can be found in environmental politics and policymaking in most countries, and some varieties of the theory do outline an explanatory framework for understanding *why* states are embarking on the route of ecological modernization when addressing environmental problems (Christoff 2006). However, an important shortcoming of the EMT is that it is not trying to understand cross-national variations in environmental governance per se; it is not at heart a comparative theory. Instead, EMT is often described as a mix between a normative theory, a set of policy prescriptions, and a specific political and administrative process taking place in advanced economies. This has meant that studies in the EMT school have tended to be interested in taking stock of how far the process of ecological modernization has progressed in different countries, rather than in understanding the causes of this variation. Another shortcoming is that the EMT framework—although it tries to move away from the issue by assuming that it is resolved—ultimately revolves around the fundamental conflict between economic growth and environmental harm. Insofar as there are other explanatory factors involved in shaping environmental policies and management than the level of economic development, the EMT is often unable to detect them.

The ecological modernization debate overlaps with another central theme in contemporary comparative research: the emergence of new policy instruments and novel forms of environmental governance. This issue in environmental policy is also connected to a more general scholarly discussion about the move toward governance in public steering and service provisioning (Pierre 2000; Rhodes 2000). Because environmental regulation is a relatively fast-changing policy area, it is a useful object of study for those wishing to keep track of emerging forms of public governance. The findings in the literature are clear: the style and contents of environmental policymaking is changing rapidly in most developed and many developing states (Durant et al. 2004; Paehlke and Torgerson 2005). Older "command-and-control" regulatory styles are still forming the backbone of many environmental policy regimes, but a new generation of policy instruments is gaining ground. Examples of such policy instruments include public-private partnerships (Glasbergen 1998), product-labeling schemes (Boström and Klintman 2011), voluntary environmental regulation programs (Prakash and Potoski 2006), and network governance (Klijn et al. 2010). Further examples of work focusing on changes in environmental policymaking and regulation can be found in a volume edited by Jordan et al. (2003), which seeks to investigate how "new" policy instruments in the toolboxes of national environmental policy makers are emerging.

A closely related research topic concerns how processes of policy diffusion, regulatory competition, and trade are shaping policymaking in the environmental area. An edited volume by William Lafferty and James Meadowcroft represents an early collection of case studies of how the concept of sustainable development is diffused and subsequently implemented in a selection of nine highly developed countries (Lafferty and Meadowcroft 2000). Other studies have investigated the role of trade between countries as a vector for spreading policy innovations and regulatory standards; for example, see Roni Garcia-Johnson's study of the role of multinational corporations in policy diffusion in Brazil and Mexico (Garcia-Johnson 2000) and Kate O'Neill's analysis of waste trading in advanced economies (O'Neill 2000). A central issue in this area has been concerns about a "race to the bottom" in the environmental area driven by regulatory competition among countries, but most studies find no evidence of this effect (Vogel 1997). A recent effort in the neighboring topic of environmental policy diffusion in is the research based on the ENVIPOLCON data set developed by Katarina Holzinger, Christoph Knill, and Art Bas. This data has mainly been used to conduct large-N investigations of policy diffusion processes between different countries, and it has yielded some highly interesting insights into how policy instruments and regulations are disseminated throughout the sample of OECD nations. Holzinger et al. (2008) find a strong and increasing trend toward the homogenization of policy portfolios and show that the main driving forces behind this development are trade and contacts with other countries mediated by international organizations (see also Delmas 2002; Tews et al. 2003).

The role of civil society, citizens, and social movements in environmental politics and policy—in particular as perceived agents behind policy change—has been the topic for a number of comparative studies. An early forerunner in this field is Russell Dalton's *The Green Rainbow*, in which he analyzes environmental movements in Western Europe (Dalton 1994). The volume *Environmental Protest in Western Europe*, edited by Christopher Rootes, consists of case studies of environmental activism and protests in seven countries and one region in Europe. The study finds no ground for worries about and increasing institutionalization of environmental social movements, and also no evidence to suggest that national protest targets had been displaced by targets on the European Union (EU) level (Rootes 2003). In a study entitled *Green States and Social Movements*, John Dryzek and colleagues set out to explore the role of social movements in the rise of what they call the "Green state" in the United States, United Kingdom, Germany, and Norway. The transformation toward a fully developed green state has not been completed anywhere, but Germany is judged to come closest to the ideal type (Dryzek et al. 2003). Through their foundation in social movement theory, this work represents a rare example of a study that seeks to analyze the driving forces behind environmental politics and policy in a more general comparative and theoretically guided perspective.

Several chapters in this volume seek to understand different aspects of the development of environmental governance in a comparative perspective. Susan Baker and Katarina Eckerberg investigate the changing but nonetheless central role of the state in environmental governance in a qualitative analysis of 27 European countries. Thomas Sommerer examines policy-diffusion processes in the environmental area using longitudinal data from 24 countries and distinguishes unexpected and shifting patterns of laggards and pioneers among countries. And Michele Micheletti, Dietlind Stolle, and Daniel Berlin focus on the emergence of norms of sustainable citizenship and elaborate on the consequences of this new form of citizenship for environmental policymaking.

Determinants of Environmental Performance

Research within a second theme in comparative environmental politics focuses on uncovering the causes of the environmental performance of states. This question has primarily been addressed using large-N studies in which indicators of environmental quality are regressed against a set of sociopolitical explanatory variables. Pioneering works in this tradition are papers by Markus Crepaz (1995), Detlef Jahn (1998), and Lyle Scruggs (1999), which tried to explain variations in point-source emission reductions among OECD countries. These early studies found that consensual democracies and corporatist labor market structures seemed to be associated with a better environmental record, possibly as a consequence of having a more developed set of institutions for negotiation and compromise between business interests and state actors. Findings from these early studies (in particular the effect of neo-corporatism) have been contested by later studies (Scruggs 2001; Neumayer 2003; Scruggs 2003; and Walti 2004), and the applicability of the results is yet to be determined for diffuse-source emissions and natural resource management.

A subtheme of studies on environmental performance has been a focus on the importance of good governance in combating environmental degradation. Here, the guiding notion has been that countries with higher levels of institutional quality (in the sense of low levels of corruption and public governance in adherence to the principles of rule of law) can be expected to facilitate collaboration between societal actors seeking to address environmental problems (Robbins 2000). Some studies have indicated the existence of such an effect (Esty and Porter 2005; Pellegrini and Gerlagh 2006), whereas others fail to find any such linkages (Grafton and Knowles 2004; Duit et al. 2009). One reason for these mixed results can be, as Barrett et al. (2006) point out, that institutional quality, in addition to being helpful when states are trying to protect the environment, is a prerequisite for effective resource utilization. Conversely, corruption might reduce the effectiveness of preservation policies, but it is also possible that corruption precludes the establishment of large-scale and effective resource utilization projects and thereby reduces society's total environmental impact.

Another subsection of environmental performance studies investigates the role of states as actors on a global arena (Neumayer 2002; Mitchell 2003; Roberts et al. 2004; Bernhagen 2008). The main object of study in this area is to identify those institutional and structural factors that determine state participation in, and compliance with, international environmental treaties. Findings from large-N studies indicate that a country's place in the world market, the viability of its civil society, the strength of environmental parties and nongovernmental organizations (NGOs), and the power of business interests are important determinants of a country's participation in international accords (Roberts et al. 2004; Bernhagen 2008). A similar picture emerges from the edited volume *Engaging Countries: Strengthening Compliance with International Accords* consisting of case studies of ratification, compliance, and implementation of five different international accords in eight countries. Despite substantial variations between countries and accords, the editors nevertheless find clear signs of an overall trend toward increased compliance and implementation: Countries seem to be more willing and able to implement and comply with international treaties over time. Cross-country variations in this trend are primarily explained by administrative capacities, political cultures, economic development, and the degree of exposure to international trade (Weiss and Jacobson 2000).

Environmental performance is currently one of the better-studied topics of comparative environmental politics. But despite a fairly substantial number of studies on this topic, a coherent understanding of what causes better environmental performance is yet to emerge. As Daniel Fiorino concludes in a recent survey of the research field, we still do not understand why some countries seem to do so much more and are so much better at protecting their own environments, as well as global environments (Fiorino 2011). A perennial problem confronting students of environmental performance has been that of finding valid indicators of performance (Neumayer 2004; the chapter by Meadowcroft in this volume). Earlier studies tended to use estimates of point-source emissions as proxies for environmental performance, which are often reliable but of limited usefulness in capturing a state's overall environmental performance. Later studies have tried to use a more diversified set of performance indicators, thereby reducing reliability as well as comparability. A related problem has been that the policy process itself has been largely black-boxed, in the sense that most studies have in effect been designed to search for correlational patterns between political structures (i.e., democracy, electoral systems, and corruption) and environmental outcomes. This design format has been a consequence of a lack of comparative data on policy instruments, which has meant that the effect of policy outputs connecting structures and environmental outcomes are largely unknown. However, scholars recently have started to look at factors influencing environmental policy outputs. With the help of freshly collected data on environmental policy instruments and policy stringency, Knill et al. (2010) were able to identify a robust correlation between the scope of environmental regulation and a more pro-environmental position of governmental parties. Such studies, which try to take into account political processes and policy dynamics in explaining policy outputs, constitute an important step toward a more complete model of environmental performance.

Four of the chapters in this volume add to the environmental performance literature. James Meadowcroft engages with the central issue of measuring environmental performance of states and offers novel suggestions for how to develop a more meaningful set of performance indicators. Roger Karapin looks at the issue of environmental performance through the perspective of renewable energy policies, and asks why German wind energy policy has been more successful than its US counterpart. By adopting a study design not commonly used in environmental performance studies—a comparative and longitudinal case study—Karapin's analysis delivers fresh insights about the policy-level mechanisms that have an important influence on the outcome of environmental policymaking. In their chapter, Christoph Knill, Susumu Shikano, and Jale Tosun employ an item response estimation technique and present detailed data on environmental policies to uncover factors that can account for cross-national patterns of diffusion and adoption of environmental policy. Finally, Detlef Jahn takes a theoretical point of departure in green political theory when investigating the environmental performance of different types of green states.

Comparative Natural Resource Management

Studies of natural resource management make up one of the largest, and perhaps also one the most fruitful, subdivisions within environmental social science. Consequently, knowledge about factors underlying successful and failed cases of natural resource management is relatively well developed. The groundbreaking work of Elinor Ostrom on understanding the role of institutions in achieving sustainable governance of common pool resources is a centerpiece in this genre (Ostrom 1990; Ostrom 2005), but several additional studies have focused on issues related to natural resource management. One such issue is forest management, which has been investigated both in large-N studies (Midlarsky 1998; Bates and Rudel 2000) and through case studies (Cashore et al. 2004; Espach 2009). However, unlike what has been the case in studies of environmental performance and policymaking, the field of natural resource management contain relatively few studies that are using a comparative approach (Agrawal 2012). Most research efforts in the field have focused on uncovering local-scale dynamics between society and nature, usually by analyzing how stakeholders and resource users regulate, harvest, and manage small-scale natural resources. Addressing such issues often requires that researchers pay close attention to resource characteristics and patterns of resource utilization, which has meant that the large majority of studies in this area have been carried out using a case study design. This design format, together with the need for careful collection of local-level data, have meant that questions about the effect of the surrounding political system on local and regional levels of environmental management initiatives and institutions have been pushed to the back of the research agenda (Ribot 2006). Most scholars would agree that general characteristics of the political system, such as levels of decentralization, corruption, democracy, and social capital, are likely to have a large impact on the possibility of achieving self-organization and local institution building for sustainable resource usage, but these are rarely addressed in a systematic fashion by Common Pool Resource studies (see, however, Andersson and van Laerhoven 2007) or in studies of stakeholder involvement (Brown 2009). As one of a very small number of ways of gauging the effect of such sociopolitical factors on local environmental management efforts, institution building, and policymaking, a comparative approach offers a valuable alternative for moving the research on this topic forward. Three of the studies in this volume adopt a comparative approach to natural resource management, and are thereby able to integrate state behavior and characteristics into their analysis of local-level resource management. Martin Sjöstedt's analysis of relationship between the rulers and the ruled in fisheries in South Africa, Namibia, and Angola seeks to analyze the role of corruption and trustworthiness of the state in the regulation of common pool resources. Krister Andersson et al. use longitudinal data and satellite imagery from 300 local governments in three Latin American countries (Bolivia, Guatemala, and Peru) to investigate the role of decentralization on local forestry management. Finally, Andreas Duit and Ola Hall employ survey and satellite data from 143 nature reserves (Man and the Biosphere Areas) in 55 countries in a study of the role of corruption and protection of political rights on stakeholder participation in nature conservation projects.

The Structure of the Book

Cross-cutting the common theme of rediscovering the state in environmental governance through a comparative perspective are three sections in this book that correspond to the three subdivisions of comparative environmental governance identified above. The first section tackles the core issue of understanding the environmental performance of states in a comparative perspective. This section contains chapters by James Meadowcroft; Christoph Knill, Susumu Shikano, and Jale Tosun; Detlef Jahn; and Roger Karapin. The second section turns to the structure and style of environmental governance. The chapters by Susan Baker and Katarina Eckerberg; Michele Micheletti, Dietlind Stolle, and Daniel Berlin; and Thomas Sommerer apply a comparative lens to issues in environmental policymaking, such as the role of the state in regional-level environmental governance, processes of policy diffusion and change, and the rise of sustainable citizenship. The third section, with chapters by Krister Andersson et al., Andreas Duit and Ola Hall, and Martin Sjöstedt, examines how political and institutional contexts affect natural resource management in a comparative perspective. The concluding chapter, by Andreas Duit, brings together the findings in the individual chapters and suggests that one way of understanding the role of the state in environmental governance is through the notion of an emerging ecostate.

References

Agrawal, Arun. 2012. Local Institutions and the Governance of Forest Commons. In *Comparative Environmental Politics: Theory, Practice, and Prospects*, ed. Paul F. Steinberg and Stacy D. VanDeveer, 313–340. Cambridge, MA: MIT Press.

Agrawal, Arun, and Clark C. Gibson. 2001. *Communities and the Environment: Ethnicity, Gender, and the State in Community-Based Conservation*. New Brunswick, N.J.: Rutgers University Press.

Andersson, Krister, and Frank van Laerhoven. 2007. From Local Strongman to Facilitator: Institutional Incentives for Participatory Municipal Governance in Latin America. *Comparative Political Studies*, 40(9): 1085–1111.

Bäckstrand, Karin. 2004. Scientisation vs. Civic Expertise in Environmental Governance: Eco-feminist, Eco-modern and Post-modern Responses. *Environmental Politics* 13 (4): 695–714.

Bäckstrand, Karin. 2008. Accountability of Networked Climate Governance: The Rise of Transnational Climate Partnerships. *Global Environmental Politics*, 8(3): 74–102.

Barrett, Christopher B., Clark C. Gibson, Barak Hoffman, and Mathew D. McCubbins. 2006. The Complex Links between Governance and Biodiversity. *Conservation Biology*, 20(5): 1358–1366.

Barry, John, and Robyn Eckersley. 2005. An Introduction To Reinstating the State. In *The State and the Global Ecological Crisis*, ed. J. Barry and R. Eckersley. Cambridge, MA: MIT Press.

Bates, Diane, and Thomas K. Rudel. 2000. The Political Ecology of Conserving Tropical Rain Forests: A Cross-National Analysis. *Society & Natural Resources,* 13(7): 619–634.

Bernhagen, Patrick. 2008. Business and International Environmental Agreements: Domestic Sources of Participation and Compliance by Advanced Industrialized Democracies. *Global Environmental Politics,* 8:78–110.

Biermann, Frank, and Klaus Dingwerth. 2004. Global Environmental Change and the Nation State. *Global Environmental Politics* 4 (1): 1–22.

Boström, Magnus, and Mikael Klintman. 2011. *Eco-Standards, Product Labelling, and Green Consumerism.* Basingstoke, UK: Palgrave Macmillan.

Brown, Katrina. 2009. Human Development and Environmental Governance: A Reality Check. In *Governing Sustainability*, ed. W. Neil Adger and Andrew Jordan, 32–52. Cambridge: Cambridge University Press.

Buttel, Frederick H. 2004. The Treadmill of Production. *Organization & Environment,* 17(3): 323–336.

Cashore, Ben W., Graeme Auld, and Deanna Newsom. 2004. *Governing through Markets: Forest Certification and the Emergence of Non-State Authority.* New Haven: Yale University Press.

Christoff, Peter. 2006. Ecological Modernization, Ecological Modernities. In *Contemporary Environmental Politics. From Margins to Mainstream*, ed. Piers H. G. Stephens with John Barry and Andrew Dobson, 179–200. New York: Routledge.

Crepaz, Markus. 1995. Explaining National Variations of Air Pollution Levels: Political Institutions and Their Impact on Environmental Policy-making. *Environmental Politics* 4(3): 391–414.

Dalton, Russell J. 1994. *The Green Rainbow. Environmental Groups in Western Europe.* New Haven: Yale University Press.

Dalton, Russell J. 2005. The Greening of the Globe? Cross-National Levels of Environmental Group Membership. *Environmental Politics* 14(4): 441–459.

Delmas, Magali A. 2002. The Diffusion of Environmental Management Standards in Europe and in the United States: An Institutional Perspective. *Policy Sciences* 35(1): 91–119.

Desai, Uday. 1998. *Ecological Policy and Politics in Developing Countries: Economic Growth, Democracy, and Environment.* Albany: State University of New York Press.

Desai, Uday, ed. 2002. *Environmental Politics and Policy in Industrialized Countries.* Cambridge, MA: MIT Press.

Dryzek, John S., David Downes, Christian Hunold, David Schlosberg, and Hans-Kristian Hernes. 2003. *Green States and Social Movements. Environmentalism in the United States, United Kingdom, Germany, and Norway.* Oxford: Oxford University Press.

Dryzek, John S., and Simon Niemeyer. 2008. Discursive Representation. *American Political Science Review* 102(4): 481–493.

Duit, Andreas, Ola Hall, Grzegorz Mikusinski, and Per Angelstam. 2009. Saving the Woodpeckers: Social Capital, Governance, and Policy Performance. *Journal of Environment & Development* 18(1): 42–61.

Durant, Robert F., Daniel J. Fiorino, and Rosemary O'Leary, eds. 2004. *Environmental Governance Reconsidered. Challenges, Choices, and Opportunities.* Cambridge, MA: MIT Press.

Enloe, Cynthia H. 1975. *The Politics of Pollution in a Comparative Perspective. Ecology and Power in Four Nations.* New York: David McKay Company.

Espach, Ralph H. 2009. *Private Environmental Regimes in Developing Countries: Globally Sown, Locally Grown.* Basingstoke, UK: Palgrave MacMillan.

Esping-Andersen, Gøsta. 1990. *The Three Worlds of Welfare Capitalism.* Cambridge, UK: Polity Press.

Esty, Daniel C., and Michael E. Porter. 2005. National Environmental Performance: An Empirical Analysis of Policy Results and Determinants. *Environment and Development Economics* 10(4): 391–434.

Fiorino, Daniel. 2011. Explaining National Environmental Performance: Approaches, Evidence, and Implications. *Policy Sciences* 44 (4): 367–389.

Garcia-Johnson, Ronie. 2000. *Exporting Environmentalism: US Multinational Chemical Corporations in Brazil and Mexico.* Cambridge, MA: MIT Press.

Glasbergen, Pieter. 1998. *Co-operative Environmental Governance: Public-Private Agreements as a Policy Strategy.* Berlin: Springer.

Grafton, Quentin R., and Stephen Knowles. 2004. Social Capital and National Environmental Performance: A Cross-Sectional Analysis. *Journal of Environment & Development* 13(4): 336–370.

Hall, Peter A., and David Soskice. 2003. *Varieties of Capitalism: The Institutional Foundations of Comparative Advantage.* Oxford: Oxford University Press.

Hanf, Kenneth, and Alf-Inge Jansen, eds. 1998. *Governance and Environment in Western Europe. Politics, Policy, and Administration.* Harlow, UK: Longman.

Holling, C. S., and Meffe, Gary K. 1996. Command and Control and the Pathology of Natural Resource Management. *Conservation Biology* 10 (2): 328–337.

Holzinger, Katharina, Christoph Knill, and Arts Bas. 2008. *Environmental Policy Convergence in Europe: The Impact of International Institutions and Trade.* Cambridge: Cambridge University Press.

Jahn, Detlef. 1998. Environmental Performance and Policy Regimes: Explaining Variations in 18 OECD-Countries. *Policy Sciences* 31(2): 107–131.

Jänicke, Martin, and Helmut Weidner, eds. 1997. *National Environmental Policies. A Comparative Study of Capacity Building.* Berlin: Springer.

John, Peter, Sarah Cotterill, Alice Moseley, Liz Richardson, Gerry Stoker, Hanhua Liu, Graham Smith, Corinne Wales, and Hisako Nomura. 2011. *Nudge, Nudge, Think, Think: Using Experiments to Change Civic Behaviour.* London: Bloomsbury Publishing PLC.

Jordan, Andrew, Wurtzel, Ruediger K.W., and Anthony R. Zito, eds. 2003. *New Instruments of Environmental Governance? National Experiences and Prospects.* London: Frank Cass Publishers.

Klijn, Erik-Hans, Bram Steijn, and Jurian Edelenbos. 2010. The Impact of Network Management on Outcomes in Governance Networks. *Public Administration* 88(4): 1063–1082.

Knill, Christoph, Marc Debus, and Stephan Heichel. 2010. Do Parties Matter in Internationalised Policy Areas? The Impact of Political Parties on Environmental Policy Outputs in 18 OECD Countries, 1970–2000. *European Journal of Political Research* 49(3): 301–336.

Lafferty, William, and James Meadowcroft, eds. 2000. *Implementing Sustainable Development. Strategies and Initiatives in High-Consumption Societies.* Oxford: Oxford University Press.

Lichbach, Mark Irving, and Alan S. Zuckerman, eds. 1997. *Comparative Politics: Rationality, Culture, and Structure.* Cambridge: Cambridge University Press.

Lijphart, Arend. 1999. *Patterns of Democracy. Government Forms and Performance in Thirty-Six Countries.* New Haven: Yale University Press.

Lundqvist, Lennart J. 1980. *The Hare and the Tortoise: Clean Air Policies in the United States and Sweden.* Ann Arbor: University of Michigan Press.

Meadowcroft, James. 2012. Greening the State. In *Comparative Environmental Politics: Theory, Practice, and Prospects*, ed. Paul F. Steinberg and Stacy D. VanDeveer, 63–87. Cambridge, MA: MIT Press.

Meyer, John W., David J. Frank, Ann Hironaka, Evan Schofer, and Nancy B. Tuma. 1997. The Structuring of a World Environmental Regime, 1870–1990. *International Organization* 51(4): 623–651.

Midlarsky, Manus I. 1998. Democracy and the Environment: An Empirical Assessment. *Journal of Peace Research* 35(3): 341–361.

Mitchell, Ronald B. 2003. International Environmental Agreements: A Survey of Their Features, Formation, and Effects. *Annual Review of Environment and Resources* 28:249–261.

Mol, Arthur P. J, David A. Sonnenfeld, and Gert Spaargaren, eds. 2009. *The Ecological Modernisation Reader: Environmental Reform in Theory and Practice.* London: Routledge.

Mol, Arthur P. J., and Gert Spaargaren. 2000. Ecological Modernisation Theory in Debate: A Review. In *Ecological Modernisation Around the World: Perspectives and Critical Debate*, ed. Arthur P. J. Mol and Gert Spaargaren, 17–49. London: Frank Cass Publishers.

Moore, Barrington, Jr. 1966. *Social Origins of Dictatorship and Democracy: Lord and Peasant in the Making of the Modern World.* Boston: Beacon Press.

Neumayer, Eric. 2002. Do Democracies Exhibit Stronger International Environmental Commitment? A Cross-Country Analysis. *Journal of Peace Research* 39(2): 139–164.

Neumayer, Eric. 2003. Are Left-Wing Party Strength and Corporatism Good for the Environment? Evidence from Panel Analysis of Air Pollution in OECD Countries. *Ecological Economics* 45:203–220.

Neumayer, Eric. 2004. Indicators of Sustainability. In *International Yearbook of Environmental and Resource Economics 2004/05*, ed. Thomas Tietenberg and Henk Folmer, 139–188. Cheltenham, UK: Edward Elgar.

O'Neill, Kate. 2000. *Waste Trading among Rich Nations: Building a New Theory of Environmental Regulation*. Cambridge, MA: MIT Press.

Ostrom, Elinor. 1990. *Governing the Commons: The Evolution of Institutions for Collective Action*. Cambridge: Cambridge University Press.

Ostrom, Elinor. 2005. *Understanding Institutional Diversity*. Princeton: Princeton University Press.

Paehlke, Robert, and Douglas Torgerson. 2005. *Managing Leviathan: Environmental Politics and the Administrative State*. Toronto: University of Toronto Press.

Paterson, Matthew. 2009. Global Governance for Sustainable Capitalism? The Political Economy of Global Environmental Governance. In *Governing Sustainability*, ed. Neil W. Adger and Andrew Jordan, 99–122. Cambridge: Cambridge University Press.

Pellegrini, Lorenzo, and Reyer Gerlagh. 2006. Corruption, Democracy, and Environmental Policy: An Empirical Contribution to the Debate. *Journal of Environment & Development* 15(3): 332–354.

Pierre, Jon, ed. 2000. *Debating Governance—Authority, Steering, and Democracy*. Oxford: Oxford University Press.

Prakash, Aseem, and Matthew Potoski. 2006. *The Voluntary Environmentalists: Green Clubs, ISO 14001, and Voluntary Environmental Regulations*. Cambridge, MA: Cambridge University Press.

Rhodes, R. A. W. 2000. Governance and Public Administration. In *Debating Governance. Authority, Steering, and Democracy*, ed. Jon Pierre, 54–90. Oxford: Oxford University Press.

Ribot, Jesse. 2006. Choose Democracy: Environmentalists' Socio-political Responsibility. *Global Environmental Change* 16:115–119.

Robbins, Paul. 2000. The Rotten Institution: Corruption in Natural Resource Management. *Political Geography* 19(4): 423–443.

Roberts, J. Timmons, Bradley C. Parks, and Alexis A. Vásquez. 2004. Who Ratifies Environmental Treaties and Why? Institutionalism, Structuralism, and Participation by 192 Nations in 22 Treaties. *Global Environmental Politics* 4:22–64.

Rootes, Christopher, ed. 2003. *Environmental Protest in Western Europe*. Oxford: Oxford University Press.

Scruggs, D. L. 2003. *Sustaining Abundance. Environmental Performance in Industrial Democracies*. Cambridge, MA: Cambridge University Press.

Scruggs, Lyle. 1999. Institutions and Environmental Performance in Seventeen Western Democracies. *British Journal of Political Science* 29(1): 1–13.

Scruggs, Lyle. 2001. Is There Really a Link between Neo-corporatism and Environmental Performance? Updated Evidence and New Data for the 1980s and 1990s. *British Journal of Political Science* 31:686–692.

Skocpol, Theda. 1979. *States and Social Revolutions: A Comparative Analysis of France, Russia, and China*. New York: Cambridge University Press.

Spaargaren, Gert, and Arthur P. J. Mol. 2008. Greening Global Consumption: Redefining Politics and Authority. *Global Environmental Change* 18(3): 350–359.

Steinberg, Paul F., and Stacy D. VanDeveer. 2012. Comparative Environmental Politics in a Global World. In *Comparative Environmental Politics. Theory, Practice and Prospects*, ed. P. F. Steinberg and S. D. VanDeveer. Cambridge, MA: MIT Press.

Tews, Kerstin, Per-Olof Busch, and Helge Jörgens. 2003. The Diffusion of New Environmental Policy Instruments. *European Journal of Political Research* 42(4): 569–600.

Underdal, Arild. 2010. Complexity and Challenges of Long-Term Environmental Governance. *Global Environmental Change* 20(3): 386–393.

Vogel, David. 1997. *Trading Up: Consumer and Environmental Regulation in a Global Economy*. Cambridge, MA: Harvard University Press.

Walti, Sonja. 2004. How Multilevel Structures Affect Environmental Policy in Industrialized Countries. *European Journal of Political Research* 43(4): 599–634.

Weidner, Helmut, and Martin Jänicke, eds. 2002. *Capacity Building in National Environmental Policy. A Comparative Study of 17 Countries*. Berlin: Springer.

Weiss, Edith B., and Harold K. Jacobson. 2000. Assessing the Record and Designing Strategies to Engage Countries. In *Engaging Countries: Strengthening Compliance with International Environmental Accords*, ed. Edith B. Weiss and Harold K. Jacobson, 551–555. Cambridge, MA: MIT Press.

Young, Oran R., ed. 1999. *The Effectiveness of International Environmental Regimes: Causal Connections and Behavioral Mechanisms*. Cambridge, MA: MIT Press.

I

Understanding Environmental Performance

2

Comparing Environmental Performance

James Meadowcroft

This chapter addresses an important theme in the study of environmental governance: the comparative assessment of environmental performance. It considers conceptual and methodological issues involved in the attempt to compare environmental performance from one period to another and from one jurisdiction to another. The problem is of interest to students of comparative environmental politics who seek to understand how the societal management of environmental issues varies across space and time, as well as the explanations for these diverse outcomes. But the issue is of practical and not just academic concern: on the one hand, an appreciation of comparative performance can contribute to understanding factors that influence the success or failure of environmental stewardship, and hence to improving policy design; and on the other, there is a link to accountability, as comparative insights can help in judging the work of responsible political authorities.

Since the birth of modern environmental governance at the end of the 1960s, efforts have been made to build up reliable data about the scale of human impacts on the natural world, the state of the environment, and the efficacy of policy approaches. Particularly over the past twenty years, considerable ingenuity has been applied to developing an array of performance measures that allow observers to draw conclusions about how well we are doing with respect to the environment and sustainability. Yet despite the proliferation of assessment instruments, it is not evident that we have come to understand a lot more about comparative performance. This is one of the themes that will be explored as this chapter unfolds.

The discussion is divided into three parts: first, a general consideration of the challenge of assessing comparative environmental performance; second, a brief examination of two well-known efforts to provide a comprehensive assessment of national environmental performance; and finally, some reflections on this experience. Although the discussion is framed largely in terms of comparisons between countries, similar issues apply to performance assessment of large subnational units (provinces, states), as well as to cities and municipalities.

Assessing Environmental Performance

The basic notion of assessing environmental performance is relatively straightforward: it refers to the evaluation of societal attainment with relation to environmental matters. The concept of "performance" here is results oriented, drawing attention to the record of accomplishment. Performance assessment requires a mapping of observed outcomes and an analysis of the significance of these outcomes (Jahn 1998; Duit 2005). It implies choices about the issues that are worth tracking and the techniques that will be used to measure them. And it entails the deployment of some standard against which performance is to be gauged. The *comparative* assessment of environmental performance sets the evaluative task in the context of multiple jurisdictions and/or time frames.

Comparative performance assessment can focus on specific policies, instruments, and laws, or on the record of organizations such as business corporations or international agencies. However, the discussion here is concerned principally with the performance of governments and the regional collectivities to which they are linked—cities, provinces, cantons, countries, and so on. The environmental scope of assessment exercises can be narrow or wide, and can be defined by concern with specific pollutants (such as mercury or nitrates), the deterioration of environmental media (water, air, or land), the pace and character of resource extraction (such as forestry, fisheries, or mining); large-scale environmental problems (such as acid rain, hazardous waste, climate change, or biodiversity loss), or more sweeping categories such as "environmental quality," "environmental pressures," or "sustainability."

A number of points should be made at the outset about this understanding of environmental performance evaluation. First, and most obviously, it is focused on judgments about *human* practices. In this context, the term *environmental performance* does not refer to the performance of the environment itself (for example, the capacity of an ecosystem to rebound from outside stress), but rather to the performance of human institutions in managing the environmental domain. Measurements of differences in environmental conditions that stem *entirely* from natural circumstances do not speak directly to the issue of performance. However, all societies must come to terms with the specific circumstances that nature presents to them (biophysical endowments, ecosystems, and so on); and the extent to which institutions successfully do so lies at the core of environmental performance.

Second, environmental performance assessment is an explicitly evaluative task. It involves the collection and manipulation of data, the design of measurement instruments, and the application of statistical techniques. But above all, it implies *judgments* about the quality of environment performance: it is not just concerned with establishing "more or less," but ultimately with assessing "better or worse," "adequate or inadequate," and so on. In other words, it has an irreducibly *normative* character. Requisite decisions about criteria, standards, and significance are inevitably value laden, as are conclusions about relative or absolute performance (cf chapter 4, "The Three Worlds of Environmental Politics," by Detlef Jahn, in this volume).

Third, it necessarily involves attention to *causal interconnections,* to tracking outcomes, and also to the developments that led to those outcomes. The word *performance* implies a definite course of human conduct with environmental consequences: it involves acts and omissions and carries an implicit notion of responsibility for that conduct. Of course, environmental outcomes can be influenced by natural phenomena as well as by human intervention, by developments outside a jurisdiction as well as by those from within, and by contingent circumstances as well as by policy action and inaction. And if one wishes to arrive at a serious evaluation of performance, as opposed to just a listing of outcomes, such issues and many others must be untangled. As Andreas Duit has argued, environmental performance sets the analytical focus "on understanding how political and social actors both cause and remedy environmental degradation" (Duit 2005, p. 13).

Fourth, environmental performance is a *multidimensional* concept. The environment is not a single "thing," but a complex of dynamic systems and relationships that interact with societies though varied mechanisms and are understood by human actors in different ways. Environmental "performance" can be conceptualized in terms of practices of environmental governance, or impacts on the biosphere, or the way these interrelate with conditions of societal life (the economy, public welfare, and so on). In the policy performance literature, it is common to distinguish the evaluation of policy "outputs" (for example, adoption of a particular policy such as a carbon tax) and policy "outcomes" (the broader consequences of policy implementation) (Fiorino 2011), but what counts as outputs and as outcomes can be framed in many ways. Moreover, the criteria invoked to assess performance are also plural, most obviously including "effectiveness"—which can be approached in terms of the attainment of self-set goals or of externally defined standards. But processes and outcomes can also be assessed in terms of their "fairness," "efficiency," "balance of costs and benefits," and so on. And these too are applied in discussions of environmental performance (Mickwitz 2006).

Finally, the evaluation of environmental performance is necessarily *provisional*. Judgments made today may be revisited in the future. Our knowledge of environmental phenomena remains limited; in many domains, causal mechanisms are only partially understood; new phenomena remain to be discovered; and many impacts take decades to become manifest. Solutions that look promising today ultimately may prove to have been illusory, so assessments of environmental performance are likely to change. Some, which look poor by today's benchmarks, may turn out to have been not so bad when all the circumstances are considered. But it is more likely that current performances will be judged more harshly in the future.

Political interest in environmental performance evaluation closely shadowed the emergence of modern environmental governance. From the late 1960s, governments established new institutions (agencies and ministries, expert advisory panels, framework laws, and national regulatory regimes) to manage environmental burdens (Jänicke and Weidner 1997; Hanf and Jansen 1998; van Tatenhove, Arts, and Leroy 2001). As the severity of environmental problems has become better appreciated (involving a wider range of issues that are more deeply embedded in the current social fabric), as the proportion of societal resources devoted to environmental governance has grown (density of regulation, attention of economic and political actors, investment in remedial technologies, and so on), and as the environment has become increasingly central to governmental legitimacy (Meadowcroft 2012), so interest in performance evaluation has grown. Moreover, continuing arguments over the effectiveness and economic costs of environmental regulation, as well as a political climate that emphasizes "evidence-based policy making," have encouraged greater attention to performance evaluation (Mickwitz 2006).

Over the past four decades, significant efforts have been invested in gathering data and developing informational frameworks to assist environmental performance evaluation. This has ranged from initiatives by the Organisation for Economic Cooperation and Development (OECD) to collect comparative environmental statistics and pioneer the pressure/state/response framework for environmental indicators (OECD 2003,

2008); to work to integrate environmental dimensions into the system of national accounts through a "green GNP," or a set of satellite accounts (Pierce and Warford 1993; World Bank 2010); and to assess sustainability through metrics of income or energy or ecological footprinting and through comprehensive indicator sets and aggregated indexes. Not only is there a huge amount of literature on each of these initiatives (and many others), but increasingly there have been studies that compare different high-level indexes and contrast their results (Böhringera and Jochem 2007; Mayer 2008; Singh et al. 2009; Pillarisetti and van den Bergh 2010; van de Kerk and Manuel 2010).

The evolution of performance metrics has been driven by a number of factors, including (1) changing perceptions of the most pressing environmental issues, (2) the shift in dominant environmental paradigms, and (3) competing visions of the underlying character of environmental dilemmas. Disciplinary influences also have been at work, with economists typically emphasizing the integration of environmental considerations into economic calculi, geographers and natural scientists stressing the monitoring of physical systems, and political scientists taking up issues such as the formulation of abatement targets and the adoption of policy instruments. Early interest in metrics for the pollution of air and water (ambient quality and pollutant loadings), waste management (volumes, types, disposal mechanisms, and recycling rates), resource consumption (nonrenewable and biological resources), and protected areas was supplemented with work on energy, acidification, and eutrophication. The 1980s witnessed the emergence of global issues such as ozone depletion, climate change, and biodiversity. From the 1990s, increasing emphasis was placed on assessing "sustainability," "environmental policy integration," and "decoupling," as well as the uptake of specificeasures (for example, environmental tax reform, "negotiated agreements," national sustainable development plans, and so on). More recently, there has been interest in monitoring the impacts of trade flows and topics such as embodied carbon and water (Kejun et al. 2008; Atkinson et al. 2010; RAE 2010).

Several interrelated tensions are evident in ongoing efforts to develop metrics to assess comparative environmental performance. Among the most important are those between *specialized and aggregated assessments*, between *physical and monetary accounting*, and between *the lens of the environment and sustainable development*. The complexity and diversity of environmental issues underlie the first tension: the problem is the extent to which it is possible to produce meaningful synthetic metrics that sum developments across a range of environmental issues. Even when tracking a single pollutant, the measurement and aggregation of data pose challenges: namely, how, when, and where can measurements be taken that will give a useful indication of the significance of an air contaminant in a city, region, or country? But the problem becomes more acute when dealing with compound issues, such as how to estimate "air pollution" or "air quality" when they relate to the presence of sulfur oxides (SOx) and nitrogen oxides (NOx), lead, particulates (of various sizes), ground-level ozone, and so on. Specialized measures give a more accurate portrait of multiple dimensions of an issue. And yet for many purposes—particularly attracting the attention of politicians, officials, and the media and communicating with the public—simplified and synthesized results may be more effective. After all, most people want to know the answers to broad questions such as: Is air pollution improving or deteriorating? Is air quality better here or there? Is the existing policy regime working or not?

The classic strategies for dealing with this problem are either (1) to select a headline indicator (perhaps a particularly revealing figure that tracks the most prevalent, persistent, or toxic contaminant, one that is likely to reflect the situation as a whole or to lead others in time); or (2) to devise a composite index that somehow combines measures of different dimensions of the problem. In the first case, there is the challenge of choosing just one revealing indicator; in the second, there is the challenge of selecting a range of measures with plausible weightings.

Dilemmas of this sort exist in every sphere of environmental evaluation (water quality, forest management, species at risk, climate, waste management, and so on), but they get progressively more serious as panoptic ambition rises, and one needs to capture ever more complex compound realities; for example, moving from air pollution, to pollution (in general), or to an even broader perspectives on environmental impacts or environmental sustainability and sustainable development.

When discussing the advantages of headline indicators or compound metrics, reference is often made to the importance of economic statistics—particularly gross domestic product (GDP) growth statistics—in policy discussions. Can we not devise one or more environmental indicators that will have a similar impact? In this context, it is worth noting that GDP is not an aggregate indicator: although the statistical processes and assumptions behind it are complex, GDP growth attempts to measure just one thing—the change in value of the total domestic economic output. In modern societies, GDP growth is linked with economic goods such as employment, living standards, and so on. So keeping GDP growth strong matters to governments, businesses, and individuals, and it is no wonder that people pay attention to this measure of the scale of economic activity. Of course, some critics (e.g., Stiglitz et al. 2009) say that GDP growth is not well correlated with increases in real social welfare, and that it should not dominate political discussion.

All this suggests reasons why it has proven so difficult to find an "environmental equivalent" of GDP. Society/nature interactions are complex and fragmented, so they are even less amenable to reduction to a single metric. No single measure comes anywhere close to capturing the many ways that environments are important to us; and the complex normative assumptions that underpin the composition of aggregate indicators make agreement on any one construct difficult. Moreover, in the environmental domain, the large-scale impacts of negative phenomena may take decades to become manifest, so the day-to-day relevance of trends revealed by an environmental indicator appear less salient to the audience than does an economic statistic such as GDP growth.

Yet even if one resists the lure of compound indexes and remains content with clusters of indicators for each relevant domain, the problems of establishing priorities and combining judgments about discrete areas into more comprehensive performance evaluations do not go away. There are so many diverse environmental concerns that some must be given more attention, and performance over different key subareas must be reconciled qualitatively—if not quantitatively—to present a synthesis of results.

The second tension relates to the relative merits of tracking environmental performance in physical or in monetary terms: do we focus on material units that relate to a biophysical process (such as parts per million of CO_2 in the atmosphere, kilograms of coal burned, or hectares of forests cleared) or on the financial dimensions of these physical processes (e.g., the value of resources harvested or left standing, the economic implications of pollution, and the price of remedial technology). A monetized approach has many advantages. It allows comparisons in a single metric. It facilitates integration with well-developed systems for tracking economic activity. Above all, because economic decisions (by business, governments, and consumers) are largely responsible for environmental destruction, prioritizing the economic value of resources and the financial consequences of environmental degradation (lost property values, health costs, and so on), "translates" environmental concerns into the metric used in everyday economic decisions—thereby making these impacts "real" for economic agents. Quite a bit of work has gone into such approaches, ranging from integrating environmental dimensions into cost/benefit analyses (to evaluate projects, programs, or policies) to proposals for "greening" the system of national accounts, and the development of economic indexes of sustainability such as the Genuine Savings Index (World Bank 2009). Whatever the precise approach, the point is that environmental performance is calculated in economic terms: by the economic benefits and costs to human societies. One problem with this approach is that many environmental goods and services are unpriced—there is no real market for them (consider, for example, the benefits that forests confer by absorbing air pollution or preventing flooding)—and increasingly sophisticated (and contested) techniques must be used to assign them economic values (Garrod and Willis 2000). Price information can often mislead about the underlying physical condition of the resource. In fisheries, for example, harvest income may rise even as the underlying stock is being depleted. More fundamentally, the disadvantage is that information about the real state of physical systems (i.e., about stocks and flows) is lost in the 'monetary conversion. The alternative is to track environmental issues in physical terms that relate most closely to the underlying biophysical processes. So, for environmental purposes, one might monitor timber resources by quantitative and qualitative measures of forest cover rather than by an economic assessment of the value of the resource. Yet there are also problems with the physical approach: there are many physical parameters of interest; making comparisons across problem areas is difficult; and interpreting their social significance can be a challenge. Historically, the tendency has been to track species of particular interest (fish, game, trees, and so on) and to focus management on these stocks, typically with deleterious long-term consequences for ecosystems as a whole (Gunderson and Holling 2002).

The third tension reflects the extent to which environmental assessment remains focused on purely environmental issues (including *environmental* sustainability, narrowly construed as an estimation of whether the environment can continue to support human activities and communities), or whether it reaches outward to connect explicitly with sustainable development (or sustainability writ large). In the first case, emphasis is placed on environmental pressures and conditions. Social circumstances are still part of the story—after all, it is people whose activities are disrupting environmental processes—but it is the environmental consequences (including the consequences for human health and welfare) to which attention is directed, and broader economic and societal implications are left to other assessment practices. In the second case, an attempt is made to integrate environment explicitly with social and economic dimensions: environment is evaluated as part of a more general effort to gauge societal sustainability. This is done in two main ways: either by presenting the environment alongside the other dimensions of sustainability, with separate indicators relating to the economic, social, and environmental issues; or by combining these elements into some form of synthetic appraisal of progress toward, or away from, sustainability as a whole.

Restricting the assessment to environmental issues makes the task more manageable. But the argument is that a true understanding of its meaning can be gained only by placing this information in the context of a broader vision of what is going on in society. On the other hand, almost any social, economic, or environmental trend can be linked in some way to sustainable development: so the assessment task for sustainability "writ large" becomes daunting indeed. A comprehensive approach can point toward the parallel presentation of dozens, perhaps hundreds, of indicators trying to capture important societal developments: see, for example, the indicator sets developed by the United Nations Commission on Sustainable Development (UNCSD; DESA 2007). One version of the UK sustainable development indicators included social issues such as rates of teenage pregnancy and car theft. Such exercises have the virtue of comprehensiveness, but they run a serious risk of losing sight of the forest for the trees. It is simply not clear what these broad sets of indicators really "indicate." But the aggregation of indicators into a compound economic, social, and environmental metric is even more problematic: such a compound index necessarily homogenizes information and assumes tradeoffs because any given overall rating can be composed of many different combinations of discrete scores on the underlying variables.

Two Recent Evaluation Approaches

To illustrate the complexity of making general judgments about environmental performance (to say nothing of sustainable development), I will examine briefly two of the widest-known approaches to measuring comparative environmental performance. Over the past decade, the concept of the "environmental footprint" has been taken up by government bodies, business, and the environmental movement. It can be calculated for countries, cities, households, and even individuals. This scalability and the powerful metaphor of measuring the human "footprint" on the planet have contributed to its success. It is an attempt to evaluate the biological resources required to support the production/consumption patterns of existing human societies. This then can be compared to available "biocapacity" to determine the extent to which a community is living within its ecological means. The footprint is presented in "global hectares per capita"—the area of "average" land and water required to provide the organic resources and waste assimilation services to support a particular community. This "ecological footprint of consumption" is the sum of six elements: cropland, grazing land, forests, fishing grounds, built-up land, and the carbon footprint. The per capita totals for different communities can then be contrasted. Thus, in 2007, France had an ecological footprint of 5.0 global hectares per capita, a bio-capacity of 3.0 global hectares per capita, and an ecological deficit of 2.0 hectares per capita. For Brazil, the figures were as follows: footprint 2.9; bio-capacity 9.0; and ecological reserve 6.1. Sometimes the results are presented in terms of the number of Earths that would be required to support humankind were everyone to live like the citizens of a particular country. For example, 4.3 Earths would be needed if we all lived like the average Canadian. Globally, footprint analysis suggests that by 1980 the total human footprint already had surpassed the existing planetary bio-capacity.

Pioneered by Mathis Wackernagel and William Rees in the mid-1990s, the ecological footprint has been subject to widespread debate (Wackernagel and Rees 1996; Rees 2000). Critics (van den Bergh and Verbruggen 1999; Fiala 2008; Opschoor 2000) refer to its "static character," neglect of major environmental impacts, assumption of "autarky," and the problematic accounting of carbon emissions. The first complaint is that the approach merely offers a snapshot of current conditions: technological optimists note a failure to acknowledge the potential for innovation, while more ecologically minded critics point to a neglect of the risks from environmental discontinuities and emphasize the importance of tracking the *resilience* of communities and ecosystems. The ecological footprint does focus on conditions at one moment in time, but by regularly repeating the studies, a more dynamic view can be created. Indeed, the Global Footprint Network is dedicated to this precise task. Footprint proponents also would argue that technological advances are incorporated as they are actually *deployed* into exiting production/consumption processes. There is no assumption that further progress is impossible, but the analysis focuses on the current state of play. Its purpose is to determine the extent to which *current* practices are outstripping available bio-capacity. Some indication of social and ecological vulnerability is provided by this measure of the (excess) demands being placed on available bio-resources. And of course, one indicator cannot be expected to track everything.

A second charge is that the ecological footprint misses a lot. Water flows are ignored; there is no direct way to incorporate toxic emissions;

there is no direct accounting of soil or nonbiological resources essential to production, such as aggregates and minerals; and biodiversity issues are not addressed explicitly. Still, some of these items are included indirectly (for example, the land requirements and carbon emissions associated with mining and the decline in bio-capacity caused by pollution). Moreover, proponents argue that the existence of such additional impacts simply reinforce the essential message of the footprint—that we are already exceeding what the planet can sustain.

Another objection is that the footprint seems to imply that each human political unit (i.e., a city, province, or country) should live on its own bio-resources. But cities have always depended on a large hinterland from which to draw sustenance. And is it really surprising that a small, densely populated country like the Netherlands requires biological resources from beyond its borders? Trade allows populations to exchange with others in order to gain access to resources beyond what is locally available. To be fair, the ecological footprint has no necessary normative bias against trade; but it does highlight the uneven international pattern of biological resource appropriation and the fact that *overall,* we are demanding more biological resources than the planet can supply. Yet this does mean that as an assessment of *comparative* performance, its utility is limited.

Perhaps the most significant issue relates to carbon emissions. The ecological footprint handles this by adding to the total footprint a forested area sufficient to draw down carbon released from fossil sources. This is consistent with the idea of assessing the land area required to provide resources and absorb all wastes, and is meant as a bio-resource accounting convention—a way to track our dependence on fossil energy derived ultimately from biological processes—and *not* as a practical solution to climate change. (No community actually burns fossil fuels for energy and then plants trees to soak up the carbon.) The trees would not hold the carbon forever, and on a global level, annual emissions are far too high for biological sinks to manage, as the ecological footprint demonstrates. But this convention is not without difficulties. In the first place, there are continuing debates about the circumstances in which real forests serve as long-term carbon sinks: so one could argue about the quantity and location of the land necessary to accomplish this notational offsetting. Moreover, other accounting approaches are possible: for example, one could take the land area required to produce bio-energy crops adequate to displace the fossil fuels that were actually consumed. And the Global Footprint Network coyly suggests another alternative: estimating the land that would be needed to allow the long-term regeneration of the fossil fuels being consumed. The inclusion of this "carbon footprint" is not trivial. While it may amount to only 5 to 10 percent of the total footprint of poorer countries in Africa, it represents *more than half* the footprint of many developed states. The 2007 accounts, for example, show that for Sweden, 46 percent of the recorded footprint was produced by carbon emissions; for Germany, the figure is 53 percent; and for the United States, it is 70 percent. Without this item, the footprints of the richest countries would drop dramatically. For instance, Germany's ecological deficit would shrink from 3.2 to 0.8 hectares per person; China's ecological deficit of 1.2 hectares per person would be eliminated altogether; and the United States would move from an ecological deficit of 4.1 to an ecological reserve of 1.47 hectares per capita. Globally, the result would be a shift from a 2007 deficit of 0.9 hectares per person to a reserve of 0.54 hectares per person.

The underlying difficulty is the attempt to represent everything in a single metric—even when that metric is something as basic as biologically productive area. In this context, it is not clear that we gain a lot by converting carbon emissions into land areas and "keeping the books" in hectares.

Scientific assessments have documented the risks of climate change; and we know that to avoid serious problems, we virtually need to eliminate greenhouse gas (GHG) emissions. So why not track progress in metrics that are closer to the problem—and are more transparent—such as absolute CO_2 emissions, CO_2 emissions per capita, and CO_2 emissions per unit of GDP?

The ecological footprint is a useful tool for illustrating the extent to which current consumption and production patterns exceed the long-term bio-capacity at the local or global level. Even if the calculations are off by a wide margin, they suggest a definite trend, where population growth and economic activity threaten to swamp the world's bio-potential. They show how much our civilization relies on bio-capacity accumulated from previous epochs (for the carbon emissions that must be absorbed are being released from fossil fuels created by earlier "bio-capacity"). And they also suggest that climate change is far from the whole story: even without the carbon footprint, we are pushing the existing bio-envelope. With respect to discriminating *among* countries, however, the ecological footprint appears less valuable. Above all, it does not give policymakers a very sensitive tool to guide their response to the environmental issue.

Another recent initiative is the Environmental Performance Index (EPI), developed by researchers at the Yale Center for Environmental Law and Policy and the Center for International Earth Science Information Network at Columbia University (Emerson et al. 2010). The 2010 version is the result of a decade of work that included a pilot Environmental Sustainability Index and earlier iterations of the EPI. The 2010 EPI tracks the performance of 163 countries in the attainment of environmental policy goals. It deals with two basic themes: "Environmental Health," which assesses environmental stressors to human health; and "Ecosystem Vitality," which gauges "ecosystem health and natural resource management." For each of the 25 indicators incorporated in the index, performance is assessed with relation to attainment of a policy target. These targets are defined by international treaties or agreements (where these exist) or developed from standards set by international organizations, governments, and expert judgment. On the Environmental Health side, half the weighting is allocated to the "environmental burden of health" (25 percent of the total EPI), with the remaining half split between water and air pollution impacts on humans (12.5 percent of the total EPI each). On the Ecosystem Vitality side, half the weighting goes to climate change (25 percent of the total EPI), with the remainder split among six themes: agriculture, fisheries, forestry, biodiversity and habitat, water, and air pollution (4.2 percent of the total EPI each). Five indicators are used to generate the results for Environmental Health, and twenty for the more diverse Ecosystem Vitality.

The top performers in this index were Iceland, Switzerland, Costa Rica, Sweden, and Norway. The lowest-ranked countries were Togo, Angola, Mauritania, the Central African Republic, and Sierra Leone. OECD countries generally performed well (four of them are in the top five, most are in the top third, and all are in the top half), buoyed up by their strong performance on Environmental Health (the result of effective heath care systems and established environmental policy institutions). But performance on the Ecosystem Vitality component was more uneven. The larger developed economies—Germany (ranked 17), Japan (20), and especially the United States (61)—were pulled down by their performance in the areas of air pollution and climate change. The Least Developed Countries (LDCs) fared poorly, reflecting their lack of resources to meet basic needs. Brazil, China, and India were ranked 62, 121, and 123, respectively.

So how useful is the EPI as a measure of environmental performance? The index has been laboriously constructed by a team that explicitly seeks policy relevance and is committed to continuous improvement. It includes many important environmental issues and takes great care to make explicit its methodology, data sources, and normative assumptions. Its creators acknowledge serious gaps in the data for many countries. A detailed sensitivity analysis (involving over 300 simulations), conducted by the Joint Research Centre of the European Commission in Ispra, Italy, argues that EPI results for the majority of countries are reasonably robust: that is to say, the EPI rank lies within the confidence interval for the median rank and this confidence interval is less than twenty ranks in size (Saisana and Saltelli 2010). However, this was not the case for sixty countries (including Japan, Spain, the United Kingdom, and the United States) and "any inference on those country's rank should be formulated with great caution" (p. 3). Moreover, the analysis showed that overall results were driven mainly by Environmental Health. Correlation between the EPI and the climate change indicators, for example, was almost nonexistent.

The EPI indicators track many different sorts of things, combining data on the signature of international agreements with GHG emissions. On some indicators (such as forestry), there is little discrimination, as many countries get full marks: but in light of the wide spread and continuing criticism of international forestry practices this appears implausible.

Assessing Comparative Assessment

These two examples point to the difficulties and complexities associated with attempts to derive an overall measure of environmental performance, to say nothing of a more comprehensive assessment of sustainable development. Several issues stand out. First, because "the environment" is not a single thing and human societies interrelate with their natural surroundings on multiple levels, the effort to reduce (or to dramatically simplify) these dimensions seems destined to obscure as much as it illuminates. While such exercises are suggestive, they do not indicate firm conclusions. Second, because there are enormous socioeconomic differences between the most highly developed states and the LDCs, it is not clear what is really gained by assessing them on a single environmental continuum. Perhaps (at least for the foreseeable future) we should limit our ambitions to mapping performance in more specific domains and across more similar jurisdictions. If we consider specific areas, such as resource management (fisheries or forests, for example), nature protection, the management of SOx emissions, disposal of toxic wastes, and so on, and focus comparisons on countries at similar stages of economic development, things will be more manageable. Of course, even here, underlying differences in ecological endowments, land area and population, and historically inherited economic structures are important. Indeed, states' juridical status in international law is just about the only dimension on which they are truly equal. Still, reasonable performance comparisons are possible. Students of comparative politics make such claims all the time in other areas (such as economic or social policy); and in the environmental realm, too, we are used to seeing judgments that support claims, such as "Germany has performed better than the United Kingdom in promoting renewable energy" or "Canada uses more water per capita than any other OECD state." Of course, even these types of judgments require nuance—for although Germany has the greatest market penetration of photovoltaics, that achievement has cost the German taxpayer dearly. So is environmental performance only about environmental results (say, the quantity of nuclear or coal-fired generation displaced) or it is also about costs? And while Canada uses a lot of water, it also has the highest per capita water resources in the OECD, and much of the recorded usage is for once-through cooling at power stations (no cooling towers here): so does the high per capita water "consumption" really matter?

Climate change is one domain where the inherent complexity of performance assessment is readily manifest. But what does it mean to assess comparative national performance in relation to climate change? What are suitable indicators of comparative performance on this topic? Let us consider issues surrounding some of the possibilities.

One place to start is with *national emissions*. Here we could focus on the absolute GHG emissions of states, where those with lower emissions are judged to have performed well compared to those with higher emissions. Of course, states vary greatly with respect to population, level of economic development, natural resource endowments, and industrial structure, so an approach focused on absolute emissions seems in some ways to be more about contingency than performance. Turning from absolute to per capita emissions puts the population issue to one side; it also accords with notions of equity—because there is no a priori justification for one group of humans to be allowed to do more of an activity that harms the common weal than another. Tracking emissions per unit of GDP (the carbon intensity of output) reveals the carbon efficiency of the economy, and factoring in population gives the carbon intensity of per capita output. Yet there are many ways to challenge such an approach. One could argue, for example, that there is no reason to privilege current emissions. Climate change is caused by the atmospheric accumulation of GHGs and the saturation of sinks, so cumulative emissions are a more accurate indication of the national contribution to the problem and thus a preferred measure of environmental performance. Yet it seems unfair to count emissions (and start allocating responsibility) before climate change was even recognized as a problem. Countries burned their fossil fuels in "good faith." So perhaps we should only begin counting cumulative emissions from the point that the world recognized climate change to be a significant issue—for instance, the 1988 Toronto Conference on the Changing Atmosphere, or the 1992 signing of the United Nations Framework Convention on Climate Change (UNFCCC). Alternatively, one could argue that it is not the emissions time frame, but the focus on releases from national territory that is the problem. In a globalized world, production and consumption may be separated by thousands of miles, but it does not matter where GHGs enter the atmosphere. So goods produced in China to maintain the lifestyles of Western consumers should really count as carbon emissions from the developed states. Otherwise, a country's performance can be improved simply by buying carbon intensive goods from abroad. On this reading, the United Kingdom's GHG emission reductions during the Kyoto Protocol period largely evaporate, swamped by foreign goods imports. But could not Australia then claim that the emissions from its coal mining should be added to China's total—for this is the destination of its fossil energy exports?

Another approach would focus on *the emissions trajectory*: has the emission growth rate slowed? Have emissions stabilized or declined? Or is this true in particular sectors, or for particular gases? Here, performance is judged as superior if the curve has been bent further. Again, per capita and per unit GDP variants could be applied. But emissions can decline for many reasons other than climate policy: the UK "dash for gas" (which resulted partially from Prime Minister Margaret Thatcher's antipathy toward the coal miners) and structural changes following German unification accounted for at least half the emissions reductions in these European climate change leaders. On the other hand, Canada's GHG emissions increased because of the "bad luck" of an oil and gas boom. So the change in emissions does not necessarily capture "performance" without a great deal of supplementary analysis and explanation.

The *policy response* is the other obvious touchstone for performance assessment. What have governments done (or not done) to address this environmental problem? Assessing comparative performance on the basis of the policy response focuses attention on the seriousness of governmental engagement with the problem. Comparisons can be made in terms of the design and implementation of regulatory frameworks and policy instruments. One could examine the signature and ratification of international agreements, the establishment of carbon pricing, research and development (R&D) expenditure on low carbon technologies, public education around climate risks and GHG mitigation, and so on. Better performance is here associated with doing more, taking more vigorous or sustained action, and being willing to devote substantial societal resources to address the problem (Duit 2005). And yet in many cases, these signs may be ambiguous. Consider the signature and ratification of international agreements. They testify to a willingness to accept an international consensus, but little more. And a higher climate R&D budget may indicate a commitment to act, but it may also indicate almost the reverse, with the decision to commit resources to R&D acting as a substitute for regulatory action. And this has actually been the case with many of the states (the United States, Australia, Canada) that committed large expenditures to support carbon capture and storage (CCS) while having difficulty with the Kyoto Protocol and binding GHG reduction targets (Meadowcroft and Langhelle 2010). In general, more policy action seems better than less. But if more action does not actually lead to substantive improvements (verifiable emissions reductions), is it really a better performance than one achieved by other jurisdictions that achieve a similar result by doing less?

So far, we have considered emissions and policy initiatives, which could be classed as "pressures" and "responses" within the OECD indicator schema. But what about the other two elements in that framework: "state" and "drivers"? In this context, *"state"* could be associated with the condition of the climate system: perhaps the concentration of CO_2 in the atmosphere, the temperature increase above preindustrial levels, the transformation of regional and local climate patterns, sea level increase, the scale of impacts on the biosphere, or human health, settlements, livelihoods, and so on. Three sorts of difficulty affect these factors as performance measures in the climate case. First, there is the time lag: impacts of elevated GHGs take many years to feed through to the climate system. Second, scientific uncertainties separating "signals" from "noise" in these complex coupled systems are challenging, so unambiguous results are hard to achieve. Third, and most important, the global nature of the climate system makes it impossible to link national behavior to national effects. The changing "state" of the climate system is driven by the behavior of all nations; and the different climate that each one receives is the products of the behavior of the whole group. So climate "state" gives little purchase to define national performance—unless, of course, one approaches the issue through the lens of differential responsibility; but that drives one back toward emissions.

So that leaves *"drivers"*: these are underling currents that give rise to increased pressures. These can be defined in different ways that are more or less directly related to the immediate pressure. Increased fossil fuel usage is a relatively direct "driving force"; beyond that, we might define increased mobility, urbanization, and changing social structure (leading to more households); and at a still more general level lie increases in population and material consumption. At first glance, many of these "driving forces" appear to be implausible elements for assessing state performance with respect to climate change, especially when they are cast at an increasingly general level. They have many impacts (besides climate change), and are rooted deep in societal economic, social, and political processes. A state cannot alter at will its industrial or social structure, nor can it shift its demographic trajectory. And yet this argument weakens over time: over the course of several decades, states can change their economic and social structures or alter their population dynamics. Whereas in the short term, "driving forces" appear as natural constants which policymakers must live with, in the longer term it is possible to act on them, either (a) by slowing or reversing the trend or (b) by cutting linkage to undesirable pressures. So, as we think about managing climate change mitigation over the course of half a century and more, policies to reduce such drivers or their impacts may acquire significance for performance assessment. In other words, measures of material consumption (or its impact) and population growth and levels (or its impact) may be able to tell us something about national *performance* in relation to climate change policy.

In reality, of course, we may be interested in using data on all these sorts of approaches to evaluate performance. But this example shows the complexity and contestability of judgments about environmental performance. It points to the importance of defining the standards on which judgments are made: in reference to a self-defined target (individual or collective), to a ranking exercise (in relation to others in a group), or to an authoritatively defined target based on scientific assessment. Above all, it shows that performance judgments require the assessment of complex causal interconnections, process tracking, and the development of an analysis of what happened and why. This suggests reasons for the difficulties in understanding environmental performance merely through quantitative indexes, even with generous footnotes explaining data inconsistencies. And it may point to why it is so difficult to develop large-N analysis that shows more than trivial results.

Looking Forward

Over the past few decades, enormous progress has been made in understanding the scope of environmental problems and monitoring their impacts. We know more than ever about the environmental performance of states and other large-scale political units. In practical terms, a great contribution has been made by monitoring and assessment bodies (such as the European Environment Agency) in standardizing reporting and making data available. Certainly, we now know enough to appreciate the larger phenomena: the gains that have been made in particular societies in controlling specific environmental impacts; the major differences in the way human pressures are manifest in the richest and the poorest parts of the world; the continuing burden that even the most environmentally concerned societies are placing on local and global environments; the significant performance gaps between leaders and laggards even among developed states; and the rapid increase in global pressures as large developing countries move toward Northern production and consumption patterns. We have a general idea about which countries have taken environmental protection most seriously and which lag behind in their political/economic peer groups (consider chapter 3, "Explaining Environmental Policy Adoptions: A Comparative Analysis of Policy Developments in 24 OECD Countries," chapter 4, and chapter 6, "Early Bird or Copycat, Leader or Laggard? A Comparison of Cross-National Patterns of Environmental Policy Change" in this book).

On the other hand, we do not have a very good grasp of the underlying causes of these performance differences and similarities. Detailed case studies and small group comparisons have teased out a variety of significant factors that relate to ideas, interests, and institutions. But so far, large-scale quantitative studies have not been enormously illuminating. There are tantalizing leads: for example, it seems that democratic political cultures that accept an active role for the state in social life are more willing to embrace substantial environmental commitments; and those with corporatist decision frameworks and more generous welfare state models also tend to be environmental leaders (Jahn 1998; Scruggs 2003; and chapter 4 of this volume). Fiorino (2011) provides an excellent survey of the literature on national environmental performance, reviewing research that explores linkages to (1) economic development (for example, Dasgupta et al. 2002; Esty and Porter 2005); (2) democratic governance (Li and Reuveny 2006); (3) political institutions (Scruggs 2003; Harrison and Sundstrom 2010); and (4) societal capacity and policy diffusion (Jänicke and Weidner 1997; Jörgens 2004). Yet despite a growing volume of work, a sophisticated theoretical understanding of comparative environmental performance remains elusive (Duit 2005; Fiorino 2011).

There are several obvious obstacles to further progress. First, there are weaknesses in data. There is uneven and inconsistent geographical coverage and many topic areas remain neglected (Srebotnjak 2007). Time series data is often unavailable, and budget cuts and periodic reorganizations of government departments affect consistency. Thus, for example, no robust data on state environmental expenditure exists, so we don't even know what governments are spending on environmental issues. And, of course, data on environmental impacts lags decades behind the deployment of the practices and technologies that are suspected of causing harm. Second, there are serious conceptual challenges about defining environmental performance: what issues should count, how are they interrelated, and how important are the different elements? These confusions mean that there is often little agreement over what is being measured: Different results relate to different phenomena, and the quest for more general insights is frustrated (Neumayer 2004; Böhringer and Jochem 2007; Mayer 2008; Fiorino 2011). Third, social/ecological linkages are complex and varied, and they result from co-evolutionary interactions: this makes the identification of simple causal mechanisms difficult. Many results are overdetermined, and the huge array of environmental contextual issues makes almost every case appear *sui generis*.

Nevertheless, there are several promising avenues to pursue. First, if one is to consider overall environmental performance (as opposed to performance in relation to specific topics, such as waste management, SOx emissions, water quality, and so on), it may be more productive to disaggregate the concept of "environmental performance" into several distinct dimensions. One suggestion would be to consider four interdependent areas:

1. *Environmental governance*. This relates to institutions and practices of governance, democracy, and participation, the effectiveness of policy, and the choice of policy instruments.

It deals with questions such as: What mechanisms are in place to safeguard the environment and prevent harm to humans and ecosystems? Are they effective? Are they enforced? What do they cost? To what extent is there conscious societal deliberation about environmental issues? How important is public education about the environment? Is scientific research a priority? What paradigms guide environmental policy? Is environmental policy integrated with other policy areas? Has the polity engaged with the imperative of decarbonization? What is the level of "capacity" for environmental policy (Jänicke and Weidner 1997)?

2. *Quality of the lived environment.* This relates to the environmental quality experienced directly by human beings. It deals with exposure to toxic chemicals, the nature of the built environment, and impacts on human health and well-being (for example, from pollutants, traffic, and noise). It deals with questions such as: What are the impacts of air and water quality on human health and activities? What is the provision of parks and green spaces? How do humans experience nature aesthetically and culturally? How does the environment relate to "quality of life"? In other words, how does the environment affect the lives of citizens?

3. *Ecosystems and natural resources.* This concerns the health of ecosystems and the management of resources on the national territory. It relates to the vitality of ecosystems and the extent to which they can support the long-term development of the community. Key questions include: What is the state of natural ecosystems, including those that provide resources for the community? How are these managed and protected? Are resources managed in a sustainable fashion? What measures exist to protect species and habitats?

4. *Contribution to global environmental issues.* The problem here is to track the interconnection between national behavior and global environmental issues. It relates to the import and export of environmental burdens and the contribution (or otherwise) to international efforts to manage environmental problems. Important questions include: What is the net contribution to global environmental burdens? What pressures does the national community impose on global ecosystems? What impacts does it receive from other states? What sort of contribution is being made to cooperate to manage these issues? What principles guide the state's international environmental behavior?

While there is some overlap among these issues, they also relate to somewhat different understandings of what environmental performance is all about. The first is concerned with the way a country is organized to manage its behavior vis-à-vis the environment; the second focuses on the lived experience of the environment; the third relates to ecosystems and resources; and the fourth deals with international dimensions. A country that performs well under one heading may not do so well on another: for example, it might invest much effort in protecting human health, while it remains a net contributor to global environmental burdens (such as climate change or pressures on global fish stocks). In contrast to the EPI discussed earlier in this chapter, this approach disaggregates environmental performance into four (rather than two) major dimensions; it does not aim to produce a compound score for each dimension, or indeed for performance as a whole; and it is more qualitative and open textured.

A second point is that whatever the dimension that is under investigation, it is important to ground environmental performance evaluation in an understanding of interlinked environmental and social processes. This includes the use of ecosystem approaches to track how human practices interact with ecological systems. In addition, it implies the invocation of knowledge from the natural and social sciences to help define standards against which performance can be measured. Of course, it is important to understand performance in relation to self-defined goals, the existing values of a community, the promises of political leaders, or an estimate of the "politically possible." But is also important to understand performance in relation to natural limits, "tipping points," and potential consequences, as identified by ecological perspectives.

Finally, in light of unambiguous evidence that the *scale* of the human impacts on the natural world are continuing to increase, particular attention should be paid to the analysis of the linkages between economic activity and environmental burdens. A substantial "decoupling" of continuing economic activity from environmental impacts appears to be a critical element in meeting the environmental challenge. Indeed, if economic growth cannot be severed from ever-increasing physical pressures on global ecosystems, then sooner or later, that growth itself will have to cease. So an analysis of decoupling (OECD 2002, UNEP 2011)—of when and where it occurs, of the conditions under which it is possible, and the tracing of international linkages and the untangling of diverse impacts—is essential to environmental performance assessment that really matters. Of course, this issue is rather complex and nuanced, but is that not always the case with comparative investigation?

References

Alfsen, Knut H., and Mads Greaker. 2007. From Natural Resources and Environmental Accounting to Construction of Indicators for Sustainable Development. *Ecological Economics* 61:600–610.

Atici, C. 2009. Pollution without Subsidy? What Is the Environmental Performance Index Overlooking? *Ecological Economics* 68:1903–1907.

Atkinson, Giles, Kirk Hamilton, Giovanni Ruta, and Dominique Van Der Mensbrugghe. 2010. Trade in 'Virtual Carbon': Empirical Results and Implications for Policy. Background Paper to the 2010 World Development Report, World Bank. http://www.iadb.org/intal/intalcdi/PE/2010/04559.pdf.

Böhringer, C., and P. Jochem. 2007. Measuring the Immeasurable—A Survey of Sustainability Indices. *Ecological Economics* 63:1–8.

Dasgupta, Susmita, Benoit Laplante, Hua Wang, and David Wheeler. 2002. Confronting the Environmental Kuznets Curve. *Journal of Economic Perspectives* 16(1): 147–168.

DESA. 2007. *Indicators of Sustainable Development: Guidelines and Methodologies*. New York: United Nations Department of Economic and Social Affairs.

Duit, Andreas. 2005. Understanding Environmental Performance of States: An Institution-Centred Approach and Some Difficulties. QOG working paper 2005:7.

Emerson, J., D. Esty, M. Levy, C. Kim, V. Mara, A. de Sherbinin, and T. Srebotnjak. 2010. *2010 Environmental Performance Index*. New Haven: Yale Center for Environmental Law and Policy.

Esty, Daniel, and Michael Porter. 2005. National Environmental Performance: An Empirical Analysis of Policy Results and Determinants. *Environment and Development Economics* 10:391–434.

Fiala, Nathan. 2008. Measuring Sustainability: Why the Ecological Footprint is Bad Economics and Bad Environmental Science. *Ecological Economics* 67:519–525.

Fiorino, Daniel. 2011. Explaining National Environmental Policy Performance: Approaches, Evidence and Implications. *Policy Sciences*, 44:367–389.

Garrod, G., and K. Willis. 2000. *Economic Valuation of the Environment: Methods and Case Studies*. Cheltenham, UK: Edward Elgar.

Gunderson, Lance, and Buzz C. Holling, eds. 2002. *Panarchy: Understanding Transformations in Human and Natural Systems*. Washington, DC: Island Press.

Hanf, Kenneth, and Alf Inge Jansen. 1998. *Governance and Environment in Western Europe: Politics, Policy, and Administration*. London: Longman.

Harrison, Kathryn, and Lisa Sundstrom, eds. 2010. *Global Commons, Domestic Decisions: The Comparative Politics of Climate Change*. Cambridge, MA: MIT Press.

Jahn, Detlef. 1998. Environmental Performance and Policy Regimes: Explaining Variations in 18 OECD Countries. *Policy Sciences* 31:107–131.

Jänicke, Martin, and Helmut Weidner, eds. 1997. *National Environmental Policies: A Comparative Study of Capacity Building*. Berlin: Springer-Verlag.

Jörgens, Helge. 2004. Governance by Diffusion: Implementing Global Norms Through Cross-National Imitation and Learning. In *Governance for Sustainable Development: Adapting Form to Function*, ed. William Lafferty. Cheltenham, UK: Edward Elgar.

Kejun, J., A. Cosbey, and D. Murphy. 2008. *Embodied Carbon in Traded Goods. IISD Report*. International Institute for Sustainable Development.

Li, Q., and R. Reuveny. 2006. Democracy and Environmental Degradation. *International Studies Quarterly* 50:935–956.

Mayer, Audrey. 2008. Strengths and Weaknesses of Common Sustainability Indices for Multidimensional Systems. *Environment International* 34:277–291.

McGillivray, M. 1991. The Human Development Index: Yet Another Redundant Composite Development Indicator? *World Development* 19(10): 1461–1468.

Meadowcroft, James. 2012. Greening the State. In *Comparative Environmental Politics: Theory, Practice, and Prospects*, ed. Paul F. Steinberg and Stacy D. VanDeveer, 63–87. Cambridge, MA: MIT Press.

Meadowcroft, James, and Oluf Langhelle. 2010. *Caching the Carbon: The Politics and Policy of Carbon Capture and Storage*. Cheltenham, UK: Edward Elgar.

Mickwitz, P. 2006. *Environmental Policy Evaluation: Concepts and Practice*. Helsinki: Finnish Society of Sciences and Letters.

Neumayer, Eric. 2004. Indicators of Sustainability. In *International Yearbook of Environmental and Resource Economics 2004/05*, ed. Thomas Tietenberg and Henk Folmer, 139–188. Cheltenham, UK: Edward Elgar.

Niemeijer, D. 2002. Developing Indicators for Environmental Policy: Data-Driven and Theory-Driven Approaches Examined by Example. *Environmental Science & Policy* 5:91–103.

OECD. 2002. *Indicators to Measure Decoupling of Environmental Pressure and Economic Growth*. Paris: OECD.

OECD. 2003. *OECD Environmental Indicators: Development, Measurement, and Use*. Paris: OECD.

OECD. 2008. *OECD Key Environmental Indicators*. Paris: OECD.

Opschoor, H. 2000. The Ecological Footprint: Measuring Rod or Metaphor? *Ecological Economics* 32:363–365.

Pierce, D., and J. Warford. 1993. *World without End: Economics, Environment, and Sustainable Development*. Oxford: World Bank and Oxford University Press.

Pillarisetti, J., and J. van den Bergh. 2010. Sustainable Nations: What Do Aggregate Indexes Tell Us? *Environment Development and Sustainability* 12:49–62.

RAE. 2010. *Global Water Security and Engineering Perspective*. London: Royal Academy of Engineering.

Rees, W. 2000. Eco-footprint Analysis: Merits and Brickbats. *Ecological Economics* 32:371–374.

Saisana, M., and A. Saltelli. 2010. Uncertainty and Sensitivity Analysis of the 2010 Environmental Performance Index, Joint Research Centre (JRC) report— European Commission, 2010.

Scruggs, Lyle. 2003. *Sustaining Abundance: Environmental Performance in Industrial Democracies*. Cambridge: Cambridge University Press.

Singh, R., H. Murty, S. Gupta, and A. Dikshi. 2009. An Overview of Sustainability Assessment Methodologies. *Ecological Indicators* 9:189–212.

Spangenberg, Joachim H. 2002. Environmental Space and the Prism of Sustainability: Frameworks for Indicators Measuring Sustainable Development. *Ecological Indicators* 2:295–309.

Srebotnjak, Tanja. 2007. The Role of Environmental Statisticians in Environmental Policy: The Case of Performance Measurement. *Environmental Science & Policy* 10:405–418.

Stiglitz, Joseph E, Amartya Sen, and Jean-Paul Fitoussi. 2009. Report of the Commission on the Measurement of Economic Performance and Social Progress. http://www.stiglitz-sen-fitoussi.fr/documents/rapport_anglais.pdf.

UNEP. 2011. *Decoupling Natural Resource Use and Environmental Impacts From Economic Growth. Report drafted by the Decoupling Working Group of the International Resource Panel.* United Nations Environment Program.

van de Kerk, G., and A. Manuel. 2010. *Short Survey of Relevant Indexes and Sets of Indicators Concerning Development towards Sustainability.* Brussels: Northern Alliance for Sustainability (ANPED).

van den Bergh, J., and H. Verbruggen. 1999. Spatial Sustainability, Trade and Indicators: An Evaluation of the "Ecological Footprint." *Ecological Economics* 69:61–72.

van Tatenhove, Jan, Bas Arts, and Pieter Leroy, eds. 2001. *Political Modernisation and the Environment: The Renewal of Environmental Policy Arrangements.* Dordrecht: Kluwer Academic.

Wackernagel, Mathis, Chad Monfreda, Dan Moran, Paul Wermer, Steve Goldfinger, Diana Deumling, and Michael Murray. 2005. *National Footprint and Biocapaciy Accounts 2005: The Underlying Calculation Method.* Oakland: Global Footprint Network.

Wackernagel, Mathis, and William Rees. 1996. *Our Ecological Footprint: Reducing Human Impact on the Earth, Gabriola Island.* Gabriola Island, Canada: New Society Publishers.

World Bank. 2009. *Adjusted Net Saving.* Washington, DC: World Bank.

World Bank. 2010. *Environmental Valuation and Greening the National Accounts: Challenges and Initial Practical Steps.* Washington, DC: World Bank.

3

Explaining Environmental Policy Adoption: A Comparative Analysis of Policy Developments in Twenty-Four OECD Countries

Christoph Knill, Susumu Shikano, and Jale Tosun

In the age of economic globalization and international cooperation, it seems reasonable to hypothesize that there exist equal global dynamics of environmental policymaking and politics. One of the implications of such global dynamics is that policy choices in one country are increasingly influenced by previous or current choices in other countries, possibly leading to cross-national policy convergence (Bennett 1991). This basic hypothesis has been evaluated by a considerable number of empirical studies of cross-national policy diffusion (see, for example, Braun and Gilardi 2006; Elkins et al. 2006; Gilardi 2005, 2008; Gleditsch and Ward 2006; Huber 2008; Meseguer Yebra 2004, 2006; Meseguer Yebra and Gilardi 2009; Simmons and Elkins 2004). Diffusion studies typically concentrate on the adoption of policy innovations by governments. In this context, it is important to understand that the aggregate empirical phenomenon is the spread of a given policy innovation across countries (i.e., diffusion), while the analytical focus lies on the individual governments' decisions whether and when to adopt the policy innovation of interest. It is this latter aspect (i.e., the adoption of policy innovations by policy-makers in the individual countries) that diffusion studies are interested in explaining by means of a range of independent variables that take into account both the national and international arenas. In a first step, diffusion studies describe these adoption patterns. These are based on the cumulative number of countries that have adopted a given policy until a time t. In most cases, this produces an S-shaped curve, implying that the adoption rises slowly first, when relatively few governments introduce the policy. Then the curve takes off as more governments adopt it. After a while, however, most governments will have adopted it and the diffusion curve begins to level off (Gray 1973). In so doing, diffusion studies then formulate the following questions: Why do some countries adopt a certain policy while others do not? Which countries are faster in adopting a certain policy? The overarching finding of this body of literature is that such transnational diffusion processes have indeed affected governments' policy decisions (Garrett et al. 2008, 360).

Despite this consistent overall finding, the empirical testing of policy diffusion suffers from a variety of problems. In our view, following Meseguer Yebra and Gilard; (2005), the most pressing one is the fact that scholars mostly study explosive diffusion processes, which is characterized by governments swiftly adopting a given policy innovation, often promoted by an international organization, with the result that most of or even all the countries covered in the analysis end up having adopted the policy innovation in question. An example of such an explosive diffusion refers to strategies for sustainable development, which were developed and promoted by the United Nations (see, e.g., Tews et al. 2003). In most of these cases, all countries covered by the sample join in the diffusion processes at either an earlier or later point in time, thus leading to the outcome that over the observation period, all countries will eventually adopt the policy innovation in question. According to prominent authors in the field, such as Meseguer Yebra and Gilardi (2009) and Strang and Soule (1998), the fact that the event of interest occurs in every country represents a selection bias (i.e., the sample includes positive cases only where policy adoption—and therefore policy diffusion—actually occurred). The fact that negative outcomes (i.e., nonadoption) are not considered by such a research focus may lead to wrong conclusions. Clearly, the focus on the explosive spreading of policies and policy innovations in particular is justified because it lies at the heart of diffusion research (Gray 1973). Also, scholars are often confronted with severe restrictions regarding access to appropriate data. Therefore, a specific theoretical interest in policy innovations and practical constraints certainly require a concise empirical focus. At the same time, however, the use of highly specific empirical information impedes the evaluation of whether states indeed increasingly tend to adopt the same set of policies. We contend that the growing cross-country similarity across countries is the broader empirical phenomenon arising from the most basic assumption of diffusion studies—namely, whether policy choices in country are influenced by the policy choices of other countries, also known as "Galton's problem" in the literature (see Braun and Gilardi 2006; Jahn 2006). In this chapter, we systematically address the overall finding of the diffusion literature and ask whether a similarity in the states' adoption of policies still persists if we extend our empirical basis to include policies that are not commonly associated with explosive diffusion. We aim to explore the spread of environmental policies, which has been particularly troubled with measurement problems (see chapter 2, "Comparing Environmental Performance," in this volume). We base this study on the adoption of forty environmental policy measures in twenty-four countries (i.e., twenty-one European countries as well as the United States, Mexico, and Japan) in the period from 1970 to 2000. The data allows us to assess the adoption process of both swiftly and slowly diffusing policies, which we expect will reduce the abovementioned problem of selection bias. Note that chapter 6, "Early Bird or Copycat, Leader or Laggard? A Comparison of Cross-National Patterns of Environmental Policy Change," provides a more differentiated analysis of a subset of the data employed here to illustrate policy change and its sequence in more general terms instead of focusing specifically on policy adoption.

Procedure and Structure of the Research

With regard to the statistical analysis, we rely on an item response model, which gives us the opportunity to model a country's likelihood of adopting environmental policies as a function of interaction between policy item–specific and country-specific factors. Item-specific factors refer to the contagiousness of a policy and hence describe the extent to which the adoption of a given policy is more or less difficult. In contrast, we use the concept of adoption resistance to cover the country-specific factors that affect the probability that—given a certain level of contagiousness— a country actually adopts the policy in question. The item response model enables us to decompose policy adoption as an observable phenomenon into two unobservable factors: contagiousness and adoption resistance. This approach takes into account the null hypothesis that domestic policy change depends on independent governmental decision making. This nuanced conceptual perspective on policy adoption provides us with clear-cut theoretical expectations.

This chapter is structured as follows: First, we present a short overview of the most relevant empirical studies addressing policy diffusion. Next, we define our central analytical concepts and then propose an explanatory model. Subsequently, we provide detailed information about the adoption of the environmental policy measures on which this analysis rests, and proceed to test our hypotheses. Finally, we discuss our findings and point to open questions to set the stage for future research.

Contribution to the Literature

The term "diffusion" is generally defined as the socially mediated spread of policies across and within political systems, including communication and learning processes across and within populations of adopters (Rogers 2003). In the past few years, research on policy diffusion has proliferated in political science. Policies whose diffusion processes are analyzed include, for example, social and welfare policy (Jahn 2006; Strang and Chang 1993; Weyland 2005), environmental protection standards (Frank et al. 2000; Tews et al. 2003), various schemes of the International Standardization Organization (ISO; Guler et al. 2002; Prakash and Potoski 2006, 2007), deregulatory policies (Eising 2002; Gilardi 2005, 2008); neoliberal macroeconomic policies (Henisz et al. 2005; Meseguer Yebra 2004, 2006), central bank independence (Marcussen 2005; Polillo and Guillén 2005), independent regulatory agencies (Gilardi 2005, 2008; Jordana and Levi-Fauer 2005, 2006), and financing models for public services (Gilardi et al. 2009; Lee and Strang 2006).

In terms of empirical analysis, this body of literature is characterized by the use of sophisticated analytical techniques such as event history analysis and spatial econometrics. With regard to theory, most empirical studies seek to uncover the causal mechanisms underlying diffusion. In so doing, they mostly tend to concentrate on international factors triggering diffusion processes, which are derived from neoinstitutional theory. This perspective posits three distinct mechanisms of diffusion: international coercion, which results from power dynamics; normative emulation, whereby actors intensely related to each other within a social structure influence each other; and competitive mimicry, a process of social comparison stemming from the pressure to remain economically effective and efficient relative to others. DiMaggio and Powell (1991) refer to these mechanisms that are expected to lead to an increasing degree of homogeneity across units as "isomorphism."

The role of coercive power exerted by governments, international organizations, and nongovernmental organizations (NGOs) has often been explored within the context of the diffusion of neoliberal macroeconomic policies (see, e.g., Henisz et al. 2005). The number of studies explicitly focusing on mimesis is lower. For instance, Radaelli (2000) shows that the European Monetary Union can be modeled as a process of mimetic isomorphism because in the presence of uncertainty, the European Union (EU) imitated the (perceived) most successful national model of monetary policy. Research on normative emulation, on the other hand, has been broader in scope and showed that government officials and bureaucrats constantly assess policy and organizational developments in other countries (see, e.g., Guler et al. 2002; Lee and Strang 2006).

In addition to these three classic international diffusion mechanisms, the concepts of "lesson-drawing" (Rose 1991) and "learning" (Hall 1993) have received increasing scholarly attention. In a nutshell, lesson drawing and learning state that policy diffusion also may result from information about the successes or failures of policy change in other countries (see, e.g., Gilardi 2008; Gilardi et al. 2009; Weyland 2005). Meseguer Yebra (2009) presents a more refined concept, and introduces the concept of "Bayesian learning." It represents a mode of rational, experience-based learning in which governments are modeled as perfectly rational learners. They update their belief systems on the consequences of policies with all available information about policy outcomes in the past and elsewhere and choose the policy that is expected to yield the best results (see also Holzinger and Knill 2005, 2008; Holzinger et al. 2009).

Despite the considerable size of this literature, almost all empirical studies come to the same conclusion: namely, that "countries need to be explicitly studied as interdependent actors" (Braun and Gilardi 2006, 317). How stable is this finding when we adopt a more extensive empirical focus? As a matter of fact, we cannot answer this question on the basis of previous studies. Rather, we could make an "educated guess," stating that it should be less stable than generally claimed due to the existence of a selection bias. While strongly acknowledging the merits of previous empirical studies of policy diffusion, we argue that a novel conceptual perspective is needed to evaluate this apparently simple question. To accomplish this, we need to come up with a higher degree of variation in the cases selected for analysis and elaborate a theoretical model that combines country-specific factors with policy-specific ones.

Theoretical Framework and Hypotheses

The central objective of this chapter is to explain why countries vary in the extent to which they are affected by cross-national environmental policy diffusion. Why might countries resist adopting successful policies that are in place elsewhere? To answer this question, we first have to specify the concept of adoption resistance (i.e., the dependent variable of our analysis). We conceive of adoption resistance as the phenomenon that—within a given population of n countries and over a given period of time t—a country i refrains from adopting a certain policy j. In a second step, we discuss theoretical factors that might account for varying levels of national adoption resistance.

We use the concept of adoption resistance in accordance with the null hypothesis that national governments make decisions independently of each other (i.e., that no adoption takes place). We assume that varying levels in adoption resistance result from the interaction between policy item–specific factors and country-specific adoption resistance. Concerning the item-specific factors, we employ here the concept of "item contagiousness." To explain the country-specific adoption resistance, we distinguish between two sets of explanatory variables. On the one hand, we discuss aspects that determine the exposure of a country to policy developments in its environment. On the other hand, we focus on endogenous factors that constrain or facilitate the adoption of foreign policy models.

Item-Specific Factor: Contagiousness

The concept of item contagiousness refers to item-specific characteristics that facilitate or inhibit the broad adoption of a certain policy within a given population of countries over time. Using medicine as an analogy might illuminate this concept further. Similar to diseases that can be more or less infectious, policies might be more or less contagious, implying that they can be "caught" by a country with varying levels of difficulty. In our perspective, item contagiousness tells us something about the ease by which a given policy is adopted by the target population. Adoption resistance should, *ceteris paribus* (all other things being equal), be strongly affected by these item-specific characteristics; the higher the contagiousness of an item, the higher we expect the likelihood will be that a given country adopts it. Note that this is not a hypothesis, but an assumption that enables us to identify country-specific adoption resistance levels. The latter in turn will be explained by further factors, which we discuss in the next section.

Country-Specific Factors: External Exposure

Generally, we expect the adoption resistance of a country to increase the less it is interlinked with other countries of a given population. In the relevant literature, international factors play an important role to account for cross-national policy diffusion (see, e.g., Bennett 1991; Dolowitz and Marsh 2000; Drezner 2001; Gilardi 2005, 2008; Holzinger and Knill 2005, 2008; Simmons and Elkins 2004). On the one hand, these factors refer to the extent to which countries are institutionally interlinked. Here, emphasis is placed not only on diffusion emanating from the harmonization of national policies through international or supranational law, but also on the effects of transnational communication within institutionalized networks. On the other hand, regulatory competition emerging from intensifying economic integration with international markets has been identified as an important factor that drives policy diffusion (see chapter 6).

International Interlinkages
Common membership of countries in international institutions is generally seen as an important factor driving the cross-national diffusion of policies. A primary source of these effects is international harmonization. It refers to a specific outcome of international cooperation, namely to constellations in which national governments are legally required to adopt similar policies and programs as part of their obligations as members of international institutions (Holzinger and Knill 2005, 2008; Holzinger et al. 2009). International harmonization and, more generally, international cooperation presuppose the existence of interdependencies or externalities that push governments to resolve common problems through cooperation within international institutions, hence sacrificing some independence for the good of the community (Drezner 2001, 60). Once established, institutional arrangements constrain and shape domestic policy choices, even as they are constantly challenged and reformed by their member states (Martin and Simmons 1998, 743).

Second, international institutions might cause cross-national policy diffusion not only by legally binding rules and decisions, but also by nonobligatory factors. Diffusion processes can occur because of their symbolic properties and their capacity to legitimize the actions of policy makers, or simply because they come to be taken for granted as appropriate solutions to a given policy problem (Gilardi 2008, 72). Frequently interacting organizations, such as national bureaucracies, tend to develop similar structures and concepts over time. Policy diffusion results from organizations striving to increase their social legitimacy by embracing forms and practices that are valued within the broader institutional environment (DiMaggio and Powell 1991).

Moreover, the role of international institutions for nonobligatory diffusion also can be based on theories of rational policy learning. The concept of lesson drawing refers to constellations of policy transfer in which governments rationally use available experience elsewhere to solve domestic problems. According to Rose (1991), who introduced the concept, lesson-drawing is based on a voluntary process whereby government *A* learns from government *B*'s solution to a common problem. This kind of learning will be enhanced when countries meet and communicate on a regular basis within international institutions (Holzinger and Knill 2005, 2008; Holzinger et al. 2008). Finally, transnational problem solving typically occurs within transnational elite networks or "epistemic communities" (Haas 1992, 3).

Based on the above considerations, the extent to which a country is subject to adoption pressures emerging from institutional interlinkage is determined by the number of international or supranational institutions of which it is a member. We also expect that adoption pressures arise from the different international or supranational institutions' obligatory and communicative potential (Holzinger and Knill 2005, 2008; Holzinger et al. 2008). In this regard, far-reaching differences exist, in particular between the European Union and other international institutions. These considerations give way to our first hypothesis, shown next.

Hypothesis 1

The lower the score of a country's institutional membership weighted by obligatory and communicative potential, the higher its resistance to adopting environmental policies.

Economic Interlinkages
A country's exposure to its international environment is also affected by the extent to which it is economically interlinked with other countries of a given population. The more these economic exchange relations are developed, the higher the likelihood of cross-national policy diffusion. This can be traced to effects of regulatory competition between nation states (see, e.g., Holzinger 2008; Holzinger and Knill 2005, 2008; Holzinger et al. 2008; Vogel 1995). Theories of regulatory competition generally predict that countries adjust regulatory standards to cope with competitive pressures emerging from international economic integration. Regulatory competition presupposes economic integration among countries (i.e., the existence of integrated markets and free trade). The pressure to redesign domestic regulatory arrangements arises either from threats from economic actors to shift their activities elsewhere or from internal lobbying of industries, emphasizing the competitive disadvantages that the industries suffer from existing environmental regulation. This development, as the theory of regulatory competition implies, induces governments to adjust their regulatory standards to respective policy developments in other countries.

Following this basic reasoning, we expected that regulatory competition—as a consequence of economic integration—should induce governments to be aware of what other jurisdictions do. This concept would imply that countries subject to a higher degree of economic integration are more likely to follow the behavior of the other countries of their population to avoid any negative effects of inaction, leading to our second hypothesis.

Hypothesis 2

The lower a country's economic integration is, the higher its resistance is to adopting environmental policies.

Country-Specific Factors: Endogenous Factors

The likelihood that a country adopts policies that diffuse internationally is affected not only by its international exposure, but also by national conditions. In this regard, we first consider functional aspects; i.e., if and to what extent a given policy actually provides an appropriate solution in view of the country-specific problem constellation. Second, the political positions of national governments might facilitate or constrain a country's capacity for policy adoption.

Problem Pressure

In the literature on cross-national policy diffusion, the existence of parallel problems across countries is an important factor that drives the international spread of respective policy responses (see, e.g., Frank et al. 2000; Holzinger and Knill 2005, 2008; Holzinger et al. 2008). By contrast, if a country is confronted with less intense pressures, or even with a complete lack of respective pressures, we should hardly expect it to adopt a given policy, as the latter will not fit into its given problem constellation. This idea leads to our third hypothesis.

Hypothesis 3

The lower the extent of environmental problem pressure in a country, the higher its resistance to adopting environmental policies.

Political Positions

As political parties should strive for different policy goals to gain reelection, different policy choices across space and time in modern democracies should be attributable to the varying composition of governments and legislatures. This view parallels the view expressed about "green ideology" in chapter 4, "The Three Worlds of Environmental Politics," but at a lower level of abstraction. Political parties that compete for votes need to implement those policies that satisfy their voters, once they control the decision making after obtaining majorities. The objectives of office seeking and the realization of policy goals are heavily interwoven (Strøm and Müller 1999; Strøm 1990). Indeed, important research contributions demonstrated that political parties in government do deliver those policies that they promised earlier in their platforms (Budge and Laver 1993; Klingemann et al. 1994; McDonald and Budge 2005).

Hence, it can be expected that the positions of political parties in legislatures and governments should translate into policy outputs, at least to some extent. As a result, we expect that the representation of pro-environmental parties should increase the likelihood that a government adopts environmental policies that are in place elsewhere (see also chapter 5, "Wind-Power Development in Germany and the United States: Structural Factors, Multiple-Stream Convergence, and Turning Points"). By the same token, the absence of strong environmental preferences should lower a government's motivation to adopt policies in place elsewhere, as posited in our fourth hypothesis.

Hypothesis 4

The less the political parties represented in government stress the requirement of environmental protection in their election manifestos, the higher its resistance to adopting environmental policies.

Operationalization

This section is dedicated to the measurement and operationalization of our key variables. We begin by presenting the operationalization of adoption resistance, including information about country sample, time frame, and data structure. Next, we turn to the measurement of the explanatory variables.

Measurement of the Dependent Variable

In empirical terms, we focus here on the adoption of forty environmental policies. Our selection of this particular policy field is based on two considerations. First, research on environmental policy diffusion has been particularly prone to scrutinizing cases of explosive diffusion (see, e.g., Frank et al. 2000; Guler et al. 2002; Prakash and Potoski 2006, 2007; Tews et al. 2003). This stems from the fact that environmental concerns have been subject to a large number of international cooperation activities, which produced innovative policy instruments that subsequently spread across jurisdictions. Strongly related to this aspect is the second point: namely, that environmental policies are particularly susceptible to transnational diffusion because they produce negative externalities. That is, pollution that originates in one country is capable of causing damage in another country's environment (see, e.g., Young 1997). Consequently, the absence of policy diffusion in this field would represent a strong case against the overall finding of the scholarly literature that policy choices tend to become increasingly interdependent.

For measuring the dependent variable, we rely on the publicly accessible database of the ENVIPOLCON project (Holzinger et al. 2008). It provides unique data on the adoption and subsequent changes in environmental protection standards covering a wide range of environmental media, which reaches from air quality issues to general guiding principles such as sustainability or the precautionary principle. Further, the ENVIPOLCON data set covers a wide range of different policy types. In addition to classic command-and-control instruments, the database comprises information about other environmental policy instruments, such as market-based incentives and benchmarks. Nineteen of the policies are harmonized at the EU level, whereas twenty-one policy items are not affected by supranational harmonization. Finally, the environmental policy measures included in the database vary with regard to the extent and the pace by which they were adopted by the countries of the sample. All these characteristics show that there is sufficient variation in our dependent variable.

The original data has varying levels of measurement, ranging from ratio to ordinal scales. As the preservation of the differential levels of measurement would have impeded the construction of a compound dependent variable, we decided to recode the data to produce forty individual binary variables indicating whether a country has adopted the policy items between 1970 and 2000 (coded as 1) or not (coded as 0). These are then summed up to reflect the approximation of the individual countries' environmental policy arrangements to the maximum value of 40. Table 3.1 gives an overview of the policy items addressed by this study. The "EU" column indicates which of the policy items is affected by supranational harmonization through the European Union.

In this vein, the data allows us to check whether our measurements correspond to the environmental policy patterns outlined in the literature. For example, there is a literature on environmental pioneers, which helps us to evaluate the accuracy of our measurements. This strand of research argues that since the early days of environmental policy in the 1970s,

Table 3.1

List of environmental policies for the dependent variable

No.	Policy	EU	No.	Policy	EU
1	Sulfur content in gas oil	•	21	Glass reuse/recycling target	
2	Lead in gasoline	•	22	Paper reuse/recycling target	
3	Passenger cars; NOx emissions	•	23	Promotion of refillable beverage containers	
4	Passenger cars; CO emissions	•	24	Voluntary deposit system beverage containers	
5	Passenger cars; HC emissions	•	25	Noise emissions standard from trucks	•
6	Large combustion plants; SO_2 emissions	•	26	Motorway noise emissions	
7	Large combustion plants; NOx emissions	•	27	Noise level working environment	•
8	Large combustion plants; dust emissions	•	28	Electricity from renewable sources	
9	Coliforms in bathing water	•	29	Recycling construction waste	
10	Hazardous substances in detergents	•	30	Energy efficiency of refrigerators	•
11	Efficient use of water in industry		31	Electricity tax for households	
12	Industrial discharges in surface water lead		32	Heavy fuel oil levy for industry	•
13	Industrial discharges in surface water zinc		33	CO_2 emissions from heavy industry	
14	Industrial discharges in surface water copper		34	Forest protection	
15	Industrial discharges in surface water chromium		35	Eco-audit	•
16	Industrial discharges in surface water BOD		36	Environmental impact assessment	•
17	Soil protection		37	Eco-labeling	•
18	Contaminated sites policy		38	Precautionary principle: Reference in legislation	
19	Waste recovery target	•	39	Sustainability: Reference in legislation	
20	Waste landfill target	•	40	Environmental/ sustainable development plan	

Source: Holzinger et al. (2008).

there have been pioneers. On a global scale, Japan and the United States are considered pioneers for the early period (Jänicke 2005, 136; Schreurs 2003). In the European context, Austria, Denmark, Finland, Germany, the Netherlands, and Sweden have generally been perceived as environmental policy pioneers (Andersen and Liefferink 1997).

Measurement of the Explanatory Variable

To measure a country's exposure to international harmonization pressures, we follow the approach developed by Sommerer et al. (2008) and employ the variable *Membership*. This variable consists of membership data for thirty-five international institutions. These data are weighted by the obligatory potential of the individual international institutions.

Following Andonova et al. (2007), we calculate the next variable (i.e., *Trade*), on the basis of a country's volume of imports and exports divided by the gross domestic product (GDP). The data for *Trade* were taken from the electronic version of the World Bank's World Development Indicators (WDI), as well as from the CD-ROM version of the National Accounts Data provided in 2006 by the Organisation for Economic Cooperation and Development (OECD).

The next variable is intended to assess the extent to which the individual countries are confronted with environmental problem pressure. In the most ideal case, the indicator to be chosen would be directly related to the environmental medium under study; e.g., deforestation rates would represent an appropriate indicator for empirically assessing the forest-related environmental stress. In this study, however, we use variables addressing a wide variety of environmental media. While such a broad focus certainly represents one of the major strengths of our analysis, it simultaneously complicates the identification of a direct indicator of environmental stress. As a result, we employ *Energy*, which indicates the total primary energy consumption per capita, as a proxy for a very general environmental stress. Several empirical studies rely on air pollution levels as a proxy for environmental stress, but considering the broad spectrum of environmental issues and media addressed by this study—including policies on air, water, soil, waste, noise, resource protection, climate, and nature conservation as well as several general environmental policy principles such as sustainability—a more encompassing indicator is needed. In fact, it is well documented in the literature that industrial and other energy-consuming activities produce pollution (e.g., Erdal et al. 2007; Paul and Bhattacharya 2004; Tapiero 2009). In recognition of the empirical evidence showing that energy consumption and environmental pollution are intimately related, we regard this proxy suitable for this study. The data for *Energy* are provided by the International Energy Statistics provided by the US Energy Information Administration (EIA).

The variable per capita *GDP* is incorporated into the explanatory model in response to the literature on the environmental Kuznets curve. This concept stipulates that the initial increase in pollution associated with economic growth will give way to declining levels of pollution per capita as countries economically develop and demand higher environmental quality (Grossman and Krueger 1995). While the relationship between income and pollution levels was originally thought to follow an inverted *U*-shape, more recent theoretical work suggests a linear relationship (Magnani 2001). Consequently, we equally assume a linear relationship and thus do not incorporate a quadratic term into the estimation model. From this, it follows that we expect that rising income levels lead to a higher demand for a clean environment, which should entail the adoption of relevant legislation. The data for *GDP* were taken from the WDI. For a more straightforward interpretation of the numeric values, we divided *GDP* by 1 million.

The next variable, *Center of Gravity*, refers to the government's ideological position with regard to environmental policy. For our purpose, a government's center of gravity is the most appropriate measurement (Cusack 1997). The indicator is based on conceptual considerations presented by Knill et al. (2010). Their measurement allows for estimating the overall position of a government for both the left-right and the environmental policy dimensions. Because legislative decision making, and therefore the adoption of policies, entails certain time lags, it is necessary to consider more than the environmental policy position and the left-right orientation of governments at the four points in time for which we have data on the number of environmental policies adopted. We must also take into account the ideological and environmental policy orientation of the former parliaments. For this reason, we refer to the arithmetic mean of the center of gravity on the left-right and on the environmental policy dimensions for the governments that have been elected in the ten years before each of the four points in time.

Finally, we include a variable for evaluating the presence of time trends, which is not unlikely when using pooled data. The corresponding variable *Trend* takes on a positive value between 1 and 4 to account for the observation points. While this variable will not be interpreted in substantive terms, it is useful for enhancing the model's explanatory power.

Table 3.2
Summary statistics of the explanatory variables

Variable	N	Mean	SD	Min.	Max.	Sign	Source
Membership	96	37.10	21.82	3	84.03	+	Sommerer et al. (2008)
Trade		0.70	0.77	0.11	6.11	+	WDI, OECD
Energy	96	2.60	1.10	.71	5.92	+	EIA
GDP (in millions)		2.44	1.57	.24	8.13	+	WDI
Center of Gravity	96	3.39	3.20	0	12.9	+ / -	Knill et al. (2010)
Trend	96	2.5	1.12	1	4	+	World Bank

Table 3.2 provides the summary statistics of the covariates, as well as outlining their sources. It further indicates the direction of the anticipated signs. The data set is based on pooled observations for four points in time (1970, 1980, 1990, and 2000) for twenty-four countries, including the member states of the EU-15 (with the exception of Luxembourg), Norway, Switzerland, Poland, Slovakia, Hungary, Bulgaria, and Romania, as well as the United States, Mexico, and Japan. We have 96 observations for each covariate. To achieve this, we imputed missing values for a limited number of observations by fitting exponential curves to each country. Because many processes have exponential dependencies, we believe that this imputation strategy is more accurate than other alternatives, which only minimally take into account the overall structure of the data.

Causal Analysis

Analytical Clarifications
Based on our null hypothesis, we model policy adoption in terms of adoption resistance because this notion corresponds best to the idea of independent problem solving. More precisely, we conceive of adoption resistance as a latent variable, which we will model subsequently by using the covariates introduced in the theoretical framework. To obtain the latent adoption resistance of individual countries from observable data, item response models (IRMs) provide the most appropriate estimation technique (e.g., Baker and Kim 2004; Boek and Wilson 2004; van der Linden and

Hambleton, 1997). IRMs were originally developed in psychology and educational science to measure latent psychological constructs based on observed responses to different question items. A typical latent construct assessed on the basis of IRMs is intelligence. Denoting the intelligence level of respondent i by β_i and the difficulty level of item j by α_j, the probability that i gives correct answer to j can be modeled as follows:

Logit(Prob.(i gives the correct answer to j)) = γ_j ($\beta_i - \alpha_j$)   (3.1)

This equation assumes that the probability of a correct answer is given by the extent to which the degree of intelligence exceeds the difficulty of the question. Hence, the higher the degree of intelligence β_i relative to the difficulty of the question α_j is, the higher the probability that respondent i will give a correct answer. In this context, γ_j is often called the discrimination parameter as it represents the impact of the latent variables (intelligence and difficulty) on the response. If $\gamma_j = 0$, there is no relationship between the latent variables and the response category.

This logic can be easily applied for the purpose of estimating adoption resistance. To translate the IRM to our context, we can interpret the intelligence level of individuals as the country-specific "adoption resistance" and the difficulty level of policy items as "item contagiousness." To recap, the term "contagiousness" refers to item-specific characteristics that facilitate or inhibit a broad spread of the policy across a given population of countries.

Note that for our analytical purpose, the parameterization outlined above needs to be modified. For the study at hand, it is of crucial relevance to which extent the contagiousness of policy adoption exceeds a country's adoption resistance. For this reason, the order of α_j and β_i is reversed. Furthermore, we have multiple measures for individual states for different points in time. Therefore, our model specification is as follows:

Logit(Prob.(adoption of policy j in state i at year t)) = γ_j ($\alpha_j - \beta_{it}$)   (3.2)

Now we can model adoption resistance by plugging in our theoretically derived covariates. In principle, we can estimate a simple linear model because the identified latent dimension has an unlimited metric scale. However, the individual values of the dependent variable—here β_{it}—are not completely independent. Consequently, adoption resistance could be subject to clustering effects and serial correlation inside the individual states might become an issue. To remedy these problems of spatial and temporal interdependence, we parameterize adoption resistance as follows:

$$\beta_{it} = \delta_i + \delta\ X_{it} + \beta_{it\text{-}1} + e_{it} \qquad\qquad (3.3)$$

where X_{it} is the matrix of covariates, δ is the vector of regression coefficients, δ_i is the state-specific constant, and e_{it} is the error term. This corresponds to a pooled, cross-sectional time-series analysis with lagged dependent variables. All unknown parameters of equations (3.2) and (3.3), including item contagiousness and adoption resistance, are estimated simultaneously in a Bayesian setting (e.g., Gelman and Hill 2007; Mislevy 1986). As prior information, we use only diffuse uninformative distribution.

Exploratory Analysis

After having introduced the logic underlying the estimation procedure, we can now turn to the exploratory analysis. Thus, in this section, we first present details about the changing countries' resistance to adopting environmental policies in place elsewhere. Figure 3.1 presents the estimated adoption resistance (i.e., the dependent variable) of the individual states at four points in time: 1970, 1980, 1990, and 2000. As a general trend, the adoption resistance has decreased over time. The pace of the decrease in adoption resistance, however, is markedly heterogeneous. The states that decreased their resistance particularly swiftly include Spain, Greece, and Portugal, implying the existence of a regulatory "catching up" effect due to low initial regulatory levels. Denmark and Finland also rapidly decreased their resistance to adopt environmental policies. Yet, both countries display a low initial resistance level, which they simply preserved throughout the observation period. In contrast to the previous countries, Japan and the United States belong to the states with slowly decreasing adoption resistance.

For an even more straightforward impression of the dependent variable, table 3.3 presents a ranking of the countries based on their mean adoption resistance, listed from lowest to highest. The table offers at least three instructive observations. First, the most progressive country across all the observations is Sweden. For all four points in time, the country adopted the highest number (1990) or second-highest number (1970, 1980, and 2000) of environmental protection standards (for an overview, see Duit 2007). Second, the least progressive country is Romania, which was replaced by the United States only in the fourth point in time (i.e., 2000). Third, the countries' positions within the ranking are

Table 3.3
Country ranking

Rank	1970	1980	1990	2000
1	Japan	Japan	Sweden	Finland
2	Sweden	Sweden	Switzerland	Sweden
3	Netherlands	Italy	Finland	Netherlands
4	Finland	Hungary	Netherlands	Denmark
5	Hungary	Finland	Germany	Germany
6	Belgium	Netherlands	Norway	Norway
7	France	Belgium	Denmark	Austria
8	United Kingdom	Switzerland	Austria	Spain
9	Germany	Denmark	Hungary	Switzerland
10	Switzerland	United States	Japan	Japan
11	United States	Germany	France	France
12	Austria	United Kingdom	Italy	Italy
13	Italy	France	Spain	United Kingdom
14	Denmark	Norway	Belgium	Hungary
15	Norway	Austria	United Kingdom	Portugal
16	Bulgaria	Ireland	Portugal	Greece
17	Slovakia	Spain	United States	Belgium
18	Mexico	Bulgaria	Greece	Slovakia
19	Portugal	Slovakia	Ireland	Poland
20	Poland	Portugal	Poland	Mexico
21	Spain	Poland	Mexico	Ireland
22	Ireland	Greece	Bulgaria	Bulgaria
23	Romania	Mexico	Slovakia	Romania
24	Greece	Romania	Romania	United States

relatively stable, notwithstanding countries that are characterized by a more volatile development.

This brings us back to the cases discussed on the basis of figure 3.1. Of these, without question the most interesting case is Japan, as it becomes clear that it moved down in the rankings. Although it was the leader in 1970 and 1980, the country lost this position in 1990 and 2000 and dropped to the tenth rank. A similarly interesting observation can be made for the United States. While in 1970 and 1980, the country held a medium-progressive position, it fell behind all the others in 2000, bringing up the rear. Overall, the adoption patterns correspond to the findings of the literature on environmental policy pioneers, which gives us additional confidence with regard to the accuracy of the empirical data. Moreover, our finding corresponds with the decreasing role of the United States as a leading supporter of international environmental initiatives (see Kelemen and Vogel 2010).

Estimation Results

Having outlined the country-specific adoption dynamics, we now turn to the modeling of the dependent variable using covariates. To recap, our explanatory model consists of variables addressing the countries' external exposure, as well as endogenous factors conceived to affect environmental policy adoption. The item contagiousness is not included as a covariate per se, but rather as a moderating factor for evaluating the impact of the country-specific variables.

Figure 3.2 presents the estimated impact of each covariate on country-specific adoption resistance. A positive estimate of a covariate means that increase in the covariate leads to a higher adoption resistance. In terms of impact on policy adoption, the coefficient should be interpreted in a reverse way in terms of positive and negative sign because a higher adoption resistance results in a lower likelihood of policy adoption.

The estimation results yield three interesting insights. First, the model produces a clear-cut effect with the predicted negative sign for the variable *Membership*. This implies that interdependence in environmental policymaking is more likely if countries are related to one another on institutional grounds, *ceteris paribus*. The more that a country participates in communication within the context of international institutions of an obligatory potential, the higher the likelihood that this country adopts environmental policies in place elsewhere.

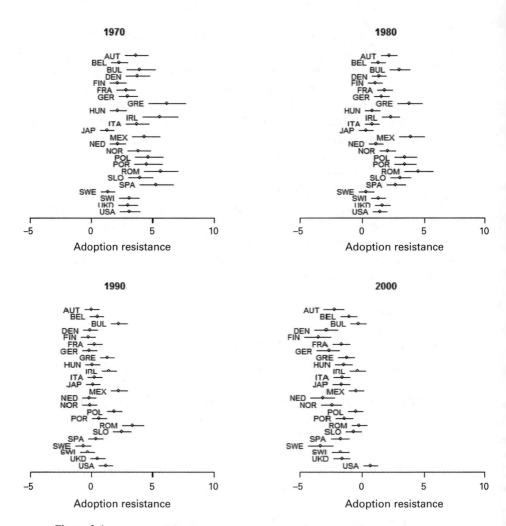

Figure 3.1

Overview of country-specific adoption resistance

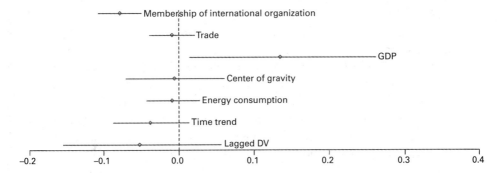

Figure 3.2

Determinants of country-specific adoption resistance; estimated impacts of country-specific covariates on adoption resistance (δ in equation [5.3]). Points are posterior means, and the lines are 90% credible intervals.

Second, the adoption resistance of environmental policies is positively affected by *GDP*, all else being equal. The result contradicts our theoretical expectation that increasing levels of income should lead to higher levels of regulatory activity. Nevertheless, this finding also somewhat supports the relevance of international interlinkages because apparently it is not the assumed public demand for stricter environmental policy but external factors that determine a government's decision on whether to adopt protection measures in place elsewhere.

Third, none of the other factors was found to have a substantive impact on the likelihood of adopting environmental policies. This finding is particularly surprising for the domestic factors because these are generally presumed work as mediators for international ones (e.g., see Braun and Gilardi 2006). Most important, the preferences of the domestic political actors (measured as *Center of Gravity*) do not seem to matter significantly for the decision to adopt environmental policies.

Furthermore, all findings reported here must be seen as particularly stable because the estimation also includes the lagged dependent variable as an additional covariate. While this specific covariate does not yield a clear-cut impact, it is important to note that the other factors' explanatory power is not absorbed by the lagged dependent variable. Considering the extensive empirical specification of our dependent variable and the theoretically derived specification of the estimation model, we can state confidently that in the case of environmental policy adoption, membership in relevant international organizations is the main determinant.

Conclusion

In this chapter, we addressed one central flaw in the analysis of transnational policy diffusion: the focus on explosive diffusion. While we acknowledge that examining policy innovations and the patterns of how they spread from one country to another lies at the heart of diffusion studies, the limited empirical perspective imposes the risk of a selection bias. Against this background, we posed the following research question: How robust are the phenomena of policy diffusion if we extend the empirical basis? To evaluate this question, we examined the adoption of forty environmental policy measures across twenty-four countries between 1970 and 2000. Based on our empirical basis, the answer to this question must be that environmental diffusion is a robust phenomenon indeed. This implies that regulatory activities in the field of environmental policy are actually interdependent, even with regular policy measures and not only instruments developed and promoted by international actors.

By decomposing policy adoption into both country- and policy-specific factors, we shed light on the main drivers of environmental policymaking. Initially, we expected that a country's institutional membership with international environmental organizations, the existence of an elevated environmental problem pressure, a high degree of economic integration, and the pro-environment preferences of political actors would contribute equally to the lowering of a country's resistance to adopting environmental protection standards in place elsewhere. The estimation results showed that the degree of institutional interlinkages of countries is the main determinant of environmental policy adoption, holding all other factors constant. While the dominance of international factors over national ones is surprising to a certain extent, it is also plausible because environmental issues are mostly of a transboundary nature, entailing the need for coordinated action.

While the findings presented in this chapter can be regarded as an important complement to the literature, our analysis also suffers from limitations. First, we concentrated on policy adoptions and did not illustrate other forms of policy change. (Fortunately, however, this perspective is covered by chapter 6 of this book.) Second, we have to acknowledge that our findings represent but an initial step for assessing the on-the-ground effects of the global dynamics of environmental politics. However, considerations of the effectiveness of environmental policy arrangements are indispensable for discussing their role for sustainable development. However, it is this dimension—commonly referred to as "environmental performance" (see chapters 4 and 5) that also poses challenges to a valid and reliable empirical assessment (see chapter 2). Thus, we encourage future research to bring together more systematically the patterns and causes of environmental policymaking and performance.

References

Andersen, Michael S., and Duncan Liefferink. 1997. *European Environmental Policy. The Pioneers*. Manchester, UK: Manchester University Press.

Andonova, Liliana B, Edward D. Mansfield, and Helen V. Milner. 2007. International Trade and Environmental Policy in the Postcommunist World. *Comparative Political Studies* 40(7): 782–807.

Baker, Frank B, and Seock-Ho Kim. 2004. *Item Response Theory: Parameter Estimation Techniques*. New York: Marcel Dekker.

Bennett, Colin J. 1991. What Is Policy Convergence and What Causes It? *British Journal of Political Science* 21(2): 215–233.

Boek, Paul de, and Mark Wilson. 2004. *Explanatory Item Response Models. A Generalized Linear and Nonlinear Approach*. Amsterdam: Springer.

Braun, Dietmar, and Fabrizio Gilardi. 2006. Taking "Galton's Problem" Seriously. Towards a Theory of Policy Diffusion. *Journal of Theoretical Politics* 18(3): 298–322.

Budge, Ian, and Michael Laver. 1993. The Policy Basis of Government Coalitions: A Comparative Investigation. *British Journal of Political Science* 23(4): 499–519.

Cusack, Thomas. 1997. Partisan Politics and Public Finance: Changes in Public Spending in the Industrialized Democracies. *Public Choice* 91(3/4): 375–395.

DiMaggio, Paul J., and Walter W. Powell. 1991. The Iron Cage Revisited. Institutionalized Isomorphism and Collective Rationality in Organizational Fields. In *The New Institutionalism in Organizational Analysis*, ed. Walter W. Powell and Paul J. DiMaggio, 63–82. Chicago: Chicago University Press.

Dolowitz, David P., and David Marsh. 2000. Learning from Abroad: The Role of Policy Transfer in Contemporary Policy-making. *Governance: An International Journal of Policy Administration and Institutions* 13(1): 5–24.

Drezner, Daniel W. 2001. Globalization and Policy Convergence. *International Studies Review* 3(1): 53–78.

Duit, Andreas. 2007. Path Dependency and Institutional Change: The Case of Industrial Emission Control in Sweden. *Public Administration* 85(4): 1097–1118.

Eising, Rainer. 2002. Policy Learning in Embedded Negotiations: Explaining EU Electricity Liberalization. *International Organization* 56(1): 85–120.

Elkins, Zachary, Andrew Guzman, and Beth A. Simmons. 2006. Competing for Capital: The Diffusion of Bilateral Investment Treaties. 1960–2000. *International Organization* 60(4): 811–846.

Erdal, Gülistan, Kemal Esengün, Hilmi Erdal, and Orhan Gündüz. 2007. Energy Use and Economical Analysis of Sugar Beet Production in Tokat Province of Turkey. *Energy* 32(1): 35–41.

Frank, David, Ann Hironaka, and Eva Schofer. 2000. Environmental Protection as a Global Institution. *American Sociological Review* 65(1): 122–127.

Garrett, Geoffrey, Frank Dobbin, and Beth A. Simmons. 2008. Conclusion. In *The Global Diffusion of Markets and Democracy*, ed. Geoffrey Garrett, Frank Dobbin, and Beth Simmons, 344–360. Cambridge: Cambridge University Press.

Gelman, Andrew, and Jennifer Hill. 2007. *Data Analysis Using Regression and Multilevel/Hierarchical Models*. Cambridge: Cambridge University Press.

Gilardi, Fabrizio. 2005. The Formal Independence of Regulators: A Comparison of 17 Countries and 7 Sectors. *Swiss Political Science Review* 11(4): 139–167.

Gilardi, Fabrizio. 2008. *Delegation in the Regulatory State: Independent Regulatory Agencies in Western Europe*. Cheltenham, UK: Edward Elgar.

Gilardi, Fabrizio, Katharina Füglister, and Stephane Luyet. 2009. Learning from Others: The Diffusion of Hospital Financing Reforms in OECD Countries. *Comparative Political Studies* 42(4): 549–573.

Gleditsch, Kristian S, and Michael D. Ward. 2006. Diffusion and the International Context of Democratization. *International Organization* 60(4): 911–933.

Gray, Virginia. 1973. Innovations in the State: A Diffusion Study. *American Political Science Review* 67(4): 1174–1185.

Grossman, Gene M, and Alan B. Krueger. 1995. Economic Growth and the Environment. *Quarterly Journal of Economics* 110(2): 353–377.

Guler, Isin, Mauro F. Guillén, and John Muir Macpherson. 2002. Global Competition, Institutions, and the Diffusion of Organizational Practices: The International Spread of ISO 9000 Quality Certificates. *Administrative Science Quarterly* 47(2): 207–232.

Haas, Peter M. 1992. Introduction: Epistemic Communities and International Policy Coordination. *International Organization* 46(1): 1–36.

Hall, Peter A. 1993. Policy Paradigms, Social Learning, and the State. The Case of Economic Policymaking in Britain. *Comparative Politics* 25(3): 275–296.

Henisz, Witold J, Bennet A. Zelner, and Mauro F. Guillén. 2005. The Worldwide Diffusion of Market-Oriented Infrastructure Reform, 1977–1999. *American Sociological Review* 70(6): 871–897.

Holzinger, Katharina. 2008. *Transnational Common Goods: Strategic Constellations, Collective Action Problems, and Multi-Level Provision*. Basingstoke, UK: Palgrave Macmillan.

Holzinger, Katharina, and Christoph Knill. 2005. Cross-National Policy Convergence: Causes, Concepts, and Empirical Findings. *Journal of European Public Policy* 12(5): 764–774.

Holzinger, Katharina, and Christoph Knill. 2008. The Interaction of Competition, Co-operation, and Communication: Theoretical Analysis of Different Sources of

Environmental Policy Convergence and Their Interaction. *Journal of Comparative Policy Analysis* 10(4): 403–425.

Holzinger, Katharina, Christoph Knill, and Bas Arts. 2008. *Environmental Policy Convergence in Europe? The Impact of International Institutions and Trade.* Cambridge: Cambridge University Press.

Huber, Joseph. 2008. Pioneer Countries and the Global Diffusion of Environmental Innovations: Theses from the Viewpoint of Ecological Modernisation Theory. *Global Environmental Change* 18(2): 360–367.

Jahn, Detlef. 2006. Globalization as "Galton's Problem": The Missing Link in the Analysis of Diffusion Patterns in Welfare State Development. *International Organization* 60(2): 401–431.

Jänicke, Martin. 2005. Trend-Setters in Environmental Policy: The Character and Role of Pioneer Countries. *European Environment* 15(2): 129–142.

Jordana, Jacint, and David Levi-Faur. 2005. The Diffusion of Regulatory Capitalism in Latin America: Sectoral and National Channels in the Making of New Order. *Annals of the American Academy of Political and Social Science* 598(1): 102–124.

Jordana, Jacint, and David Levi-Faur. 2006. Toward a Latin American Regulatory State? The Diffusion of Autonomous Regulatory Agencies across Countries and Sectors. *International Journal of Public Administration* 29(4–6): 355–366.

Kelemen, Daniel R., and David Vogel. 2010. Trading Places: The Role of the United States and the European Union in International Environmental Politics. *Comparative Political Studies* 43(4): 427–456.

Klingemann, Hans-Dieter, Richard I. Hofferbert, and Ian Budge. 1994. *Parties, Policies, and Democracy.* Boulder, CO: Westview.

Knill, Christoph, Marc Debus, and Stephan Heichel. 2010. Do Parties Matter in Internationalised Policy Areas? The Impact of Political Parties on Environmental Policy Outputs in 18 OECD Countries, 1970–2000. *European Journal of Political Research* 49(3): 301–336.

Lee, Chang K., and David Strang. 2006. The International Diffusion of Public-Sector Downsizing: Network Emulation and Theory-Driven Learning. *International Organization* 60(4): 883–909.

Magnani, Elisabetta. 2001. The Environmental Kuznets Curve: Development Path or Policy Result? *Environmental Modelling & Software* 16(2): 157–165.

Marcussen, Martin. 2005. Central Banks on the Move. *Journal of European Public Policy* 12(5): 903–923.

Martin, Lisa L, and Beth A. Simmons. 1998. Theories and Empirical Studies of International Institutions. *International Organization* 52(4): 729–757.

McDonald, Michael, and Ian Budge. 2005. *Elections, Parties, Democracy: Conferring the Median Mandate.* Oxford: Oxford University Press.

Meseguer Yebra, Covadonga. 2004. What Role for Learning? The Diffusion of Privatisation in OECD and Latin American Countries. *Journal of Public Policy* 24(3): 299–325.

Meseguer Yebra, Covadonga. 2006. Learning and Economic Policy Choices. *European Journal of Political Economy* 22(1): 156–178.

Meseguer Yebra, Covadonga. 2009. *Learning, Policy Making, and Market Reforms.* Cambridge: Cambridge University Press.

Meseguer Yebra, Covadonga, and Fabrizio Gilardi. 2009. What Is New in the Study of Diffusion? A Critical Review. *Review of International Political Economy* 16(3): 527–543.

Mislevy, Robert J. 1986. Bayes Modal Estimation in Item Response Models. *Psychometrika* 51(2): 177–195.

Paul, Shyamal, and Rabindra N. Bhattacharya. 2004. CO2 Emission from Energy Use in India: A Decomposition Analysis. *Energy Policy* 32(5): 585–593.

Polillo, Simone, and Mauro F. Guillén. 2005. Globalization Pressures and the State: The Worldwide Spread of Central Bank Independence. *American Journal of Sociology* 110(6): 1764–1802.

Prakash, Aseem, and Matthew Potoski. 2006. Racing to the Bottom? Trade, Environmental Governance, and ISO 14001. *American Journal of Political Science* 50(2): 350–364.

Prakash, Aseem, and Matthew Potoski. 2007. Investing Up: FDI and the Cross-Country Diffusion of ISO 14001 Management Systems. *International Studies Quarterly* 51(3): 723–744.

Radaelli, Claudio M. 2000. Policy Transfer in the European Union: Institutional Isomorphism as a Source of Legitimacy. *Governance: An International Journal of Policy Administration and Institutions* 13(1): 25–43.

Rogers, Everett M. 2003. *Diffusion of Innovations.* New York: Free Press.

Rose, Richard. 1991. What Is Lesson-drawing? *Journal of Public Policy* 11(1): 3–30.

Scharpf, Fritz W. 2001. Introduction: The Problem-Solving Capacity of Multi-level Governance. *Journal of European Public Policy* 4(4): 520–538.

Schreurs, Miranda A. 2002. *Environmental Politics in Japan, Germany, and the United States.* Cambridge: Cambridge University Press.

Simmons, Beth A, and Zachary Elkins. 2004. The Globalization of Liberalization: Policy Diffusion in the International Political Economy. *American Political Science Review* 98(1): 171–189.

Sommerer, Thomas, Katharina Holzinger, and Christoph Knill. 2008. The Pair Approach: What Causes Convergence of Environmental Policies. In *Environmental Policy Convergence in Europe? The Impact of International Institutions and Trade,* ed. Katharina Holzinger, Christoph Knill, and Bas Arts, 144–195. Cambridge: Cambridge University Press.

Strang, David, and Patricia M. Y. Chang. 1993. The International Labor Organization and the Welfare State: Institutional Effects on Welfare Spending, 1960–80. *International Organization* 47(2): 235–262.

Strang, David, and Sarah A. Soule. 1998. Diffusion in Organizations and Social Movements: From Hybrid Corn to Poison Pills. *Annual Review of Sociology* 24(1): 265–290.

Strøm, Kaare. 1990. A Behavioral Theory of Competitive Political Parties. *American Journal of Political Science* 34(2): 565–598.

Strøm, Kaare, and Wolfgang C. Müller. 1999. Political Parties and Hard Choices. In *Policy, Office, or Votes?* ed. Wolfgang C. Müller and Kaare Strøm, 1–35. Cambridge: Cambridge University Press.

Tapiero, Charles S. 2009. Energy Consumption and Environmental Pollution: A Stochastic Model. *IMA Journal of Management Mathematics* 20(3): 263–273.

Tews, Kerstin, Per-Olof Busch, and Helge Jörgens. 2003. The Diffusion of New Environmental Policy Instruments. *European Journal of Political Research* 42(4): 569–600.

van der Linden, Wim J., and Ronald K. Hambleton. 1997. *Handbook of Modern Item Response Theory*. New York: Springer.

Vogel, David. 1995. *Trading Up: Consumer and Environmental Regulation in the Global Economy*. Cambridge, MA: Harvard University Press.

Weyland, Kurt. 2005. Theories of Policy Diffusion: Lessons from Latin American Pension Reform. *World Politics* 57(2): 262–295.

Young, Oran R. 1997. *Global Governance Drawing Insights from the Environmental Experience*. Cambridge, MA: MIT Press.

4

The Three Worlds of Environmental Politics

Detlef Jahn

The classification of states as "green states" has received increasing attention in recent years.[1,2] To identify green states, a comparative analysis of the role of the state is most helpful because a judgment about the degree of a state's greenness can be given only by comparing environmental governance. However, it is difficult to find criteria by which to classify countries into the category of green or non-green states. Moreover, to what extent states with successful environmental policies build or possess a social structure that supports environmental success is an even more difficult question. The ability to connect policy outcomes and social structures is the strength of Gøsta Esping-Andersen's analysis of welfare states. Through analyzing social structures, he found that welfare states can be clustered into one of three groups: liberal, corporatist, and social-democratic. These clusters correspond with the level of decommodification as an outcome of welfare policies. "Decommodification" is defined as the "degree to which individuals, or families, can uphold a socially acceptable standard of living independently of market participation" (Esping-Andersen 1990, 37).

Constructing a similar typology in the field of environmental policy must consider at least two important aspects. First, it has to identify how states perform in environmental policy. Like Esping-Andersen, I identify variance in performance by comparing the *outcomes* of environmental policies in highly industrialized democracies. In this view, environmental performance is equivalent to decommodification in welfare state research. The second aspect must consider structural developments that have an environmental impact on highly industrialized countries. In this respect, not much has been done in comparative environmental research, although advances in green political theory have moved this aspect to the center of interest. In social policy, Esping-Andersen studies the social stratification of welfare states that roughly follow the distinction between major political ideologies of industrial societies: liberalism, conservatism, and socialism (Bobbio 1996). Likewise, to render this conceptualization useful for environmental research, I will need to differentiate between ideological approaches to industrial societies while incorporating a green component. For such an undertaking, I refer to the theoretical literature on green politics that distinguish between a productionist and a green paradigm of social development.

This research design reveals which states are green states and which are not, as well as distinguishing between successful and less successful environmental outcomes. On the one hand, the design identifies environmentally successful states that are in line with green ideological principles; on the other hand, it also calls out states with successful environmental performance that nevertheless adhere to productionist-oriented policy principles. Finally, it describes less environmentally successful states that are productionist. Of course, it might be possible that states attempt to follow a green paradigm but are not environmentally successful. However, that would not be a result of policies, but special and exogenous environmental circumstances.

This chapter is divided into three parts. First, I analyze the environmental performance of twenty-one countries in the Organisation of Economic Cooperation and Development (OECD). Second, I distinguish the degree to which these 21 OECD countries have structural features that are consistent with green ideology. Needless to say, all established industrialized countries have inherited the productionist paradigm, but some have come to gradually embrace more green-oriented structural features. Third, I combine the two dimensions. This analysis leads to the identification of three worlds of environmentalism among the most highly industrialized states.

Environmental Performance in Twenty-One OECD Countries

To facilitate classifying states in terms of their environmental status and achievements, I will focus on their performance in respect to key issues of environmental policy outcomes. To do so, the term "performance" must be defined. There are several aspects constituting performance in general, and environmental performance in particular. In general, the concept of performance is evaluative and has been used in political science since the 1970s (Dahl 1967; Gurr and McClelland 1971; Eckstein 1971). Evaluation can be accomplished through comparison to a preset target or baseline, or to other cases or time periods (Eckstein 1971, 8).

Concerning environmental performance, the use of set targets or base-line models is difficult because of a lack of clearly defined and universally accepted targets concerning the abatement of environmental degradation or achievements.[3] Therefore, I focus on the comparative approach. However, a relative comparison needs comparable cases (Lijphart 1975), requiring a restriction of the analysis to the twenty-one most developed and democratic OECD countries.

The concept of environmental performance is complex and multidimensional. Therefore, appropriate indicators must fulfill several requirements, which are in turn highly contested both analytically and methodologically (for a more lengthy discussion on this aspect, see chapter 2, "Comparing Environmental Performance"). In the following list, I will outline the basic features of an empirical indicator of environmental performance suitable for analyzing a longer time period with a large number of cases:

• Performance is a typical *outcome* variable. While the introduction of an environmental policy or the establishment of an environmental institution or organization may have the intention of reducing pollution, the empirical proof of its effectiveness can be measured only by the outcomes.

• Environmental problems have to be *obvious to political actors*. Political actors can react only to problems that are known to them. Regarding climate change, this problem has been debated since the mid-eighteenth century, with scientific indications of global environmental effects emerging in the 1970s (Fleming 1998). However, the political acceptance of climate change as an environmental threat did not begin to happen until the 1990s. This implies that environmental performance is "necessarily provisional," as outlined in chapter 2, meaning that judgments made today may be revised in the future.

• Environmental performance indicators must correspond to aspects that *can be influenced by political action*. Causality in political research requires this precept (Jahn forthcoming). Emissions from volcanic activity—although having substantial consequences for atmospheric emissions—cannot be included in a performance measure. It is more difficult to argue that environmental disasters should not be taken into account. Environmental disasters often occur because politicians do not introduce effective regulatory instruments. However, in this investigation, I am interested in the impact of politics on regular environmental performance; therefore, I exclude environmental disasters from the analysis. However, even if environmental performance analyzes known and more or less accepted environmental problems, there are notorious issues with data quality and comparability to consider (both over time and across cases).

• For an index of environmental performance, it is important to note that accumulated measures are difficult to use because they can be changed only in the long run. However, if we wish to identify the effects of human action, causal links can best be established when analyzing short-term changes. Therefore, performance indicators should focus *primarily on changes in outcomes* and only secondarily on levels. The latter is important nevertheless because levels define the precondition for changes and themselves have important consequences for the environment.

• Environmental performance deals with *complex* aspects and processes. There are various issues to consider, from atmospheric and water emissions, waste, and biodiversity, to gene manipulation. Thus, environmental performance is multidimensional, leading to another measurement issue: should we aim to construct a comprehensive index of environmental performance by aggregating often distinct aspects? Or should we analyze disaggregated issues to better trace causal relationships? The former approach has the advantage of being relevant for a wide range of environmental issues, as well as potentially having a stronger political signaling function like gross domestic product (GDP).[4] The disadvantage is that an aggregated index may conceal causality because specific environmental issues may demonstrate contrary developmental directions (Pillarisetti and van den Bergh 2010). This is the advantage of disaggregated environmental performance indicators. Nevertheless, it is difficult to "explain" environmental performance when various disaggregated indicators come to different conclusions.

• A performance measure must account for *other factors* that might be responsible for the outcome. This is a particular challenge for environmental performance because there might be a host of other factors, such as geographical and climatic conditions, technological development and innovations, change in economic cycles, and international pressures.

• Finally, performance measures need to be *comparable over time and across countries,* while accounting for the fact that some environmental issues are more important in one country than in another.

There are a few existing indices that consider environmental conditions in various countries. Pioneering studies in political science have been conducted by Crepaz (1995), Palmer (1997), and Jahn (1998). Crepaz analyzed some indicators of pollution but fell short of aggregating them into an environmental performance index. Palmer constructed a composite index based mainly on CO_2 emissions, fertilizer consumption, and deforestation. Jahn subsequently developed one of the first comprehensive environmental performance indices in political science. He considered air emissions, municipal waste, fertilizer consumption, hazardous waste, and protected areas. While Crepaz used several indicators of air emission levels individually, Jahn used an index that includes levels and changes as a performance index. In his study of eighteen OECD countries, Jahn concluded that the Netherlands, West Germany, Austria, and Sweden had the highest environmental performance in 1990. The United States, Italy, Canada, and Ireland were laggards in his study. Lyle Scruggs (2003), who has written what is thus far the only book-length study of environmental performance of OECD countries in macro-comparative politics, used almost all of Jahn's indicators, but he exclusively analyzed changes from 1975 to 1995. According to his index, Germany ranked first of the 17 OECD countries analyzed, followed by Sweden, Denmark, and Austria. The United States, Canada, and Spain were the countries with the lowest environmental performance, followed only by Ireland, which had the worst performance. In her very comprehensive study of performance in twenty-one established OECD countries, covering the fields of domestic security, social, economic, and environmental policy, Edeltraut Roller (2005) used the levels of air emissions, municipal waste, fertilizer consumption, and freshwater abstraction. She offered indices for 1974–1979, 1980–1984, 1985–1989, and 1990–1995, as well as an aggregated index from 1974 to 1995. For 1990–1995, she reached the conclusion that Switzerland, Sweden, and Austria performed best. Contrary to most other studies, she found that Portugal and Greece followed, ranking 4th and 5th, respectively. Over the entire time period, Portugal places second, rather surprisingly. This result is a likely consequence of the fact that Roller measured only pollution levels and failed to incorporate relative change.

All these studies used indicators that can be influenced by political actors and employed the comparable cases approach by analyzing highly industrialized democracies. Therefore, they all satisfy most of the criteria for environmental performance spelled out above. However, they fall short due to an exclusive reliance on cross-sectional analysis.[5] However, there are also environmental performance indices outside political science that I would like to introduce in this chapter—two in particular.

The first is the Environmental Performance Index (EPI), conducted by the World Economic Forum, which has been published for 2006, 2008, and 2010 (Esty et al. 2006; Emerson et al. 2010). This index is composed of various indicators measuring ecosystem vitality and environmental health. The index began in 2006 by covering 17 countries, but was expanded to 163 in 2010. That means that the index has been revised substantially and repeatedly, so "it is important to note that owing to changes in the data and methods used in 2010 . . ., the results cannot be directly compared to the 2008 or 2006 Pilot EPIs" (Emerson et al. 2010, 63). The 2010 index ranked 163 countries on twenty-five performance indicators. The top-ranked countries for environmental performance were Iceland, Switzerland, Costa Rica, Sweden, and Norway. At the bottom were Angola, Mauritania, Central African Republic, and Sierra Leone. The United States, Poland, Greece, and Belgium had the poorest environmental performance within the OECD. The EPI of the World Economic Forum uses set targets to identify country-specific environmental performance. However, it is not always clear if the targets are similarly binding on political actors. An inability to conduct analysis over time is a further drawback of this index. In fact, the measures refer to the latest available data for any given country, subsequently presenting significant difficulties in determining to which actual year the data refers. This makes causal analysis highly problematic.

The Ecological Footprint Index is the only index suitable for a multi-country, time-series analysis (Wackernagel et al. 2002). It weighs the biocapacity of a country with anthropogenic impacts on the environment (i.e., the ecological footprint). For instance, biocapacity is high in countries with considerable biodiversity and a short history of industrialization, such as New Zealand and Australia. The United Kingdom and many other European countries have low biocapacity. The relation between biocapacity and the ecological footprint is expressed by an index of ecological deficits. In other words, it expresses a situation where the footprint is greater than the biocapacity. The worldwide ecological deficit is (0.8), meaning that the ecological footprint–to-biocapacity ratio was 1.3 in 2005. This in turn suggests that humans use the resources of 1.3 worlds.

The highest ecological deficit existed in the desert oil states Qatar, Kuwait, and the United Arab Emirates. In the OECD, Japan, Spain, Belgium, Greece, and the United Kingdom had the highest ecological deficits in 2005. Countries with the best balance between biocapacity and ecological footprint were Congo and Bolivia. The leading OECD countries were Australia, Canada, Finland, and Sweden. For the twenty-one established OECD countries, the ecological footprint–to-biocapacity ratio was 2.4 in

2005. Although the Ecological Footprint Index is the only environmental indicator suited for time-series cross-section analysis from 1960 to 2009, its shortcomings make it ill suited for the present analysis. First, the methodology is not very transparent and replication is probably not possible. Second, biodiversity plays a substantial role in the index. This does not meet the criteria for performance because it deals with an aspect that is difficult to influence by political action, at least in the short or medium term.

Because all established indices contain substantial conceptual and methodological deficiencies for this analysis (see also chapter 2), I decided to develop my own EPI. Although this index may also provoke controversy concerning the indicators included and their aggregation, the guiding principle is straightforward. The indicators should have an obvious environmental meaning for the most highly advanced industrial societies (a major analytical requirement of environmental performance, as discussed above). Furthermore, the data for the indicators should be publicly available (transparency and replication) and comparable between countries and over a long time period.[6]

In concrete terms, I analyze environmental performance by referring to periodically appearing OECD data (OECD Environmental Data Compendium). The advantage of using OECD data derives from its public availability and the fact that it covers all OECD countries over an extended time period. In addition, to ensure that the indicators were comparable, the OECD homogenized the data by consulting their member states. This does not solve all problems (as can be seen by the extensive footnotes in OECD publications), but the data of the OECD comes closest to being a valid and reliable comparative data set of environmental indicators.[7]

Operationalization

To obtain a comprehensive index for the environmental performance of twenty-one OECD countries (i.e., established OECD countries with a long data record), I use fourteen indicators that allow for a time-series— cross-sectional analysis from 1980 until 2005. To take into account those factors that are not a result of political action, I controlled for climate and structural changes.[8] To discover the latent dimensions of the fourteen indicators, I conducted a principal component analysis of the pollution and environmental abatement levels over all the years, extracting three factors that yield a distinct pattern.[9] For the most part, the variables fit well into the factor model. However, the variable *nuclear waste* is problematic

Table 4.1

Dimensions of environmental performance in twenty-one OECD countries

Variable	Factor 1	Factor 2	Factor 3	Uniqueness
Sulfur emissions	0.7653			0.2635
Nitrogen emissions	0.9516			0.0941
VOC* emissions	0.8442			0.2318
Carbon monoxide emissions	0.9206			0.1301
Carbon dioxide emissions	0.8277			0.2658
Municipal waste	0.5213	0.4892		0.4848
Nuclear waste	0.3285			0.8333
Freshwater abstraction	0.7044			0.4474
Glass recycling		0.7669		0.3008
Paper recycling		0.8123		0.3006
Connection to sewage		0.7777		0.3601
Fertilizer consumption			0.5735	0.6025
River pollution			0.7251	0.4299
Lake pollution			0.6549	0.5275

Explanation: Principal component analysis with orthogonal varimax rotation. A cutoff point of < 0.328 was chosen so that each variable had at least one loading on one of the three factors.
*Volatile organic compounds

due to its high uniqueness. The results of the rotated factor analysis are presented in table 4.1.

The first factor gathers variables that represent *general environmental contamination*. This factor had particularly high loadings for all air emission indicators, yet other environmentally harmful indicators loaded on the first factor as well. Among these were fresh water abstraction, nuclear waste, and municipal waste. The second factor can best be interpreted as an *environmental relief factor*. In addition to the connection to water purification plants, paper and glass recycling explicitly loaded on this factor. The last factor captures *water pollution*, which primarily comprises the contamination of rivers and lakes. The variable measuring the use of fertilizer fits into this factor well, as fertilizer-intensive agriculture results in water pollution. Therefore, the third factor combines both the indicator for the cause of contamination (use of fertilizer) and the indicator for water pollution in rivers and lakes.

For the purpose of this chapter, I create a composite indicator of environmental performance (EPI = Environmental Performance Index). I create this index by adding the general and water pollution performance indicators, then subtracting the environmental relief figure from this number. The question of how to treat the level of pollution in relation to changes, therefore, is a matter of concern. Is a country a better performer when it reduces its environmental impact, even if the reduction is from a very high initial level (Scruggs's approach)? Alternatively, is better performance best illustrated by the level of environmental degradation (Roller's approach)? As outlined above, authors have dealt with this issue very differently, and one can find good reasons for all approaches. I will follow my own research tradition (Jahn 1998) by combining level and change. I give change twice the weight of level because change is the basis of the concept of performance. To do this, I standardized both variables (level and change) between 0 and 100.

This EPI concludes that between 2001 and 2005, Sweden was the best-performing country. Sweden was firmly established in the top group of countries and was the leading country for most of the period covered in this analysis. Other countries in the top group were Norway, Switzerland, Germany, Finland, and Austria. Like Sweden, these countries performed well on environmental indices over the whole period of analysis. Further countries with an above-average balance sheet of environmental performance were Japan, Denmark, and Italy. On the other hand, Greece, Belgium, Australia, the United States, and Canada sat at the bottom of the league.

The overall index suggests no clear and uniform developments (see figure 4.1). A clear positive trend is most obvious in Sweden, Switzerland, and Germany. Finland and Norway are late comers because their positive balance sheet started in the 1990s. The positive trend in Ireland and the United Kingdom started even later, in the late 1990s or early 2000s. Austria was a leader until the late 1990s, but thereafter environmental performance decreased. Positive trends also emerged in Denmark, Belgium, the Netherlands, Canada, and the United States, although they rose from a much lower starting point. In Japan, Italy, and Spain, the trend was also positive but much weaker than the others. Australia, Greece, Portugal, and New Zealand lagged in environmental performance. New Zealand was in fact the only country with a negative trend over the period of analysis. In the early 1980s, New Zealand belonged to the top group of

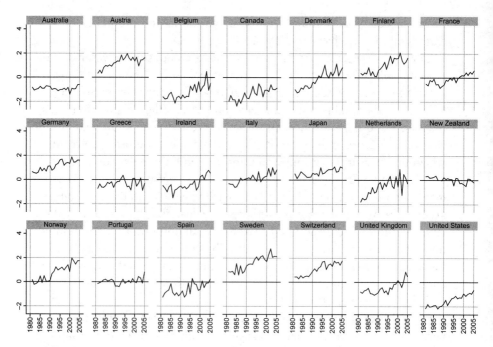

Figure 4.1

Environmental performance in twenty-one OECD countries (1980–2005)

environmental performers; at the beginning of the new millennium, it fell into the bottom third.

I conduct a validity check by comparing the index developed here to the other indices mentioned above. In addition, I compare it to the Environmental Sustainability Index (ESI) of the World Economic Forum (Esty et al. 2005), which measures the state of the environment in several countries and is comparable to the Ecological Footprint Index.[10] These two indices do not aim to measure performance rather, they aim to grasp the state of the environment in every country, meaning that the index developed here should not correlate highly with them. Table 4.2 reports the correlation coefficients for all the indices.[11]

The results show that the EPI developed in this chapter correlates to a significant degree with all other indices that focus on environmental performance. This is particularly true regarding the EPIs of Scruggs, Jahn (1998), Roller, and the World Economic Forum. Only Palmer's EPI has a lower correlation. However, his index seems to be more a "state of the environment index" than an EPI. EPIs also correlate highly among each

Table 4.2
Correlation between various environmental performance indices

	Footprint	Economic Forum ESI	Economic Forum EPI	Scruggs EPI	Roller EPI	Palmer EPI	Jahn (1998) EPI
EPI (new)	0.23	0.59	0.82*	0.63	0.75*	0.14	0.70*
Footprint		0.35	0.20	0.11	0.58	0.48	0.28
Economic Forum ESI			0.53	0.36	0.21	0.68	0.28
Economic Forum EPI				0.23	0.31	0.51	0.08
Scruggs EPI					0.29	−0.25	0.81*
Roller EPI						−0.12	0.48
Palmer EPI							0.41

Explanation: *Pearson'r with Bonferroni-adjusted significance levels of < 0.05 or less.

other, except for Roller's. It also shows that the two indices measuring the state of the environment do not correlate with the performance indices to any significant degree. This correlation matrix provides evidence that the index developed here measures what it is supposed to measure: environmental performance in twenty-one OECD countries.

Environmental Regimes in Twenty-One OECD Countries

I turn now to the second aspect of a green state, which is structural foundations. These structural foundations will be called "environmental regimes." In Esping-Andersen's (1990) analysis of the welfare state, regimes are the complex of legal and organizational features mediating the interwoven relation between the state and the economy. He measures this by looking at policy outcomes (such as social spending, salience of means-testing, or emphasis on private social insurance). As a result, he identifies three welfare state regimes that correspond to established political ideologies: liberal, conservative, and social democratic. According to Esping-Andersen, each of the eighteen OECD countries that he examined fulfills all this criteria, but only to various degrees. The Nordic states and the Netherlands are social democratic welfare states, continental European states are conservative welfare states, and the Anglo-Saxon countries, as well as Switzerland and Japan, are liberal.[12]

Connecting political ideologies to patterns or regimes in environmental policy requires a distinction that illuminates the extent that the green dimension is expressed in policy outcomes. This requires identifying the principles of a green ideology and contrasting these with the principles of current industrial societies. I do this in two steps. First, I identify the basic concepts that help us to distinguish between green and non-green regime types. Second, I offer more details about how to operationalize these concepts for the purpose of this study.

Because "green ideology" is a rather new concept, it is more difficult to identify its legal and organizational features than the classical ideologies of liberalism, conservatism, and socialism. Dryzek et al. (2003) use output variables and see environmentally related taxes as an indicator of the degree of a green state. However, in the real world, it is difficult to estimate the goal of taxes. I therefore refrain from using taxes and state spending as indicators of a green state. In contrast, I use concepts and indicators that are structurally grounded and that reproduce or change the relationship between the economy and the state (on the one hand) and the environment (on the other). To achieve this goal, I review the literature on green political theory to identify the distinctive features of a green position in contrast to a non-green position.

The literature on the ecological development of highly industrialized societies distinguishes between the different "social paradigms" that these societies pursue. "A social paradigm incorporates beliefs about how the world works physically, socially, economically, and politically" (Milbrath 1989; see also Cotgrove 1982). Unfortunately, when it comes to empirical research, it is very difficult to find indicators that identify a green state or distinguish the "new environmental" from the "dominant" paradigm. In recent years, however, there have been some attempts to develop theoretical concepts suitable for collecting data that distinguishes a green from a non-green state (Dryzek et al. 2003; Eckersley 2004; Duit 2009). Naturally, "it hardly need be said that as yet there is no green state in these terms" (Dryzek et al. 2003, 165). Thus, a measure that captures highly industrialized countries' degrees of greenness is a more acceptable approach to estimate if some states are on the way to becoming greener. For this analysis, I distinguish green states, which incorporate some elements of the new environmental paradigm, from productionist states, which abide by the dominant paradigm.

The term "productionist" elucidates the character of a state oriented toward the dominant paradigm, in which nature is used to produce goods (Milbrath 1989, 120). In other words, production and consumption are the key objectives of states where the dominant paradigm of productionism is hegemonic. However, it would be misleading to place economic growth at the center of the productionist label. We simply cannot say that states with no or little growth are green states or that green states do not seek growth. So far, no state has developed a strategy in which the idea of economic growth is subordinated to the goal of environmental protection.

Therefore, sustainable development, which was introduced by the World Commission on Environment and Development (WCED 1987), is the most far-reaching concept of greenness at present. The concept of sustainable development attempts to combine both economic growth and environmental protection. In this strategy, some have seen a reorientation, a "new politics of pollution" (Weale 1992). The basic assumption of the concept of sustainable development is that social and economic development "meets the needs of the present without compromising the ability of future generations to meet their own needs." To fill this concept with empirical substance is a controversial matter. One simple indicator of sustainable development is that economic growth is decoupled from environmental pollution. Furthermore, there are some studies that analyze to what extent industrial democracies have developed and implemented strategies for sustainable development. In their comprehensive study, Lafferty and Meadowcroft (2000) conclude that the Netherlands, Norway, and Sweden enthusiastically follow sustainable development strategies, while the United States is not interested. In their four-country comparison, Dryzek et al. (2003) conclude that Germany most closely approximated the status of a "green state," followed by Norway, Great Britain, and the United States.

To assess where states are located on the green/productionist spectrum, indicators of policy outcomes that capture the essence of this dimension are needed. Andrew Dobson (1995, 88–99) considers consumption to be a starting point for the analysis of green states when suggesting that "consumption implies depletion implies production implies waste or pollution" (Dobson 1995, 88); see also Lindberg (1977); Paehlke (1989); Goodin (1992); Neumayer (2003b); and Eder (2009). Three aspects are particularly important indicators of the structure of a green state: energy consumption, the way that energy is produced, and the priority given to alternative modes of transportation.

Energy consumption and energy policy are areas in which we can identify the two paradigms most clearly according to these approaches. Therefore, most studies aiming to identify a change of paradigm focus on the energy policy and outcomes of highly industrialized societies. Lindberg identifies the basic principle of productionism as the "energy syndrome" that guides industrialized societies (for alternatives, see also Lovins 1977). High consumption is unfavorable to a green state because it degrades the environment. Therefore, most argue that abundant energy consumption is the basic principle of a productionist state. A green state would not simply use energy more efficiently to allow for increased consumption, but also seeks to decrease the overall level of consumption. "A low-energy strategy means a low-consumption economy; we can do more with less, but we'd be better off doing less with less" (Porritt 1984, 174). Even though measuring the complex relationship between energy use and consumption is beyond the scope of this chapter, I use energy consumption as the first indicator to distinguish between green and productionist states. The higher the per capita energy consumption, the more productionist we can consider the state to be.

Aside from consumption rates, the means by which energy is produced is another highly controversial aspect in the debate on modern societies' environmental development. The distinction between nuclear energy and alternative energy sources, such as wind and solar, is used as a watershed between the two principles of societal development (Kitschelt 1983; 1984). "Our questions about whether we should use nuclear power showed some of the largest differences between DSP [Dominant Social Paradigm] supporters and NEP [New Environmental Paradigm] supporters" (Milbrath 1989, 126). The following quote brings the role of nuclear power to the forefront when it states what nuclear energy means for advocates of the new environmental paradigm:

Nuclear power stations in particular have come to have a deep symbolic significance: centralized, technologically complex and hazardous, and reinforcing all those trends in society which environmentalists most fear and dislike—the increasing domination of experts, threatening the freedom of the individual, and reinforcing totalitarian tendencies. Opposition to nuclear power is seen for many as a key issue on which to take a stand against the further advance of an alliance between state power and commercial interests. For the objectors, the material advantages from nuclear power cannot justify the risks involved. (Cotgrove and Duff 1980, 338).

As an indicator, I use the proportion of energy obtained from nuclear power, in relation to the proportion obtained from wind/solar energy (see also chapter 5, "Wind-Power Development in Germany and the United States: Structural Factors, Multiple-Stream Convergence, and Turning Points"). When nuclear energy outweighs wind and solar energy, a state is more productionist. If the proportion of wind and solar energy is higher, a state is greener.

Transportation is another aspect illustrating which of the two principles of environmental development guides societal organization (Dobson 1995, 103–104). Because the right to mobility is considered essential in liberal societies, the private automobile has emerged as a basic feature of industrial society. However, not only is mobility a basic right, but the private automobile itself is often considered a basic right, especially in productionist societies. The automobile is not only a means of transportation but also a fetish, a symbol of individual identity. In contrast, public transportation is a basic cornerstone of a green state. Collective transport is seen as environmentally friendly compared to individual automobile traffic. This suggests that the ratio of automobile traffic to public transportation can be used as a third indicator of the distinction between green and productionist states. To capture this dimension, I use the ratio of people and freight transported by road versus rail traffic as a proxy for cars/trucks versus collective transport.[13] Very similar principles apply to the relationship between road and rail when considering freight transportation.

Tractor-trailers (trucks) offer individual flexibility for businesses, while more collective freight trains offer more environmental efficiency. I therefore include the ratio of freight tonnage transported by road versus rail in the transportation indicator along the same lines.

In summation, I use three indicators to measure policy outcomes of highly industrialized societies on the green/growth dimension. A green state would (1) have low and decreasing energy consumption, (2) favor solar and wind energy over nuclear energy, and (3) favor rail transport over road transport. In a productionist society, these relationships are reversed. In the following section, I discuss the operationalization of these three indicators.

Operationalization

To obtain a valid indicator for energy consumption, I use the energy consumption per GDP unit (in US dollars) to control for economic conditions. Furthermore, because energy consumption is highly dependent on the annual winter temperature, I weight energy consumption by the Heating and Cooling Degree Months (as discussed in Jahn 2013b). The energy consumption index is finally standardized between 0 and 1. In these terms, in 2005, Australia, the United States, and New Zealand have the highest energy consumption (lowest scores), although—perhaps with the exception of New Zealand—they increased their energy efficiency. Countries with the highest energy efficiency are Switzerland, Denmark, Norway, Ireland, and Austria. From 1980 to 2005, energy efficiency increased the most in Ireland, Denmark, the United States, Canada, Great Britain, and Finland, although sometimes from very high levels, as in the case of the United States and Canada. Energy consumption increased in Italy, Switzerland, Spain, Greece, Norway, and Portugal.

For the energy mix index, I calculated the relationship between alternative or regenerative energy sources (i.e., wind and solar energy) and nuclear energy. I standardized the empirical scores over all countries and years for both indicators between 0 and 1, which means that I weighted alternative energy sources around 15 times higher than nuclear energy.[14] The final scale was again standardized between 0 and 1, meaning that a country without alternative and nuclear energy scores 0.5 on this index. If nuclear energy dominates alternative energy sources, the score is below 0.5; and in the reverse case, it is above 0.5. Only very few countries follow a green path with respect to their energy mix. The most apparent of these was Denmark. Since 2000, Denmark has increased its wind energy substantially; and because it did not use nuclear energy at all, the energy mix score was very high. Other countries that use alternative energy to a significant degree without using nuclear energy were Austria, Greece, Ireland, New Zealand, and Portugal. In all these countries, the index score was above 0.5. In Australia, Italy, the Netherlands, and Norway, the score was also above 0.5 because all these countries did not use nuclear energy—although they scored only slightly above 0.5 because they used alternative energy to a marginal degree. Countries with a substantial share of nuclear energy were France, Sweden, Belgium, Switzerland, Finland, and Canada. The United States and Great Britain also used nuclear energy, but only to a moderate degree. However, all these countries used very little wind and solar energy, so their score was well below 0.5. Finally, there were countries with a high share of nuclear *and* alternative energy. This is true for Germany and Spain, where nuclear and alternative energy were almost equally prevalent in 2005 according to my calculation. Japan also used both nuclear and alternative energy, but here, the share of nuclear energy clearly dominated.

Over time, Denmark has increased its use of alternative energy substantially and was clearly leading in this regard. This is especially obvious when considering the rapid increase in alternative energy resources (especially wind energy), observable since the late 1990s. In Spain, Greece, and Portugal, solar energy increased its share in the energy mix. On the other side, Japan, Finland, Sweden, Belgium, and France, have seen an increase in the share of nuclear energy in the last 25 years and have not developed alternative energy resources to a significant degree.

The third indicator of a green or productionist regime refers to the road/rail transport mix. The data refers to the transport of both goods and persons. For the transport of goods, I use the unit tons of freight multiplied by kilometers. For transport of persons, I use the number of persons traveling multiplied by kilometers. Both indices were standardized with the empirical scores over all countries and years separately, which results in weighting freight transport at 1.09 in relation to the transport of persons. Both indices, transport of freight and persons, contained extreme values, so I used the logarithm for this data. The index is 0.5 when both rail and road transport have equal shares; it is above 0.5 when rail transport dominates, and below 0.5 when road transport dominates. The index of freight transport in territorially large countries is a special situation because there are huge transport volumes transported by so-called mega trains. Therefore, these countries receive a high value for the rail transport of goods. This is particularly true for Australia, Canada, and the United States. However, the decision to transport goods by train in these countries is not primarily motivated by ecological concerns. In contrast to European countries, railway equipment in these countries is normally old and consists of heavily polluting diesel trains. Nevertheless, in Australia, the railway network has expanded in recent decades, which affects the transport of goods by train. In 2004, almost 3,000 kilometers of the railway track from Adelaide to Darwin were improved to normal-size rails, which made it possible to run a new, long-distance train (called The Ghan). The share of rail transport for persons is radically different in Canada and the United States. Because the traffic in goods is traditionally high and over long distances, and because the use of railways is not directed by ecological concerns but pragmatic concerns, I weighted the transport of persons double that of the transportation of goods.

In total, Austria, Germany, Australia, Sweden, and Switzerland had a "green" transport mix, where rail transport is strong compared to road transport. Road transport dominated unchallenged in the United States, Ireland, Greece, and New Zealand.

Finally, I summed up the three indicators of energy consumption, energy mix, and rail versus road transport to an index of environmental regimes. This index is composed in a similar way as the EPI, meaning that I weighted the changes twice as much as the level. The index then was standardized between 0 and 1. Table 4.3 shows the results for 1996–2005.

Germany, Switzerland, Austria, and Denmark diverge from the productionist path, at least to some degree. In these countries, structural developments accord with green ideology. Although these countries are not classified as green states, they most clearly exhibit the features that green theoretical approaches attribute to an environmental paradigm shift. The most productionist countries are Japan, Belgium, the United States, France, Greece, and New Zealand.

The Three Worlds of Environmentalism

In the last part of this chapter, I combine the two aspects of environmental development to answer two questions:

• To what extent is environmental performance connected to environmental regimes?

• Are there distinct patterns of development?

The indices in the analysis incorporate both level and change, using an average of ten years. I use such a long time frame because structural changes occur only over long periods of time. The bivariate correlation

Table 4.3

Environmental regimes in twenty-one OECD countries (1996–2005)

Country	Environmental Regime Index (0 = Productionist; 1 = Green)	Energy Consumption (0 = high; low= 1)	Energy Mix (0 = dominance of nuclear power; 1 = dominance of alternative energy)	Transport Mix (0 = dominance of road transport; 1 = dominance of rail transport)
Denmark	1	0.87	0.81	0.62
Austria	0.57	0.83	0.55	0.77
Germany	0.57	0.69	0.41	0.79
Switzerland	0.54	0.88	0.31	0.94
Norway	0.45	1	0.5	0.65
Spain	0.44	0.39	0.44	0.62
Finland	0.42	0.69	0.3	0.76
Sweden	0.39	0.67	0.09	0.86
United Kingdom	0.38	0.62	0.39	0.55
Canada	0.37	0.37	0.4	0.64
Netherlands	0.35	0.57	0.51	0.57
Italy	0.34	0.66	0.51	0.64
Ireland	0.33	0.82	0.53	0.34
Australia	0.31	0	0.51	0.82
Portugal	0.31	0.4	0.53	0.59
Japan	0.21	0.45	0.35	0.66
Belgium	0.18	0.38	0.25	0.67
United States	0.13	0.19	0.41	0.46
France	0.07	0.47	0.01	0.7
Greece	0.06	0.41	0.59	0.03
New Zealand	0	0.14	0.51	0.11

between both variables is 0.37 over the whole period of analysis. However, because environmental policy is a new field and the cleavage structures are not as strongly institutionalized as other ideological dimensions, it is not as clear in this situation as it is in social policy, for instance, whether structure determines policy or whether policy decisions still shape structure. To shed some light on this question, I conducted regression analyses with different time lags, using both environmental policy performance and environmental regimes as the dependent variables. The strongest correlation is after one year, when we use environmental performance as the dependent variable and environmental regime as the independent variable. This suggests that sociostructural factors have a significant impact on environmental performance. They actually explain 13 percent of the variance. Reversing the causal relationship shows that environmental performance affects environmental regimes as well. However, here the time lag is five years and the explanatory power is 11 percent. This fact shows that structure is reshaped by policy decisions in the medium to long term.

These results make sense from an agency-structure perspective: policy (defined as agency) is determined by structure (regimes) in the short run. However, persistent changes in policy, and therefore outcomes, affect the structure in the long run. In this respect, the results of this study relate to the classic debate about agency and structure from Karl Marx, Emile Durkheim, and Max Weber.

To identify patterns, I cross-tabulated the two variables and conducted a cluster analysis.[15] I divided the scatterplot into sections that correspond to the identified clusters of the analysis. The results are shown in figure 4.2.

The scatterplot shows that there are three worlds of environmentalism. This interpretation is also confirmed by the cluster analysis, which identifies two main clusters: countries with high environmental performance and countries with less successful environmental performance. Within these two clusters are subclusters. The environmentally successful countries of Austria, Germany, and Switzerland constitute the most homogeneous cluster in the whole analysis. Norway, Finland, and Sweden are clearly divided from this homogeneous group of countries. Denmark, because of its very isolated position, constitutes its own cluster. This constellation can be interpreted to suggest that Austria, Germany, Switzerland, and Denmark constitute the first world of environmental policy, which combines a high score in environmental performance and an environmental regime. Norway, Finland, and Sweden, in contrast, form a group

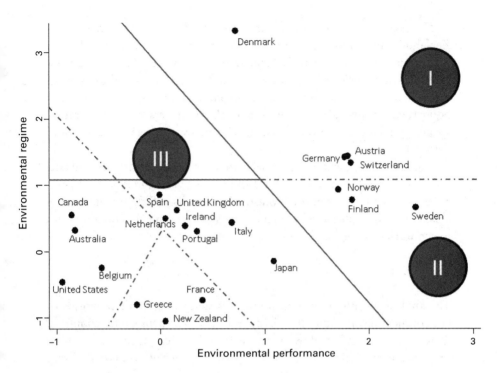

Figure 4.2

The three worlds of environmentalism (1996–2005)

of countries with a high environmental performance but a productionist policy regime. The country with the most successful environmental
performance in the last decade is Sweden. However, this success did not
translate into (nor was it motivated by) a decoupling from productionist
development. Denmark, in contrast, initiated many steps in the direction
of a green state. However, its environmental performance is just slightly
above average.

The environmentally less successful states can also be divided into several groups. Here, the clearest dividing lines are among three groups: first,
Canada, Australia, Belgium, and the United States; second, Greece, New
Zealand, and France; and third, the other countries, including the Netherlands and the United Kingdom.

The first group of countries has the lowest environmental performance
and adheres firmly to productionist development. All these countries except Belgium have a large territory and belong to the "New Anglo-Saxon
World" countries.

The second group constitutes the most productionist countries. New Zealand is the most productionist country in the sample, but Greece and France are also extremely productionist countries. All the countries in this cluster, perhaps with the exception of France, also show a rather low environmental performance. In fact, these two subclusters can be merged because all the countries in these clusters combine a rather strong attachment to productionism with below-average environmental performance.

The last group includes countries with a moderate environmental performance (only Japan and Italy perform above average in this group) and a clear attachment to productionism. Japan falls a little bit out of this cluster, which implies that it might have some features in common with the countries in the second cluster.

Conclusion

The analysis in this chapter demonstrates that there are three worlds of environmentalism among the highly industrialized democratic countries. Most countries still follow a productionist developmental trajectory. However, environmentally successful performance can take two different avenues: countries diverging from a productionist paradigm to a certain degree on the one hand, and continuing adherence to productionism on the other. Countries without success in environmental performance are all productionist.

Unlike with welfare states, these groupings do not correspond as well into geographical categories or into families of nations. However, some patterns are obvious. Countries with a green and successful environmental performance style are the three "German-speaking countries" of Austria, Switzerland, and Germany. The Nordic countries—with the clear exception of Denmark—belong in the group of environmentally successful states that follow the spirit of productionism. There are some indications that Japan also falls in this category. This dividing line between green and productionist development has also been identified in studies comparing the ecological discourses of political actors and the mass media in Sweden and Germany (Jahn 1999).

The less environmentally successful countries with productionist development fall into several groups. First, there is the group of countries with a large territory and low population density (Australia, New Zealand, Canada, and the United States). All these countries belong to the non-European, Anglo-Saxon family of nations. The European,

Anglo-Saxon countries (namely, the United Kingdom and Ireland) are placed with most of the Mediterranean countries (Italy, Spain, and Portugal) and the Netherlands into a group of productionist countries with slightly below-average environmental performance. In this group, a big surprise is the Netherlands. In the past, the Netherlands was often a leading nation in both environmental performance and the development of a green state. However, in the last decade, the Netherlands lost this position and now finds itself among other less successful productionist countries. This conclusion results from the fact that development in both dimensions fails to meet the criteria for environmental success and a green state.

The Netherlands is also a good example for a country where policy and outcome do not move in the same direction. The Netherlands undoubtedly belongs to the group of countries that perform very well when it comes to the introduction of environmental institutions and policies (Holzinger et al. 2008; see also chapter 3, "Explaining Environmental Policy Adoption: A Comparative Analysis of Policy Developments in Twenty-Four OECD Countries," and chapter 6, "Early Bird or Copycat, Leader or Laggard? A Comparison of Cross-National Patterns of Environmental Policy Change"). However, this leading role in institution building has only a moderate effect on outcomes. This is particularly true for the last decade. The gap between institution building and policy on the one hand, and outcomes on the other, indicates that more research is necessary to reveal the causal mechanisms between environmental policy and institution building and environmental performance.

The analysis in this chapter shows that there is a similar divide of countries in the field of environmental policy as there is in welfare state research. The major conclusion is that ideology matters in policy research. As social democrats in (government) power have been able to establish a comprehensive welfare state, green ideology goes with less productionist development. However, because no country with a green environmental regime has a record of poor environmental performance, green ideology is the most effective element in enhancing environmental performance. However, good environmental performance also can emerge within the productionist logic of societal development, as exemplified by most Nordic states. The contrast emerges between large-scale technology, such as nuclear energy or high energy consumption, and small-scale technology, such as solar or wind energy. This divide was identified in political theory twenty or thirty years ago, and it has become a guiding principle in the field of environmental policy in the twenty-first century.

Notes

1. The chapter shows results of "Environmental Pollution as a Global Phenomenon," a research project sponsored by the German Research Foundation (DFG) and conducted by me. I wish to thank Kathrin Deadlow, Bertram Welker, Stefanie Korte, and Christoph Oberst for data collection. I thank Douglas Voigt for editing the chapter, which was written during the time I was a permanent fellow at Kolleg-Forschergruppe (KFG) in "The Transformative Power of Europe" program at the Free University Berlin, directed by Tanja A. Börzel and Thomas Risse.

2. Martin Jänicke (2008) summarizes the debate on the "Eco-State" (*Umweltstaat*), which has its roots in the 1980s (Kloepfer 1989). This debate is extensive, increasing in volume in the late 1990s and into the 2000s (Dente 1998; Mol and Buttel 2002; Dryzek et al. 2003; Eckersley 2004; Christoff 2005; Meadowcroft 2005; Spaargaren et al. 2006; Eisner 2007; Duit 2009). Most of these studies focus on the institutionalization of ecological principles in various areas of polity and policies. However, even if some draw parallels between the development of the welfare state and the Eco-State (see Meadowcroft 2005), they do not refer explicitly to the programmatic claims of green ideology, such as I do in this chapter.

3. This is even true for climate change policy, where political actors introduced set targets. However, these targets were disputed strongly and, over time, revised frequently (Gupta 2010).

4. Some also claim that environmental performance should be of strategic and political use, just like GDP (Jesinghaus 2012). Although this is an important consideration, practical politics alone certainly cannot guide scientific analysis.

5. Crepaz pools two time points. However, he does not really conduct a time-series—cross-sectional analysis. Roller (2005) compares four time periods; however, she does not conduct a pooled data analysis. This was done later for individual indicators of air emissions by Eric Neumayer (2003a).

6. These requirements exclude indicators, such as biodiversity, where there is no reliable data on a regular basis over a longer time period and comparable across countries. This, of course, implies that we may reach different results once other data are available, subsequently implying that other aspects of environmental performance become available for measurement. However, this problem is inherent in a concept of environmental performance as described above because every index is provisional. It is also inherent in any scientific research because analytical tools improve when more data become available.

7. In this context, it is necessary to note that the more important environmental data have become to society and politics, the more difficult it has been for the OECD to compile a comparable data set. In short, country administrations are increasingly afraid of negative sanctions if their countries perform poorly compared to others. This even led to an interruption after 2004 in the regular, biannual rhythm of publication that existed from 1987 to 2004 (interviews with OECD officials; for further information on environmental performance reports, see Lehtonen 2007). Currently, there is no longer any regularly appearing publication on environmental data by the OECD, and the latest data refer to 2005–2006.

8. This was done by using the predicted values of a regression analysis that included the Heating and Cooling Degree Months for each year and country (see Jahn 2013a,b) and the significant industrial sectors. This applied to all indicators except recycling rates and connection to wastewater treatment systems. Even if this procedure is only a rough adjustment, it is better than using the raw data. An ideal analysis must control for other factors as well. However, such an analysis would need specific analytical models, which exceed the scope of this chapter.

9. I use an orthogonal varimax rotation here, although other kinds of rotation reach very similar results. The first factor explains 51 percent of the variance, the second 23 percent, and the third almost 16 percent. Overall, the factor analysis explains more than 90 percent of the variance. The eigenvalues of the factors amount to 4.95, 2.26, and 1.51.

10. According to the ESI index, Canada, Iceland, Sweden, Uruguay, Norway, and Finland are leading. Germany ranks thirty-first, just before Namibia and Russia. At the end of the list are North Korea, Taiwan, Turkmenistan, and Iraq. At the end of the list of the OECD countries are the United Kingdom, Italy, Greece, Spain, and Belgium.

11. In conducting the correlation analysis, if an index had more time points and periods than the other to which it was compared, then the closest years between the indicators were used. For this EPI, I mainly used the 1991–1995 period. To calculate significance levels, I applied the Bonferroni adjustment, which is a method used to address the problem of multiple comparisons.

12. The classification of the United Kingdom, Ireland, and New Zealand is ambiguous in these terms.

13. Data exists only for this rough distinction. There is no comparable data over time for public versus private automobile transport in urban areas, despite the fact that the daily use of private bicycles in urban areas is a mainstay of green ideology, further complicating the measurement.

14. This measure of the index is based on a comparison of the countries included in this study and does not necessarily reflect the real relationship between the use of alternative and nuclear energy. This method of weighting leads to an overemphasis on alternative energy. However, in the framework of this analysis, it is feasible to do so because the use of the raw data would have concealed the role of alternative energy sources. The decision, however, leads to a bias in favor of a green state—meaning that I may overestimate the degree of greenness of highly industrialized states.

15. I applied a cluster analysis with both variables, environmental performance and environmental regimes, for the period 1996–2005. I used the within-groups linkage method. Here, the dissimilarity between cluster A and cluster B is represented by the average of all the possible distances between the cases within a single new cluster determined by combining cluster A and cluster B. Furthermore, I used the Chebyshev distance, which is the maximum absolute difference between a pair of cases on any one of the two or more dimensions that are being used to define distance. The Chebyshev distance is an effective method for hierarchical clustering with a single linkage criterion.

References

Bobbio, Norberto. 1996. *Left and Right: The Significance of a Political Distinction*. Cambridge, UK: Polity Press.

Christoff, Peter. 2005. Out of Chaos, a Shining Star? Toward a Typology of Green States. In *The State and the Global Ecological Crisis*, ed. John Barry and Robyn Eckersley, 25–52. Cambridge, MA: MIT Press.

Cotgrove, Stephen F. 1982. *Catastrophe or Cornucopia: The Environment, Politics, and the Future*. Chichester, UK: Wiley.

Cotgrove, Stephen F., and Andrew Duff. 1980. Environmentalism, Middle-Class Radicalism, and Politics. *Sociological Review* 28(2): 333–351.

Crepaz, Markus M. L. 1995. Explaining National Variations in Air Pollution Levels: Political Institutions and Their Impact on Environmental Policy Making. *Environmental Politics* 4:391–414.

Dahl, Robert A. 1967. The Evaluation of Political Systems. In *Contemporary Political Science: Toward Empirical Theory*, ed. De Soola Pool Ithiel, 166–181. New York: McGraw-Hill.

Dente, Bruno. 1998. "Towards Sustainability: Instruments and Institutions for the Ecological State." Paper presented at the Fifth Workshop of the Concerted Action, "The Ecological State," Florence, Italy, European University Institute, May 25, 1998.

Dobson, Andrew. 1995. *Green Political Thought: An Introduction*. London: Routledge.

Dryzek, John S., David Downes, Christian Hunold, David Schlosberg, and Hans-Kristian Hernes. 2003. *Green States and Social Movements: Environmentalism in the United States, United Kingdom, Germany, and Norway*. Oxford: Oxford University Press.

Duit, Andreas. 2009. "The Ecological State: Cross-National Patterns of Environmental Governance Regimes." Paper presented at ECPR General Conference, Potsdam, Germany; September 10–12, 2009.

Eckersley, Robyn. 2004. *The Green State: Rethinking Democracy and Sovereignty*. Cambridge, MA: MIT Press.

Eckstein, Harry, ed. 1971. *The Evaluation of Political Performance: Problems and Dimensions*. Beverly Hills, CA: Sage.

Eder, Klaus. 2009. *Social Construction of Nature: A Sociology of Ecological Enlightenment*. London: Sage.

Eisner, Marc Allen. 2007. *Governing the Environment. The Transformation of Environmental Regulation*. Boulder, CO: Lynne Rienner Publishers.

Emerson, Jay, Daniel C. Esty, Marc A. Levy, Christine Kim, Valentina Mara, Alex de Sherbinin, Tanja Srebotnjak, and Malanding Jaiteh. 2010. *2010 Environmental Performance Index*. New Haven, CT: Yale Center for Environmental Law and Policy.

OECD Environmental Data Compendium. 2004. [Issued biennially from 1989–2005.] OECD: Paris.

Esping-Andersen, Gøsta. 1990. *The Three Worlds of Welfare Capitalism*. Princeton, NJ: Princeton University Press.

Esty, Daniel C., Marc A. Levy, Tanja Srebotnjak, and Alex de Sherbinin. 2005. *2005 Environmental Sustainability Index: Benchmarking National Environmental Stewardship*. New Haven, CT: Yale Center for Environmental Law and Policy.

Esty, Daniel C., Marc A. Levy, Tanja Srebotnjak, Alex de Sherbinin, Christine Kim, and B. Anderson. 2006. *Pilot 2006 Environmental Performance Index*. New Haven, CT: Yale Center for Environmental Law and Policy.

Fleming, James Rodger. 1998. *Historical Perspectives on Climate Change*. Oxford: Oxford University Press.

Goodin, Robert E. 1992. *Green Political Theory*. Cambridge: Polity Press.

Gupta, Joyeeta. 2010. A History of International Climate Change Policy. *Wiley Interdisciplinary Reviews: Climate Change* 1(5): 636–653.

Gurr, Ted Robert, and Muriel McClelland. 1971. *Political Performance: A Twelve-Nation Study*. Beverly Hills, CA: Sage.

Holzinger, Katharina, Christoph Knill, and Arts Bas. 2008. *Environmental Policy Convergence in Europe. The Impact of International Institutions and Trade*. Oxford: Oxford University Press.

Jahn, Detlef. 1998. Environmental Performance and Policy Regimes: Explaining Variations in 18 OECD-Countries. *Policy Sciences* 31:107–131.

Jahn, Detlef. 1999. The Mobilization of Ecological World Views in a Post-Corporatist Order. In *New Perspectives on Environmental Policy in Europe*, ed. Ute Collier, Gokhan Orhan, and Marcel Wissenburg, 129–155. Aldershot, UK: Ashgate.

Jahn, Detlef. 2013a. Temperature Per Capita: Measuring Annual Heating and Cooling Degrees in 21 OECD Countries from 1960–2005. *Greifswald Comparative Politics—Working Paper Series*, No. 1/2013.

Jahn, Detlef. 2013b. The Impact of Climate on Atmospheric Emissions: Constructing an Index of Heating Degrees for 21 OECD Countries. *Weather Climate and Society* 5(2): 97–111.

Jahn, Detlef. Forthcoming. *Agenda Setter and Veto Player: Environmental Performance and Politics in Highly Globalized and Industrialized Democracies*. Book manuscript.

Jänicke, Martin. 2008. Schritte auf dem Weg zum "Umweltstaat" – Umweltintegration am Beispiel Deutschlands. In *Megatrend Umweltinnovation. Zur ökologischen Modernisierung von Wirtschaft und Staat*, ed. Martin Jänicke, 146–169. München: Oekom.

Jesinghaus, Jochen. 2012 Measuring European Environmental Policy Performance. *Ecological Indicators* 17:29–37.

Kitschelt, Herbert. 1983. *Politik und Energie: Energie-Technologiepolitiken in den USA, der Bundesrepublik Deutschland, Frankreich, und Schweden.* Frankfurt am Main: Campus.

Kitschelt, Herbert. 1984. *Der ökologische Diskurs. Eine Analyse von Gesellschaftskonzeptionen in der Energiedebatte.* Frankfurt am Main: Campus.

Kloepfer, Michael, ed. 1989. *Umweltstaat.* Berlin: Springer-Verlag.

Lafferty, William M., and James R. Meadowcroft. 2000. *Implementing Sustainable Development. Strategies and Initiatives in High- Consumption Societies.* Oxford: Oxford University Press.

Lehtonen, Markku. 2007. Environmental Policy Integration Through OECD Peer Reviews: Integrating the Economy with the Environment or the Environment with the Economy? *Environmental Politics* 16(1): 15–35.

Lijphart, Arend. 1975. The Comparable Case Strategy in Comparative Research. *Comparative Political Studies* 8:158–177.

Lindberg, Leon N. 1977. *The Energy Syndrome: Comparing National Responses to the Energy Crisis.* Lanham, MD: Lexington Books.

Lovins, Amory. 1977. *Soft Energy Path: Towards a Durable Peace.* Cambridge, MA: Ballinger Publishing Company.

Meadowcroft, James R. 2005. From Welfare State to Ecostate. In *The State and the Global Ecological Crisis*, ed. John Barry and Robyn Eckersley, 3–24. Cambridge, MA: MIT Press.

Milbrath, Lester W. 1989. *Envisioning a Sustainable Society: Learning Our Way Out.* Albany: State University of New York Press.

Mol, Arthur P. J., and Frederick H. Buttel, eds. 2002. *The Environmental State under Pressure.* Oxford, UK: Elsevier Science.

Neumayer, Eric. 2003a. Are Left-Wing Party Strength and Corporatism Good for the Environment? Evidence from Panel Analysis of Air Pollution in OECD Countries. *Ecological Economics* 45:203–220.

Neumayer, Eric. 2003b. *Weak Versus Strong Sustainability. Exploring the Limits of Two Opposing Paradigms.* Cheltenham, UK: Edward Elgar.

Paehlke, Robert C. 1989. *Environmentalism and the Future of Progressive Politics.* New Haven, CT: Yale University Press.

Palmer, Monte. 1997. *Political Development: Dilemmas and Challenges.* Itasca, IL: Peacock.

Pillarisetti, J. Ram, and Jeroen C. J. M. van den Bergh. 2010. Sustainable Nations: What Do Aggregate Indexes Tell Us? *Environment Development and Sustainability* 12:49–62.

Porritt, Jonathon. 1984. *Seeing Green: The Politics of Ecology Explained.* Oxford, UK: Blackwell.

Roller, Edeltraut. 2005. *The Performance of Democracies: Political Institutions and Public Policy.* Oxford: Oxford University Press.

Scruggs, Lyle. 2003. *Sustaining Abundance. Environmental Performance in Industrial Democracies*. Cambridge: Cambridge University Press.

Spaargaren, Gert, Arthur P. J. Mol, and Frederick H. Buttel, eds. 2006. *Governing Environmental Flows: Global Challenges to Social Theory*. Cambridge, MA: MIT Press.

Wackernagel, Mathis, Monfreda Chad, and Diana Deumling. 2002. *Ecological Footprints of Nations. November 2002 Update*. Oakland, CA: Redefining Progress.

WCED, The World Commission on Environment and Development. 1987. *Our Common Future*. [Brundtland Report] Oxford: Oxford University Press.

Weale, Albert. 1992. *The New Politics of Pollution*. Manchester, UK: Manchester University Press.

5

Wind-Power Development in Germany and the United States: Structural Factors, Multiple-Stream Convergence, and Turning Points

Roger Karapin

Renewable energy is seen as a solution to problems of anthropogenic climate change, air pollution, resource scarcity, and dependence on energy imports (Vasi 2009; IPCC 2012). Of the various forms of renewable energy, wind-generated electricity has a unique set of advantages, which makes its potential environmental benefits especially large. Wind power produces relatively low levels of environmental damage over its life cycle (like solar),[1] relies on relatively mature technology, and already comprises a nontrivial share of energy production (like hydro and biomass). It is also based on a potentially enormous natural resource and has been growing rapidly in many industrialized and developing countries (Greene et al. 2010). Global installed capacity of wind power rose an average of 25 percent a year from 2001 to 2012, to a total of 282,000 megawatts (MW), and wind power now generates about 3 percent of global electricity consumption (according to data from World Wind Energy Association and BP 2012).[2]

Although common experiences with the oil crises of the 1970s and more recent concerns about global warming have motivated most industrialized countries to adopt wind-power development policies, they vary greatly in the extent to which they have developed this renewable-energy source successfully. The key indicator used in this chapter for description and causal explanation is the share of a country's total electricity generation or consumption that is produced from wind, because this describes the extent of wind-power development while controlling for the size of the country's economy and energy sector.[3] By focusing on an outcome measure, this chapter concerns environmental performance; by choosing a specialized outcome measure that is directly linked to improvements in environmental quality, it avoids the problems with aggregated indices noted in chapter 2, "Comparing Environmental Performance," in this volume. In terms of electricity generated from wind as a share of total national electricity consumption, the leading countries in the world in 2012 were Denmark (30 percent), Portugal (20 percent), Spain (18 percent), Ireland (15 percent), and Germany (8 percent). Laggards among the industrialized democracies, according to this metric, include Italy (4 percent), the United States (3.5 percent), Canada (3 percent), Australia (3 percent), and Japan (0.5 percent) (IEA 2012, 5).

Questions, Methods, and Case Selection

The central question that this chapter addresses is whether these differences among countries are simply due to structural constraints (i.e., factors that political actors cannot alter significantly in the short or medium term), or whether to a large degree they are due to political processes that wind-power advocates could initiate, influence, or exploit. This question lends itself to a comparative approach to use countries' contrasting experiences as a basis for explanation and theory development. A cross-national comparison is appropriate because wind-power policies are largely made by national governments, although laggard European countries (not included here) may be affected by the renewable-energy policies of the European Union (EU), such as its 2001 directive setting national targets. At the same time, the role of subnational governments in federal systems must be taken into account where it is of major importance for a country's wind-power development.

In this chapter, I will focus on two country cases to examine a large number of process variables, which are difficult to include in large-N studies because of the unavailability of reliable, comparable information on those variables across many cases. Thus, I aim to correct for the bias of large-N studies toward structural factors (cf. Karapin 2012). A small-N approach also facilitates longitudinal analysis, and the comparative analyses in this chapter will include cross-temporal comparisons within each country, in addition to comparisons between them. To aid theory development further, I will assess multiple theories rather than merely providing support for one approach. This approach helps guard against assuming that one's favorite theory is correct and makes it possible to discover where different theories are each applicable (Sabatier 2007, 330; Zahariadis 2007, 86–87).

Germany and the United States comprise a useful comparison for addressing this chapter's central question and for building theory, for several reasons. They currently represent a leading and a laggard country, respectively, in terms of wind-power development as a share of total electricity generation, and hence provide a clear contrast to be explained, even though this does not mean that their relative positions are necessarily fixed, as I will explain below. More generally, the two countries represent what Jahn terms the first and third worlds of the environmental state, respectively; and in fact, they occupy extreme positions in that typology (see chapter 4, "The Three Worlds of Environmental Politics," esp. figure 4.2 and table 4.3, in this volume). At the same time, in absolute terms, both the United States and Germany are very significant wind-power producers in the global context. Because of the large size of its energy sector, the small share of wind power generated in the United States still translates into a large absolute volume. In terms of absolute wind-power capacity, the United States is in second place globally, while Germany is in third. China leads in total capacity, although its wind sector provided only 2.0 percent of its national electricity demand in 2012. Together, the United States and Germany account for about 32 percent of global capacity, so explaining these two country cases is of inherent importance for understanding the politics of wind power worldwide (IEA 2013, 5-6).

There is another, perhaps more important, reason for comparing these two countries. Plausible structural explanations have been advanced to explain the differences between their wind sectors, so comparing them provides a good test of the structural theories. Given the structural differences between the two countries, was it inevitable that Germany would increase its wind share so much more than the United States? Or did political and other processes, which could have unfolded differently, play a significant role? If so, under what conditions did actors have significant scope for influencing these outcomes?

The following section describes two contrasting theoretical perspectives that will be assessed in this chapter, one based on socioeconomic structures and political institutions, and the other on the interaction between problem and political streams. Next, I briefly describe the differing wind-power outcomes and policies in Germany and the United States. The rest of the chapter explains the differences between the two countries in a series of analyses that examine structures first, and then processes. These analyses go beyond a static comparison by distinguishing three phases: US leadership (1978–1993); Germany's surge into a growing leadership role (1993–2004); and the beginning of catch-up by the United States (2004–present). The conclusion to the chapter discusses how certain combinations of structures and processes drove the turning points between these phases, and considers the implications for theories of environmental outcomes and for multiple-streams theory.

Theoretical Perspectives

Implicit in many causal analyses of environmental policies and outcomes is the question of how much difference actors such as government officials, environmental organizations, political parties, or coalitions of actors have made and could make in the future. To bring this question into sharp relief, it is useful to contrast a structural theory of environmental performance with the multiple-streams theory of policymaking.[4]

A Structural Approach: Socioeconomic Structures and Political Institutions

Major works on environmental outcomes and on climate or energy policies argue for the importance of structural features; i.e., those that are basically unchanging over the medium term (e.g., Kitschelt 1986; Jänicke et al. 1996; Paterson 1996; Jahn 1998; Neumayer 2003; Scruggs 2003). Some studies of renewable energy also take this approach (e.g., Huang et al. 2007; Keller 2010; Toke et al. 2008). These literatures examine a wide range of structural factors, such as wind resources, fossil-fuel endowments, export dependence, the openness of policymaking institutions, electoral systems, the strength of implementation institutions, planning systems, and the strength of landscape protection organizations. These factors are too numerous to treat adequately in this chapter, so to provide the best test of the structural theory, here I will describe four factors that seem best able to explain the differences between the United States and Germany (Keller 2010). While these factors refer to conditions that may change somewhat over time, the rate of any change is very slow, so they can be seen as structures with basically stable effects on these two countries from the 1970s to the present.

First, countries' energy policies are influenced by their fossil-fuel endowments and industries. Those with less domestic production of fossil fuels, and therefore greater dependence on energy imports, are more likely to develop policies that promote renewable energy at the expense of fossil fuels. Their fossil-fuel-sector lobbies are relatively weak, and thus less able to block renewable-energy policies. Moreover, such countries have a national interest in developing domestic energy sources to improve their balance of payments and reduce dependence on oil and natural-gas imports, because these have been subject to price volatility and supply insecurity (Paterson 1996, 80; Keller 2010, 4741). Second, countries dependent on manufacturing and on exports have an interest in promoting industries that would create new manufacturing jobs, in part to meet international demand. Because wind power is a much more labor-intensive form of energy production than coal or natural gas, it is in such countries' interests to promote it over fossil fuels (Gipe 1991, 763).

Third, the nature of the electoral system affects a political system's openness to innovations in environmental and energy policy. In countries with proportional representation, ecological or left-libertarian political parties were more likely to become established by the early 1990s than in countries where electoral rules strongly favor major parties at the expense of minor parties, such as the United States (Willey 1998). Green parties have been strong advocates of renewable-energy policies and may influence policy through participation in government, by exerting competitive pressure on other parties, and by playing crucial roles in advocacy coalitions (Neumayer 2003, 205, 218–219; Karapin 2012, 60–61).

Fourth, the nature of the interest-group system is also seen to bear on environmental policies, including renewable energy. Pluralist interest-group systems are marked by the fragmentation of business interests and an adversarial relationship between business and government, and hence relatively unstable or uncertain environmental policies. Neocorporatist systems are characterized by a relatively centralized and concentrated representation of interests, cooperative relations between business and government, and more stable policies that are more smoothly implemented by economic actors (Crepaz 1995; Jahn 1998, 119–120, 125; Scruggs 2003, 133–161).

A Process Approach: Stream Convergence, Policy Windows, and Advocacy Coalitions

By contrast, most of the cross-national literature on wind power, as well as some key works specifically on Germany and on the United States, argue for the importance of short-term, relatively steerable processes. These include the positions and strategies of economic, nongovernmental, and state actors, mobilization by environmental movements, advocacy-coalition formation, issue framing, policy feedback, market formation, election results, and policy design features (e.g., Kraft and Axelrod 1984; Bird et al. 2005; Jacobsson and Lauber 2006; Swisher and Porter 2006; Gan et al. 2007; Walz 2007; Sovacool 2008; Stenzel and Frenzel 2008; Laird and Stefes 2009; Portman et al. 2009; Szarka 2007; Vasi 2009). The multiple-streams theory of agenda setting and policymaking can subsume many of these variables into a coherent framework (Kingdon 2003; Zahariadis 2007). In this theory, an issue is most likely to reach the decision-making agenda when intense problem perception, viable policy solutions, and strong political commitment converge. Each of these is affected both favorably and adversely by streams of events and other processes, which are partly but not completely independent of each other. When problem and politics streams converge to produce both the perception of a severe problem and strong political commitment to address it, a policy window is created. This window may be exploited by policy entrepreneurs who promote particular solutions, which already have been generated and tested in their own complex, slow-moving stream (Kingdon 2003, 15–18, 203). Similar to theories of critical junctures (Collier and Collier 1991) and of punctuated equilibrium (Baumgartner and Jones 1993), the theory of multiple streams attempts to locate and explain moments of unusual openness, when policies that are usually stable may rapidly change.

Here, I will focus on the problem and politics streams. Doing so will simplify the comparative analysis and provide a somewhat tougher test of this process theory because policy streams are those most amenable to influence by political actors. Problem streams concern which problems are seen as most important and hence get onto policy makers' decision-making agendas. A problem stream is altered significantly through extraordinary focusing events (such as environmental disasters), dramatic new information about environmental conditions, strong feedback from existing policies, or all three. The politics stream concerns the people who hold power, their ideological or value-based commitments, and the political constraints that they anticipate. This stream is affected by election results, changes in government leadership positions, shifts in public opinion, and mobilization by organized groups that affect their balance of power (Kingdon 2003, 197–198).

The politics stream requires further elaboration, partly to capture more fully the influences on decision making as opposed to agenda setting. Hence, I will use certain concepts from the advocacy-coalition framework, which is largely consistent with and complementary to multiple-streams theory.[5] Among other things,[6] advocacy-coalition theory analyzes the changing balance of power between a coalition of advocates for a policy direction and a coalition of their opponents (Sabatier 1988; Sabatier and Weible 2007). Both kinds of coalition can draw on specialists in a variety of government and private organizations, including government agencies, parties, parliaments, interest groups, nonprofit organizations, social-movement organizations, research institutions, and media outlets (Jänicke 2005, 138; Jost and Jacob 2004; Watanabe 2011). The composition, political resources, and constraints of advocacy and opposing coalitions can be influenced by elements of the problems and politics streams, such as environmental crises, other socioeconomic developments, shifts in public opinion, changes in the governing coalition, or policies in other subsystems (Sabatier and Weible 2007, 202).

The combination of these two theories offers to explain the development of renewable-energy policy. When problem and policy streams converge, they open a policy window. At such times, if the advocacy coalition is more powerful and mobilizes more energetically than the opposition coalition, a stronger policy will result.

Testing and Interrelating Structural and Process Theories

Although structural and process theories often appear as rivals in explanations of environmental outcomes, in fact they can be complementary. Few studies of renewable energy examine both structures and processes,[7] and fewer still analyze the relative weights of these causes or how they relate to each other.[8] Perhaps without intending to, studies that examine only structural factors imply that these comprise a sufficient explanation, while studies that are limited to an analysis of processes imply that the latter are relatively unaffected by structures.

In the rest of this chapter, I try to contribute to our understanding of the relative causal weights of structures and processes and how they fit together in producing environmental outcomes. First, the structural theory described above will be tested for congruence between the putative causes and the outcomes (e.g., whether Germany has a high degree of reliance on manufacturing that would predispose it to renewable energy more than the United States). Then, in the case studies, the multiple-streams theory will be tested through process tracing to try to produce causal chains of mechanisms that stretch from processes that generate the problem and politics streams (such as the 1970s oil crisis and election results in each country), to the creation of policy windows at times of stream convergence, and on to policies and outcomes.

In addition, there are several ways to test these theories against each other, which will be employed in the empirical sections of the chapter and discussed in its conclusion. Are the stable outcomes predicted by the structural theory actually produced, in the form of a relatively invariable gap between wind-power development in Germany and the United States? Or, alternatively, do the relative positions of the two countries vary over time in ways that can be explained by process variables? Are the mechanisms identified by the process theory, such as the development of a coalition of wind-power advocates, simply determined by structural factors, such as electoral institutions that promote ecological parties; or do they depend on extraneous and contingent factors such as election results or the timing of events? Finally, are structural features, such as the shape of the interest-group system, themselves influenced by the processes studied? By using each theory to critique the other, I aim to improve our understanding of the scope and limits of each and of how they complement each other.

Wind-Power Outcomes and Policies

Both Germany and the United States recently have experienced rapid growth in wind power, but the takeoff occurred later in the United States, and hence its wind share of electricity generation was still less than half of Germany's in 2012 (see the two black lines in figure 5.1).[9] In Germany, wind power began to grow rapidly in 1991, sustained an average growth rate of 39 percent per year over the 1994–2004 period, and reached 16,600 MW in 2004. Although growth slowed markedly since then, it still averaged 8 percent per year during the 2004–2012 period (the solid gray line in the figure). In 2012, wind power generation had reached

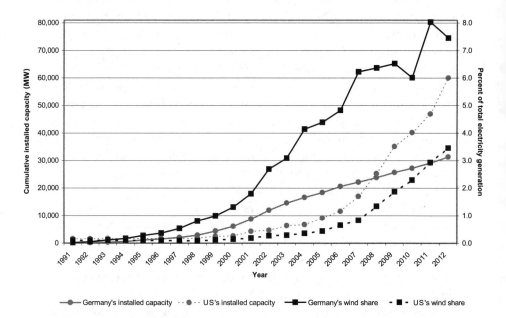

Figure 5.1

Cumulative installed wind power capacity and wind shares of electricity generation in Germany and the United States, 1991–2012

46,000 gigawatt hours (GWh), or 7.4 percent of total electricity generation in Germany (the solid black line).

By contrast, in the United States, growth in wind power has been uneven. Installed capacity grew rapidly, at 34 percent a year, over the 1983–1989 period, but it reached only 1,400 MW in 1989. Then, capacity grew very slowly for the next decade, rising only 3 percent per year, from 1989 to 1998. The real takeoff of wind power in the United States began in 1999, which was eight years later than in Germany. Installed capacity then grew at an average rate of 28 percent a year in the 1998–2012 period and reached 60,000 MW in 2012 (see the dashed gray line in the figure). However, because of its relatively late takeoff, wind power still made up only 1.8 percent of US electricity generation in 2009, although this had risen to 3.5 percent in 2012 (shown by the dashed black line).

It is clear that the state, including both national and subnational levels, has been crucial for the growth of wind power. The development of wind power requires government support for research and development, subsidies to level the playing field (at least partially) with energy sources that have large environmental externalities and their own government subsidies, and regulatory assistance to break into semi-monopolistic markets (Walz 2007). Not surprisingly, the leading countries have stronger wind-power development policies than the laggards (Szarka 2007, 68–86).

Two main differences in their wind-power policies go a long way toward explaining why Germany's wind share is still more than double that of the United States. First, financial incentives have been larger in Germany, although exact comparisons are difficult (Walz 2007, 78). In Germany, wind-energy producers were guaranteed 90 percent of the retail price of electricity from 1991 to 2000, and since then, about 8–9 euro cents (currently about 11–12 US cents) per kilowatt hour (KWh) (Lauber and Mez 2006, 113). In the United States, the federal production tax credit has provided a subsidy of about 2 US cents/KWh since 1992, which is only about 20 percent of the average retail electricity price. The incentives from state-level renewable portfolio standards (RPSs) currently add between 0.5 and 5.5 cents/KWh, depending on the state, and green-power-choice programs add another 1.5 cents/KWh, a subsidy that applies to about 30 percent of wind-generated electricity (Stern and Wobus 2008, 16, 20; Bird, Kreycik, and Friedman 2008, 10). Hence, total subsidies for wind power in the United States are highly variable across states and now range from 4 to 9 US cents/KWh, compared to the approximately 11 US cents/KWh offered everywhere in Germany.

Second, policies and the policy context have been much more stable in Germany than in the United States, which has created a more secure climate for investors. In Germany, the 1990 Electricity Feed-In Law guaranteed prices to producers beginning in 1991, and the only major political challenge to those rates was defeated in 1997. Then, in 2000, the Renewable Energy Sources Act guaranteed feed-in tariffs for twenty years and distributed the costs of the subsidies equally across all utilities and their customers, which reduced political opposition. Moreover, the overall policy context for renewable-energy development has been more favorable and stable in Germany, as it has included prominent, ambitious GHG reduction targets since 1990 and renewable-energy targets since 2000—12.5 percent of electricity by 2010, later increased to 30 percent by 2020 (Bechberger and Reiche 2004, 50; Lauber and Mez 2006, 110; Walz 2007, 69; Mez 2009, 386–387; Vasi 2009, 328).

By contrast, in the United States, tax credits for wind power often have been allowed to expire or have been renewed only at the last minute. National targets for renewable energy and GHG emissions also have been lacking, along with any sustained federal government interest in binding international commitments concerning the latter, all of which signal a lack of political commitment in those areas. Taken together, these factors have created much uncertainty for investors, insurers, and employees in the wind-power sector (Wiser 2007, 5; Sovacool 2008, 156–57).

Explaining these differences in policies and outcomes is the topic of the next two sections.

Structural Explanations of the Germany-US Differences

Germany and the United States differ in a number of structural features that have been advanced as explanations of their differences in environmental outcomes (Jahn 1998; Scruggs 2003), energy and climate policies (Paterson 1996), and specifically wind-power development (Keller 2010). These might explain why Germany has developed wind power much more intensively than the United States.

However, some of these explanations have little support here. For example, wind resources have been advanced as an explanation in several cross-national and cross-state studies (Bird et al. 2005; Menz and Vachon 2006; Toke, Breukers, and Wolsink 2008, 1133; Vasi 2009). But the United States does not lag Germany in wind power because the United States lacks adequate wind resources. The most recent estimate of the wind potential in the United States is 36,900,000 GWh/year, which is about nine times the country's current electricity consumption and over 250 times the current amount generated from wind (NREL 2010). Although comparable cross-national measures of wind resources are not available, the available studies show that Germany has much less wind resources than the United States (cf., e.g., Toke et al. 2008, 1132–1133).

Another possible explanation is that the United States lags in wind power because it has developed other renewable energy sources to a greater degree than Germany. This would reduce the pressure to develop wind power more aggressively that might come from any domestic environmental or energy-security interests or international climate-policy commitments. However, Germany also leads the United States in most other areas of "new renewable energy" (excluding large hydroelectric plants), including biomass, solar, and wind; the United States leads Germany only in geothermal energy. Germany leads in the total generation share of all new renewables, by 18.5 percent versus 5.4 percent, a nearly 3:1 ratio that is even larger than its lead in wind power (based on 2012 data from BMU 2013, 12; USEIA 2013, Tables 1.1, 1.1.A).

Several other structural explanations do have empirical support, although shortly I will argue that their contribution to an overall explanation is limited. First, Germany has smaller fossil-fuel endowments and greater dependence on energy imports, manufacturing, and manufacturing exports, than the United States. Germany's energy imports totaled about 74 percent of total energy consumption in 2000, compared with 25 percent for the United States, according to data from the Arbeitsgemeinschaft Energiebilanzen (AGEB) and the US Energy Information Administration (USEIA). Also, among major industrial countries, Germany has ranked at or near the top in the share of manufacturing jobs in the economy since 1970, while the United States has ranked near the bottom. In 1995, 23 percent of Germany's jobs were in manufacturing, compared with 13 percent in the United States (Pilat 2006, 6). Germany also depends much more on exports than the United States does; exports made up 30 percent of Germany's gross domestic product (GDP) in 2002, compared to only 7 percent in the United States.[10] As a result, Germany arguably has a stronger national interest than the United States in developing renewable energy because this reduces reliance on energy imports, improves its balance of payments, and increases domestic employment. Indeed, renewable-energy policy in Germany has been justified partly by reference to a government estimate that it has created 280,000 jobs (BMU 2009, 31).

Second, proportional representation allowed the Green Party to become a national force in Germany, averaging 7.5 percent of the vote in parliamentary elections since 1983, while the plurality winner-takes-all electoral system in the United States has protected both the Democrats and the Republicans against potential third parties. Hence, ecological parties have been insignificant in the United States, which reduces the pressure for innovation in energy policy. Conversely, the Greens, as the case studies will show, have been important advocates of renewable energy in Germany. Third, the competitive, pluralist interest-group system in the United States is accompanied by an adversarial relationship between government and business, while Germany has a neocorporatist system and more cooperative business-government relations in both environmental and economic policy areas (Scruggs 2003, 227–228, 233). Hence, business opposition to environmental policy is usually more muted in Germany than in the United States, where environmental deregulation was pushed by business interests and attempted by the Reagan administration in the 1980s (Kraft and Axelrod 1984).

Some Problems with the Structural Accounts

Together, these structural factors contribute to an explanation of why Germany took a leadership role in wind-power development. However, the structural explanations are also problematic in several ways. First, the structures in the United States do not present only disadvantages; it has greater wind resources than Germany, and its institutions favor innovation in energy policy in at least two ways. The US pluralist interest-group system and weak political parties make the country more open to new interests and ideas than Germany, with its moderately strong neocorporatist arrangements and programmatic parties (Kitschelt 1986, 66, 81). Also, while both countries have federalist institutions, state governments in the United States have more autonomy in renewable-energy and climate policy than do Germany's *Länder* (Rabe 2004, 16–18; Keller 2010), though this is partly due to the relative passivity of the US national government in those areas (Derthick 2010). This provides another avenue for innovation in the US system because states are relatively free to experiment with renewable-energy policies (Rabe 2006). If they have success, their actions may be imitated in other states, and the national government may come under pressure to adopt reforms, partly to make policy more uniform.

It is possible that these advantages for the United States operate at the same time as the disadvantages. Because the United States lags behind Germany in wind power and other new renewables, one could defend the structural theory by arguing that its structural disadvantages outweigh its advantages in terms of renewable-energy policy and outcomes. But in theoretical terms, it is unclear why some structural features would be more important than others. A theoretically more fruitful line of analysis starts with the fact that the two sets of structural factors cancel out or balance each other to a certain extent. This creates scope for other causal factors, such as political processes, to contribute to the outcomes.

Another, larger problem for the structural explanation of Germany's leadership over the United States is that the outcomes fit this explanation only since the early 1990s. In fact, the same structural explanation would be completely wrong for the 1970s and 1980s, when the United States actually *led* Germany in wind power, and the explanation has been losing force since 2004, when the United States began to catch up. The United States initiated the commercial use of modern wind power almost ten years before Germany did, in California (Dismukes et al. 2007). Hence, as late as 1992, the United States had over 60 percent of the world's wind-power capacity (and nine times Germany's), and the United States still led Germany in its share of wind power in total electricity generation.[11] Also, since 2004, the United States has been catching up to Germany in its wind-power share due to rapid recent growth in capacity in the United States (28 percent per year) and a slowdown in Germany's growth (to 8 percent per year).

In short, the relative positions of Germany and the United States in wind-power development are variable, not fixed. Although the United States now lags Germany's share of wind power, this was not always the case, and it might not be the case even five years from now. Indeed, if their respective post-2004 growth rates were to continue, the United States would begin to exceed Germany's share of wind power in electricity generation in 2016.[12] The variability in their relationship can be seen in figure 5.2,[13] which shows the ratio between the German and the US share of wind power in total electricity generation; a ratio above 1:1 means that Germany is leading. The ratio rose from less than 1:1 in the 1981–1992 period to a peak of almost 12:1 in 2004. Since then it has declined, to reach just over 2:1 in 2012.

Therefore, rather than simply saying that Germany leads the United States in wind power, it is more complete and accurate to say that the relative positions of the two countries have gone through three phases: US leadership (1978–1993), US stagnation and growing leadership by Germany (1993–2004), and the beginning of catch-up by the United States (2004–present). Hence, there are three turning points in the relative

Figure 5.2

Ratio between the wind share of electricity generation in Germany and the United States, 1981–2012

positions of Germany and the United States to explain, which the next section does in terms of the convergence of problem and political processes.

Process Explanations of the Germany-US Differences

The United States Takes the Lead (1978–1993)

In the 1970s, problem and politics streams converged to produce a window for innovation in energy policy in the United States, both at the federal level and in some states. The 1973 oil crisis presented large, unexpected problems, which included a quadrupling of oil prices, a five-month embargo of the United States by the Organization of Petroleum-Exporting Countries (OPEC), and gasoline rationing (Smith 2002, 24–25). Unusually strong political commitment resulted largely from the election of Jimmy Carter, who won the 1976 presidential election by only a 50–48 percent margin. Carter had a strong interest in energy policy and advocated the creation of a Department of Energy during the presidential campaign (Laird 2001, 90). During the Carter administration (1977–1981), research and development (R&D) spending for wind power sharply increased, peaking in 1980. Congress also created the cabinet-level Department of Energy, and within it, an office for renewable energy, led by an assistant secretary (Cox et al. 1991, 353; Nemet and Kammen 2007, 750; Walz 2007, 67; Laird and Stefes 2009, 2620). In addition, Carter proposed major energy legislation, and one part that survived interest-group and congressional resistance was the Public Utility Regulatory Policy Act (PURPA) of 1978. This law gave independent power producers grid access and required utilities to purchase their power at the "avoided costs" of the utility's own generation. Implementation was left to state governments, some of which set high purchase prices for renewable-sourced electricity by choosing to interpret the avoided costs under generous assumptions concerning future fossil-fuel prices (Swisher and Porter 2006, 186). Wind projects also were eligible for two federal investment tax credits adopted in 1978 and for accelerated depreciation starting in 1981 (Cox, Blumstein, and Gilbert 1991, 354). These aggressive national Carter-era policies set the stage for a boom in wind power in California during the 1980s, when the vast majority of the world's wind-power capacity was built in that state.

California's structural advantages were also necessary conditions of the boom there. The state had abundant land, some with excellent wind resources (in mountain passes), as well as many rich individuals potentially interested in investing in wind projects. The latter was important because utilities were not eligible for most federal tax credits, a feature that also encouraged the growth of independent power producers (Cox et al. 1991, 354; Gipe 1995, 30).

But the wind boom in California also would not have occurred without strong state-level political support for wind and solar power. This occurred through a key development in the politics stream: the election of Jerry Brown as governor of California (1975–1983). In his 1974 election campaign, Brown ran on the slogan "Serve the people, protect the earth, explore the universe" and advanced a strong environmental agenda. Although Brown was elected only narrowly (by 50–47 percent), his victory helped create a policy window for renewable energy in the state. Advised by renewable-energy advocates such as Amory Lovins, Brown appointed "soft-energy" proponents to the California Energy Commission and California Public Utilities Commission, and also created an Office of Alternative Technology (van Est 1999, 34–35). Staffers in the latter agency helped ensure that a 1978 solar tax-credit law would also apply to wind power. The same year, the state legislature passed the Mello Act, which set relatively ambitious targets for renewable-sourced electricity—1 percent by 1987 and 10 percent by 2000 (Righter 1996, 204–206).

Encouraged by this political support, the California Public Utilities Commission, all of whose members were Brown appointees after January 1979, went on the offensive and levied multimillion-dollar fines against two large utilities for failing to pursue alternative energy, conservation, and cogeneration (Harris and Navarro 1999, 12–13). The fines spurred the utilities to negotiate with independent power producers and the commission, which led to contracts with high purchase prices for wind power (Gipe 1995, 30). The commission also had begun mapping wind resources in 1977, providing crucial data that were not available in other states (Righter 1996, 204–206).

The convergence of federal and state politics streams led to a very generous combination of financial incentives for wind power at this time: a 15 percent federal investment tax credit for certain energy properties enacted in 1978, a generic federal investment tax credit of 10 percent, accelerated depreciation from the federal government, and a California tax credit of 25 percent for wind power (Cox, Blumstein, and Gilbert 1991, 347, 354). On top of all these tax incentives, the California Public Utilities Commission required that wind-power operators be guaranteed a price of about 7 cents/KWh in contracts that lasted at least 10 years. As a result, 50,000 individual investors provided $2 billion in capital, and companies rushed to build wind farms in California, mainly in three mountainous areas. Companies installed 1,141 MW of wind-power capacity by 1985, and California generated 87 percent of the world's wind power that year (Cox et al. 1991, 356; Righter 1996, 203–204; Harris and Navarro 1999, 14).

The United States Falters and Germany Surges Ahead (1993–2004)
In the early 1990s, Germany began to exceed the US wind share of electricity generated because wind power stagnated in the United States after 1985, while it took off in Germany after 1990. The US boom in wind energy turned into a bust because its problem and politics streams simultaneously reversed direction and created a different kind of policy window, one that conservatives used to undermine renewable energy. One factor was the large decline in fossil-fuel prices during the 1980s. After peaking in 1980, crude oil prices fell for the next six years, declining by 70 percent in real terms, and coal and natural-gas prices also fell sharply during the 1980s (according to USEIA data). The low prices sharply reduced political interest in alternative energy and made the aggressive renewable-energy strategy of states like California difficult to sustain (Sovacool 2008, 150–154).

The problem stream also shifted in the United States because the top-down approach to wind-power development that was taken in the 1970s led to technological failures in the 1980s; wind power became seen as more of a problem than a solution. US-made wind turbines broke down in field conditions in California in the early 1980s, because aerospace firms failed to anticipate problems such as insects and ice, and the extremely high subsidies led to shoddy construction (Cox et al. 1991, 349). By 1985, only 38 percent of US-built turbines were working reliably, compared to 98 percent of those built by Danish firms. Hence, another effect of the technical problems was that Danish turbine producers came to dominate the US market (Heymann 1998, 646).

Around the same time, the politics stream presented a reversal of fortune for environmental policy, including renewable-energy policy. Nationally, the election of Ronald Reagan as US president in 1980 inaugurated a shift toward environmental deregulation and drastically curtailed national support for renewable energy (Kraft and Axelrod 1984, 319; Kraft 2004, 115). Congress largely went along with Reagan's proposed cuts to wind and other renewable-energy programs. R&D spending for wind power fell by 50 percent from 1981 to 1984 and kept declining through the early 1990s. Spending for renewable energy and energy conservation fell by about 90 percent during the 1980s (Kraft 2004, 174; Nemet and Kammen 2007, 750). Federal tax credits and accelerated depreciation for wind power were allowed to expire in 1985, dealing a severe blow to the industry (Gipe 1991, 758).

California also saw a political reversal after a Republican, George Deukmejian, was elected governor in 1982. He attacked the state's renewable-energy tax credits, and although the legislature initially resisted him (van Est 1999, 55), the combination of changes in the problem and politics streams was irresistible. California stopped signing preferential contracts for wind power in 1985, gave operators with existing contracts only five years to get their projects online, and let its tax credit for wind projects expire (Swisher and Porter 2006, 187; Harris and Navarro 1999, 16). Some wind and solar companies went bankrupt, many skilled and experienced people left the industries, and wind and solar became saddled with a stigma of ineffectiveness and financial failure (Sovacool 2008, 154–155; 2009, 4502).

The problem and politics streams continued to be unfavorable for national renewable-energy policy into the 2000s. Real oil prices remained near or below their late 1980s levels until 2004, and Republicans controlled either the White House or the House of Representatives for all but two years until 2009. However, a brief spike in oil prices in 1990, the 1990–1991 Persian Gulf War, and President George H. W. Bush's interest in environmental issues helped lead Congress to pass the Energy Policy Act, which Bush signed it into law in October 1992. A broad coalition, including the natural-gas and renewable-energy industries as well as energy-efficiency advocates, helped get a national production tax credit for wind power into that legislation (Eikeland 1993, 65–68).

Nonetheless, these forces were too small and short-lived, and the 1992 Act was too weak to have much effect on wind-power development in the 1990s. The production tax credit's low level of subsidy, combined with low fossil-fuel prices, meant that wind power did not become competitive with natural gas until about 2000. Moreover, providing the incentive as a tax credit made it difficult for wind-power operators to claim it because their long-term tax liabilities were not large and predictable enough. This promoted the sale of wind-generation facilities to large corporations and hindered the development of a new set of economic actors with a strong interest in technological development and political advocacy for wind power (Swisher and Porter 2006, 188). Furthermore, Congress gave low priority to the renewal of the production tax credit, allowing it to expire for periods of several months in 1999, 2002, and 2004. The expirations in some years, and uncertainties about the credit's extension in others, created insecurity for investors and led to a boom-bust cycle in the construction of new wind projects, which reduced the long-term growth of the industry (Wiser 2007).

By contrast, commercialization of wind power in Germany did not really begin until after 1990. But Germany quickly overtook the United States in its wind-power share of electricity (in 1993) and in absolute wind-power generation (in 1997). The extremely rapid growth in Germany was largely due to the 1990 passage of the Feed-In Law, its defense in 1997, and its strengthening in 2000. And that legislative history, in turn, was due to a convergence of politics and problem streams in the late 1980s that created an extraordinarily large window for energy and climate policy.

The politics stream included the emergence of a broad, strong advocacy coalition that eventually included the Green Party, environmental organizations, research institutes, unions, religious organizations, hydroelectric producers, key elements of the Christian-Democratic and Social-Democratic parties, the Environment Ministry, and the Federal Environmental Agency. The environmental movement and its legacies contributed crucially to the advocacy coalition, in several ways. The West German environmental movement was stronger than in most industrialized countries, with a major focus on opposition to nuclear power during the 1970s and 1980s (Koopmans 1995). The movement spawned organizations such as the Eco-Institute and the solar association Förderverein Solarenergie, which in the late 1980s began to develop feed-in tariff proposals for renewable energy (Jacobsson and Lauber 2006, 263). Environmentalists also helped to found Green parties, initially at local and regional levels, in alliance with leftist activists, and these parties grew into the national Green Party by the early 1980s. Furthermore, the environmental movement influenced public opinion on nuclear power and climate change in the long term (Vasi 2009, 328). Moreover, once the Greens gained Bundestag seats beginning in 1983, the new party strongly influenced the established parties toward environmental positions. The Christian-Democratic federal government began to strictly regulate sulfur oxide (SO_X) emissions in efforts to prevent forest dieback, and the Social Democrats backed away from their support for nuclear energy (Papadakis 1989; Schreurs 1997, 153; Weidner 2002, 154).

These political developments linked to the environmental movement were peaking at the same time that a series of focusing events created major changes in the perception of environmental problems in the 1985–1986 period, in West Germany and elsewhere. First, ozone depletion rose as an international issue in 1985, when the Vienna Convention on Ozone Depleting Substances was adopted and British scientists published the first findings demonstrating the existence of the ozone hole over Antarctica. Then, in January 1986, a report by the respected German Physical Society warned of climate change and rising sea levels, which sparked much anxious media coverage and public attention in West Germany (Der Spiegel 1986; Kords 1996, 204; Beuermann 2000, 100). Finally, in April, the Chernobyl nuclear disaster sharply focused attention on environmental problems in West Germany, which was one of the countries receiving the greatest amount of radioactive fallout from that accident (Peplow 2006, 983).

These events drove changes in the politics stream, creating a window of opportunity for elites who promoted climate protection and renewable energy. In the wake of Chernobyl, public support for environmental protection reached a peak in West Germany in the late 1980s (Bang 2003, 217). The federal government created the Environment Ministry in June 1986 and a parliamentary inquiry commission on ozone and climate issues that October (Watanabe and Mez 2004, 112). The commission, which created and expressed a consensus across all the parliamentary parties, called for Germany to reduce carbon dioxide (CO_2) and methane emissions by 30 percent compared with 1987 and to adopt an electricity feed-in law to assist renewable energy (Lauber and Mez 2006, 105–106).

Bound up with these political changes was the formation of a broad advocacy coalition for renewable-energy policy. In 1989, Green and Christian-Democratic deputies entered into a highly unusual alliance to support a feed-in law for renewable-energy sources, including wind, solar, small biomass, and small hydroelectric projects. The law was resisted by the large electricity utilities, the Federal Economics Ministry, and party leaders, but it gained support from deputies in all parliamentary parties. The unanimous passage of the 1990 Feed-In Law was aided by two additional contingent events. In 1987, the electricity utilities announced that they had reached their limits in making payments for small hydroelectric producers, which drove those producers into the advocacy coalition. In fact, the head of their association, who was also the operator of a hydroelectric plant, was the Christian-Democratic deputy who led the parliamentary push for the feed-in law (Reiche 2004, 146). Also, as the unification of Germany unfolded rapidly in 1989–1990, the utilities became distracted by their takeover and restructuring of the former East German electricity sector, so they did not anticipate the Feed-In Law's likely effects on new renewables and did not strongly oppose it (Bechberger 2000, 4–5; Jacobsson and Lauber 2006, 264; Lauber and Mez 2006, 106).

Although the principle of a feed-in tariff—government sets a price at which utilities must purchase renewable-sourced electricity—had also been used in the US PURPA, as well as in Denmark's renewable energy policies, there is no evidence that this policy instrument diffused out of either of those countries to Germany. Rather, a voluntary feed-in tariff for renewable-sourced electricity had existed in Germany since 1979, in the form of an agreement among trade associations, before the 1990 Feed-In Law converted it to a regulatory measure on terms that were much more generous to wind-power producers; similar developments occurred in Denmark (Reiche 2004, 146). This suggests, in contrast to chapter 3, "Explaining Environmental Policy Adoptions: A Comparative Analysis of Policy Developments in Twenty-Four OECD Countries," in this volume, that the apparent "spread" of policy instruments across countries often may be due to approximately simultaneous invention in broadly similar contexts rather than to transferral from one country to another.

The 1990 law strengthened the advocacy coalition for renewable energy by creating a market for wind turbines and solar photovoltaic cells (and hence reducing their costs), and by spurring interest-group formation and mobilization. Hence, when large utilities and the federal government tried to reduce the feed-in subsidies in 1997, the advocacy coalition responded by holding a demonstration in Bonn that drew about 4,000 people. The coalition included wind turbine suppliers and operators, solar energy producers, metalworkers, farmers, environmental and religious organizations, and the large Equipment and Machinery Producers Association. Their mobilization helped defeat the proposed cuts in a Bundestag committee (Michaelowa 2005, 195; Jacobsson and Lauber 2006, 265; Watanabe 2009, 151–152, 160, 166, 179, 184).

The advocacy coalition reached a new level of influence after the 1998 parliamentary elections, in which the Greens received 6.7 percent of the vote and joined in a national coalition government with the Social Democrats for the first time (Weidner 2002, 182; Jänicke 2005, 138). Under that government, a series of major climate policies were enacted, including a 100,000-roof program for solar photovoltaics and the 2000 Renewable Energy Sources Act (Bechberger and Reiche 2004, 50). The advocacy coalition grew to include one of the large utilities (namely, Preussen Elektra) and officials in the environment ministry. It overcame strong opposition from the Federation of German Industry and most large utilities to help pass the Renewable Energy Sources Act in 2000 and amendments to it in 2004 (Jacobsson and Lauber 2006, 267–269; Wüstenhagen and Bilharz 2006, 1688).

The United States Begins to Catch Up (2004–Present)

The next turning point occurred in the early 2000s, when the growth of new wind installations became rapid in the United States and began to slow in Germany (see the gray lines on figure 5.1, shown earlier in this chapter). The slowdown in Germany was due in part to gradually falling feed-in tariffs for new onshore wind turbines after 2001 (Lauber and Mez 2006, 111) and the declining availability of suitable new onshore sites. The growth in the United States, which began in 1999 and accelerated in 2005, was caused by a convergence of factors in the problem and politics streams that opened policy windows in a large number of states.

First, fossil-fuel price increases made the problem of finding alternative energy sources more acute, and wind power appeared to be an increasingly viable solution. In particular, US natural gas prices at the wellhead rose sharply beginning in 2000, rising from about $2 per thousand cubic feet in the 1986–1999 period to a peak of $6–8 in the 2005–2008 period, before falling to about $4 in the next three years and to less than $3 in 2012 (as per USEIA data). Hence, the operating costs of fossil-fuel electricity plants rose from about 2.0 cents/KWh in 1998 to 3.6 cents in 2008 (according to data from table 8.2 in USEIA 2010). At the same time, the cost of wind-power generation fell by about 1.1 cents/KWh from 1998 to 2006 (Wiser and Bolinger 2008, 23–25). Given its subsidies, wind power became commercially competitive with natural-gas-fired electricity generation in the United States around 2000, when the price of natural gas passed $3.50/thousand cubic feet (Bird et al. 2003, 5). Since 2003, wind power has sold for about 4 cents/KWh on average, which is cheaper than gas-fired generation but still more expensive than coal-fired generation (Wiser and Bolinger 2008, 18–19).[14]

However, economic drivers by themselves were not enough to produce a surge in wind power because competitiveness also depends on state and federal subsidies. The approximately 4 cents/KWh charged by wind-power producers is supplemented by a 2-cent subsidy from the federal production credit, as well as by state and consumer subsidies that currently range from about 2 to 7 cents/KWh, depending on the state (Bird et al. 2008, 10; Stern and Wobus 2008, 16, 20). In this period, adequate subsidies and relatively high natural-gas prices have both been necessary for wind power to be competitive.

Although national subsidies have grown in inflation-adjusted terms since 1992, the rapid rise in supportive state-level policies that occurred from the late 1990s was crucial for wind-power development. The most effective state policies have been renewable portfolio standards and programs promoting green-power choice, which began in the late 1990s (Menz and Vachon 2006). The former are instruments that require utilities to produce, purchase, or subsidize specific, accelerating amounts of renewable-sourced electricity, while the latter are programs that encourages customers to subsidize renewables voluntarily. By 2008, RPSs had been adopted in 28 states, over half of all electricity customers had an option to buy "green power," and 2 percent of them did so (Bird et al. 2008, 1; Sovacool 2008, 158). Although already in 1978 the PURPA law had included a feed-in tariff, RPSs have remained the most important instrument for US states in the last three decades. Despite the success of feed-in tariffs in Germany, Denmark, and Spain, that policy instrument has largely failed to diffuse to the United States, where it is limited to a few recent, small local and state-level experiments (Rickerson et al. 2008; Gipe 2009).

Why did some states move strongly in favor of wind and other renewable energy? The answer does not lie in the party control of government. The legislatures of states that adopted RPSs were roughly divided between Democratic and Republican control, and 16 of the first 22 states with RPSs had Republican governors, even though Democrats are now stronger advocates for renewable energy on the national level (Rabe 2006, 6). However, there is an important structural factor that facilitates state action in the United States. Historically, state governments have regulated energy and areas related to it, such as economic development, land-use planning, and disaster planning. Hence, in the 1990s many states already possessed strong administrative capacities to regulate energy production, and their citizens accepted and even expected state action in this area. This constellation reduced the power of energy lobbies at the state level (Byrne et al. 2007, 4567). While this may have been a necessary condition, it was not sufficient, as shown by the lack of action by all but a few states before the late 1990s.

At that time, a conjunction of changes in the problem and politics streams created windows of opportunity for policy entrepreneurs in many states (Rabe 2004). First, the delayed effects of the 1992 Energy Policy Act created problems and opportunities for state-level energy policy. Spurred by the planning reviews required by the federal act, states began to open their electricity markets beginning in 1996, by promoting new kinds of power producers and retail choice for customers. The reforms also led to a lull in the construction of new power plants, and hence there was a need for new capacity by the late 1990s. Second, natural-gas prices increased sharply and became volatile beginning in June 2000, which prompted state-level concerns about energy security and diversification (Martinot, Wiser, and Hamrin 2005, 5, 19).

At the same time, two political changes occurred that also favored renewable-energy advocates. First, the United Nations Framework Convention on Climate Change (UNFCCC) process and the adoption of the Kyoto Protocol by its members in December 1997 raised expectations for climate policy measures in the United States. But because the US Senate symbolically voted 95–0 to reject the protocol in that year and President George W. Bush withdrew US support for the protocol in 2001, some US state governments responded by developing climate policies to compensate for the lack of federal leadership (Rabe 2004). Moreover, policy entrepreneurs saw an opportunity to link climate policy with other important state goals; i.e., economic development, the creation of in-state manufacturing jobs, increased energy security, and compliance with the Clean Air Act (Rabe 2006, 6–7; Peterson and Rose 2006; Byrne et al. 2007, 4567). Second, renewable energy also became very popular with the public. In a 2001 survey, 90 percent supported alternative and renewable energy development, and in 2006, 77 percent said that it should be the top priority for energy policy in the United States (Byrne et al. 2007, 4566).

How states responded to these pressures and opportunities depended on their particular circumstances and processes. Texas is a key example: if it were an independent country, it would now be the world's eighth-largest GHG emitter, according to WRI and UNFCCC data. It would also rank sixth in absolute wind-power capacity, with 12,214 MW in 2012, which was 20 percent of the US total (according to data from AWEA 2013 and IEA 2012, 6). Its share of wind-generated electricity was 7.4 percent that year, the same as Germany's (data from USEIA 2013, tables 1.6B and 1.17B). The growth of wind power in Texas has been largely the result of an aggressive RPS adopted in 1999, while George W. Bush was governor. The RPS initially required for-profit electricity retailers to install 2,000 MW in new renewable capacity within ten years; it was amended in 2005 to require a total of 5,880 MW by 2015. A strength in the law is the fact that noncompliance triggers automatic penalties (Rabe 2004, 50–1; 2006, 10–12; Schmalensee 2010, 224).

This remarkable Texas policy was adopted because of a convergence of problem and political processes in that state. Texas officials undertook a restructuring of the electricity market, aiming to deregulate markets and diversify electricity supplies, partly because the state is in an isolated position in the US electricity grid and because this energy-inefficient oil- and gas-producing state actually had become a net importer of energy in 1992. The RPS was a small component of this reform (it appeared on only one of the bill's sixty-one pages), and neither the bill nor its supporters mentioned the contentious issue of climate change. Moreover, this provision probably would not have been included were it not for a previous, unusual use of "deliberative polling" that brought customers, utility representatives, and regulators into discussions in the 1996–1998 period. The polling showed that customers wanted renewable-sourced electricity and were willing to pay $1 a month for it. Surprised by these results, officials started a pilot project that cost consumers 25 cents a month, and this project helped build political support for the 1999 RPS (Bird et al. 2003, 8–9; Rabe 2004, 56–59).

Summary and Conclusions

This comparative analysis has three main theoretical implications. First, structural theories of environmental outcomes have serious limitations in explaining renewable-energy development. They cannot account for change over time within a country case, such as the boom (1980–1985), bust (1985–1998), and boom (1998–present) in the United States, and hence they cannot explain the changing relative positions of Germany and the United States. In addition, the growth of government support for wind power in both countries, which has helped make them both increasingly "environmental states," cannot be explained by referring to structural factors. A combination of socioeconomic, political, and international processes is likely responsible (see chapter 12, "Conclusion: An Emerging Environmental State?" in this volume).

Second, combinations of structural and multiple-streams theories are most effective in explaining environmental performance. Although structural factors matter, their effects are not stable over time. Rather, which structures matter and when they do so depend on processes that drive the opening of policy windows, and hence the turning points between periods. Those processes consist of convergent developments in problems and in politics, which are contingent on focusing events and accidents of timing and hence cannot be readily derived from any set of structural features. If structures and processes are used together as competing explanations or rival hypotheses, we can see how far each is supported by the evidence and criticize each from the perspective of the other. The resulting analysis can help clarify how much scope actors have and under what conditions.

The interplay of structural factors and stream convergence can be seen in three periods in this comparison. During the period of US leadership (1978–1993), the institutional autonomy of state-level actors in US-style federalism were necessary conditions, as were the structural advantages that California enjoyed, such as strong wind resources in an area with low-intensity land use and an innovative set of investors. But those structural factors mattered only because of contingent events: the 1973 oil crisis, the elections by narrow margins of Jimmy Carter and Jerry Brown in the mid-1970s, and their administrations becoming committed to the aggressive promotion of renewable energy.

Next, in Germany's period of expanding leadership (1993–2004), its proportional-representation electoral system and cooperative business-government relations were major underlying causes of renewable-energy policies. Without them, the Greens would not have become a national-level actor capable of sparking and later extending the Feed-In Law, and the Helmut Kohl government's (1982–1998) neoliberal efforts in economic policymaking might have been extended to environmental deregulation, as occurred in the Reagan administration in the United States (Prasad 2006, ch. 3; Weidner 2002). But those institutional features came into play and helped produce the Feed-In Law only because politics and problem streams developed in certain unpredictable ways and converged in the mid- and late 1980s. These included the growth of a strong (West) German environmental movement, the establishment of the Greens, the Chernobyl accident and other extraordinary focusing events that occurred around the same time during the 1985–1986 period, and the cooperation of Greens, conservatives, and the small hydro industry in an unusual alliance.

Moreover, political processes helped change the structures that shaped renewable-energy policy in two key ways. The *combination* of the proportional electoral system and the environmental movement, not either factor by itself, led to a structural change in the West German party system with the addition of the Greens in 1983 (Kitschelt 1986, 83). In addition, the environmental movement of the 1970s and 1980s ultimately helped to reshape the neocorporatist system regarding environmental policy as new economic interest groups and environmental organizations became political players (Jänicke 1997) and the protest demonstration became an acceptable method of participation (Koopmans 1995).

The decline of new wind installations in the United States after 1987 also was shaped by a combination of structures and processes. The decline revealed certain structural weaknesses of the United States in terms of establishing a renewable-energy policy, when compared with Germany: the winner-take-all electoral system (hence no party consistently promoted renewable energy) and the adversarial relationship between business and government on environmental policy (hence a strong business backlash against environmental regulations in the late 1970s and 1980s) (Kraft 2002, 36–37; Prasad 2006, 181–184). But wind power might have continued to grow rapidly in the 1980s and 1990s were it not for a convergence of developments in the problem and politics streams that were unfavorable to renewable energy: the economic recession and Iranian hostage crisis that helped Reagan win the 1980 presidential election, the 1980s drop in oil prices, and the technical failures of many US-made wind turbines in the field.

Finally, since the late 1990s, the United States has begun to catch up with Germany because of a potent combination of structures and processes. Substantial state autonomy and capacities in energy policy made possible the adoption of RPSs, green-power-choice programs, and other supports for renewable energy. But these would not have been adopted without a conjuncture of problem and political processes favorable to the renewable sector, especially rising and volatile natural-gas prices, declining wind-power costs, climate-change concerns raised by the Kyoto process, and the successful promotion of green consumerism by environmentalists. The supportive state policies that resulted were not simply determined by natural gas prices because those began to rise only in 2000, after the trend toward RPSs and green-power options was underway in many states (Price 2002, 40).

In short, while the United States has structural disadvantages for renewable energy, both country cases show that there is more scope for actors to make a difference when certain aspects of problem and politics streams converge, for whatever reasons. Based on the cases analyzed here, this seems most likely when extraordinary focusing events direct public and elite attention to environmental problems, fossil fuel prices rise, environmental movements build public support, and environmentally oriented parties or candidates win elections. These processes not only counteract structural forces, they also shift the political terrain in ways that makes certain structures more or less relevant.

The third implication of these case comparisons is that multiple-streams theory needs to be improved in certain ways. The relation of the different streams to each other requires more theoretical attention (cf. Zahariadis 2007, 81–82). Problem and politics streams are often independent, in which case their convergence to produce policy windows is largely coincidental. But sometimes the two streams are both driven by the same forces, are manipulated by certain actors, or influence each other through positive feedback. The cases examined here illustrate some of these possibilities. The period of US growth in the 1970s was driven by the fact that the oil crisis and the elections of Brown and Carter occurred around the same time. However, the oil crisis affected not only problem perceptions, but also the political commitment to energy policy that those political leaders were able to muster. In Germany, the 1990 renewable energy law resulted partly from a similar convergence of domestic political factors (e.g., the rise of the Greens) and internationally driven problems (e.g., the ozone issue and Chernobyl). But in Germany, this stream convergence triggered two decades of positive feedback that produced strong, consistent policies for renewable energy. The positive feedback involved the broadening and mobilization of an advocacy coalition, the strengthening of renewable-energy policies and targets, the development of administrative institutions, and perceptions of growing technical and economic feasibility. The weakness of this kind of positive feedback so far in the United States has made the progress on renewable energy there more fragile.

Acknowledgments

I would like to thank Katarina Eckerberg, Andreas Duit, Christine Ingebritsen, Detlef Jahn, Krister Andersson, and the other participants at the Workshop on the Politics of Ecology, held June 28–29, 2010, in Stockholm, Sweden, as well as the anonymous reviewers for MIT Press, for their comments and suggestions on earlier versions of this chapter.

Notes

1. Sovacool 2009, 712. However, critics have emphasized its damage to the aesthetic value of landscapes, noise pollution, and harm to bird and bat populations (Sovacool 2008, 185–186).

2. Where data sources are cited, calculations are by the author.

3. An alternative would be a measure based on installed capacity. However, the wind-power generation share measure has the advantage of adjusting for differences in the capacity factors of the wind turbines, which can vary significantly across countries. Although generation shares are affected by annual differences in weather conditions, this is of little importance over the relatively long time frames studied here.

4. While the former aims to explain outcomes and the latter to explain agenda setting, both will be applied here to explain policymaking and implementation. This is reasonable because wind-power outcomes depend strongly, though not exclusively, on renewable-energy policies. Also, the causal factors in multiple-streams theory often bear directly on decision making and implementation as well as agenda setting (Zahariadis 2007, 80), and I will supplement them with some elements of advocacy-coalition theory to address policymaking more fully.

5. Although they are sometimes seen as competitors and have different emphases, the two theories have much in common and few real points of disagreement. Both hold that the usual inertia in policymaking can be overcome when sharp changes in perceived problems and in political alignments create openings for political actors to press successfully for major policy changes.

6. Advocacy-coalition theory also emphasizes belief systems and long-term changes in ideas, which are not analyzed in this chapter.

7. Among those that do, Reiche and Bechberger (2004) and Gan, Eskeland, and Kolshus (2005) assess a wide range of factors without distinguishing relative weights, and Snyder and Kaiser (2009) and Bohn and Lant (2009) largely view the processes that they consider as more influential than the structures.

8. Exceptions to the latter point include Laird and Stefes (2009) and Stefes (2010), which examine the roles of historical contingency and path dependence, and Breukers (2006), which relates structures and processes in a theory of institutional capacity building.

9. Data for figure 5.1 are from BMU (2013, 12–13); AGEB (2008, 21; 2012, 26; 2013, 28); Dismukes et al. (2007, 778); AWEA (2010; 2012a; 2012b, 2013); and USEIA (1996, table A1; 2013, tables 1.1, 1.1.A). The US wind shares for 1981–1994 are based on my estimates of wind-power generation, which are derived from known installed capacity and average capacity factors for the 1995–1999 period.

10. *CIA World Factbook*, 2003 data from http://www.nationmaster.com.

11. Dismukes et al. 2007, 778; author's calculations from the data cited for figure 5.1.

12. Based on 2004–2012 annual growth rates in both countries; data from the sources used for figure 5.1. However, the sharp decline in US natural-gas prices since 2008 is a potential problem for the further growth of the US wind industry.

13. Data for figure 5.2 are calculated from the same sources as for figure 5.1 (see note 9 above).

14. However, coal power plants have faced stricter regulation by the Obama administration, making the price comparison to natural gas the more pertinent one.

References

AGEB [Arbeitsgemeinschaft Energiebilanzen], 2008. *Energieverbrauch in Deutschland im Jahr 2007*. Berlin: AGEB.

AGEB [Arbeitsgemeinschaft Energiebilanzen], 2012. *Energieverbrauch in Deutschland im Jahr 2011*. Berlin: AGEB.

AGEB [Arbeitsgemeinschaft Energiebilanzen], 2013. *Energieverbrauch in Deutschland im Jahr 2012*. Berlin: AGEB

American Wind Energy Association (AWEA). 2010. *Year End 2009 Market Report*, January. http://awea.files.cms-plus.com/FileDownloads/pdfs/Annual-Market-Report-Year-Ending-2009%281%29.pdf.

American Wind Energy Association (AWEA). 2012a. *U.S. Wind Industry Fourth Quarter 2011 Market Report*, January. http://awea.files.cms-plus.com/FileDownloads/pdfs/4Q-2011-AWEA-Public-Market-Report_1-31.pdf.

American Wind Energy Association (AWEA). 2012b. "Industry Statistics." http://www.awea.org.

American Wind Energy Association (AWEA). 2013. "State Wind Energy Statistics: Texas," June 3. http://www.awea.org/Resources/state.aspx?ItemNumber=5183.

Bang, Guri, 2003. *Sources of Influence in Climate Change Policymaking*. Dr. Polit. thesis, Department of Political Science, University of Oslo, June.

Baumgartner, Frank, and Bryan D. Jones. 1993. *Agendas and Instabilities in American Politics*. Chicago: University of Chicago Press.

Bechberger, Mischa. 2000. "Das Erneuerbare-Energien-Gesetz (EEG)," FFU-report 00–06, Environmental Policy Research Centre, Free University of Berlin.

Bechberger, Mischa, and Danyel Reiche. 2004. Renewable Energy Policy in Germany. *Energy for Sustainable Development* 8:47–57.

Beuermann, Christiane. 2000. Germany. In *Implementing Sustainable Development*, ed. William Lafferty and James Meadowcroft, 85–111. New York: Oxford University Press.

Bird, Lori, Brian Parsons, Troy Gagliano, Matthew Brown, Ryan Wiser, and Mark Bolinger, 2003. *Policies and Market Factors Driving Wind Power Development in the United States*, National Renewable Energy Laboratory Technical Report, July.

Bird, Lori, Mark Bolinger, Troy Gagliano, Ryan Wiser, Matthew Brown, and Brian Parsons. 2005. Policies and Market Factors Driving Wind Power Development in the United States. *Energy Policy* 33:1397–1407.

Bird, Lori, Claire Kreycik, and Barry Friedman, 2008. *Green Power Marketing in the United States* (11th ed.), National Renewable Energy Laboratory Technical Report, October.

BMU [Bundesministerium für Umwelt, Naturschutz, und Reaktorsicherheit], 2009. *Erneuerbare Energien in Zahlen*. Berlin: BMU, June.

BMU [Bundesministerium für Umwelt, Naturschutz, und Reaktorsicherheit], 2013. *Entwicklung der erneuerbaren Energien in Deutschland im Jahr 2012*. Berlin: BMU, February.

Bohn, Christiane, and Christopher Lant. 2009. Welcoming the Wind? *Professional Geographer* 61:87–100.

BP. 2012. *BP Statistical Review of World Energy June 2012*, http://www.bp.com/content/dam/bp/pdf/Statistical-Review-2012/statistical_review_of_world_energy_2012.pdf.

Breukers, Sylvia. 2006. *Changing Institutional Landscapes for Wind Power*. Amsterdam: Amsterdam University Press.

Byrne, John, Kristen Hughes, Wilson Rickerson, and Lado Kurdgelashvili. 2007. American Policy Conflict in the Greenhouse. *Energy Policy* 35:4555–4573.

Collier, Ruth, and David Collier. 1991. *Shaping the Political Arena*. Princeton, NJ: Princeton University Press.

Cox, Alan, Carl Blumstein, and Richard Gilbert. 1991. Wind Power in California. In *Regulatory Choices*, ed. Richard Gilbert, 347–374. Berkeley, CA: University of California Press.

Crepaz, Markus. 1995. Explaining National Variations of Air Pollution Levels. *Environmental Politics* 4:391–414.

Derthick, Martha. 2010. Compensatory Federalism. In *Greenhouse Governance*, ed. Barry Rabe, 58–72. Washington, DC: Brookings Institution Press.

Dismukes, John, Lawrence Miller, Andrew Solacha, Sandeep Jagani, and John Bers. 2007. "Wind Energy Electrical Power Generation," *PICMET Proceedings*, Portland, Oregon, August 5–9.

Eikeland, Per Ove. 1993. *US Energy Policy in the Greenhouse, EED Report 1993/1.* Lysaker, Norway: Fritdtjof Nansen Institute.

Gan, Lin, Gunnar Eskeland, and Hans Kolshus. 2005. "Green Electricity Market Development," Center for International Climate and Environmental Research, Oslo, Norway, June.

Gan, Lin, Gunnar Eskeland, Hans Kolshus, Harald Birkeland, Sascha van Rooijen, and Mark van Wees. 2007. Green Electricity Market Development. *Energy Policy* 35:144–155.

Gipe, Paul. 1991. Wind Energy Comes of Age. *Energy Policy* 19:756–767.

Gipe, Paul. 1995. *Wind Energy Comes of Age.* New York: John Wiley.

Gipe, Paul. 2009. "Vermont Feed-in Tariffs Become Law," Grist.org, May 29. http://grist.org/politics/2009-05-29-vermont-feed-in-tariffs.

Greene, D. L., P. R. Boudreaux, D. J. Dean, W. Fulkerson, A. L. Gaddis, R. L. Graham, R. L. Graves, et al. 2010. The Importance of Advancing Technology to America's Energy Goals. *Energy Policy* 38:3886–3890.

Harris, Frank, and Peter Navarro, 1999. "Policy Options for Promoting Wind Energy Development in California: Report to the Governor and State Legislature," Graduate School of Management, University of California at Irvine, November.

Heymann, Matthias. 1998. Signs of Hubris. *Technology and Culture* 39:641–670.

Huang, Ming-Yuan, Janaki Alavalapati, Douglas Carter, and Matthew Langholtz. 2007. Is the Choice of Renewable Portfolio Standards Random? *Energy Policy* 35:5571–5575.

Intergovernmental Panel on Climate Change (IPCC). 2012. *Renewable Energy Sources and Climate Change Mitigation.* New York: Cambridge University Press.

International Energy Agency (IEA), 2013. *IEA Wind Energy Annual Report 2012.* N.p.: International Energy Agency, July.

Jacobsson, Staffan, and Volkmar Lauber. 2006. The Politics and Policy of Energy System Transformation. *Energy Policy* 34:256–276.

Jahn, Detlef. 1998. Environmental Performance and Policy Regimes. *Policy Sciences* 31:107–131.

Jänicke, Martin. 1997. The Political System's Capacity for Environmental Policy. In *National Environmental Policies,* ed. Martin Jänicke and Helmut Weidner, 1–24. New York: Springer.

Jänicke, Martin. 2005. Trend Setters in Environmental Policy. *European Environment* 15:129–142.

Jänicke, Martin, Harald Mönch, and Manfred Binder. 1996. Umweltindikatorenprofile im Industrieländervergleich. In *Umweltpolitik der Industrieländer,* ed. Martin Jänicke, 113–131. Berlin: Edition Sigma.

Jost, Gesine, and Klaus Jacob. 2004. The Climate Change Policy Network in Germany. *European Environment* 14:1–15.

Karapin, Roger. 2012. Explaining Success and Failure in Climate Policies: Developing Theory through German Case Studies. *Comparative Politics* 45:46–68.

Keller, Sarina. 2010. Sources of Difference. *Energy Policy* 38:4741–4742.

Kingdon, John. 2003. *Agendas, Alternatives, and Public Policies.* 2nd ed. New York: Longman.

Kitschelt, Herbert. 1986. Political Opportunity Structures and Political Protest. *British Journal of Political Science* 16:57–85.

Koopmans, Ruud. 1995. *Democracy from Below.* Boulder, CO: Westview Press.

Kords, Udo. 1996. Tätigkeit und Handlungsempfehlungen der beiden Klima-Enquete-Kommissionen des Deutschen Bundestages (1987–1994). In *Klimapolitik*, ed. Hans Günther Brauch, 203–214. Berlin: Springer.

Kraft, Michael. 2002. Environmental Policy and Politics in the United States. In *Environmental Politics and Policy in Industrialized Countries*, ed. Uday Desai, 29–70. Cambridge, MA: MIT Press.

Kraft, Michael. 2004. *Environmental Policy and Politics.* 3rd ed. New York: Pearson/Longman.

Kraft, Michael, and Regina Axelrod. 1984. Political Constraints on Development of Alternative Energy Sources. *Policy Studies Journal: The Journal of the Policy Studies Organization* 13:319–330.

Laird, Frank. 2001. *Solar Energy, Technology Policy, and Institutional Values.* New York: Cambridge University Press.

Laird, Frank, and Christoph Stefes. 2009. The Diverging Paths of German and United States Policies for Renewable Energy. *Energy Policy* 37:2619–2629.

Lauber, Volkmar, and Lutz Mez. 2006. Renewable Electricity Policy in Germany, 1974 to 2005. *Bulletin of Science, Technology & Society* 26:105–120.

Martinot, Eric, Ryan Wiser, and Jan Hamrin. 2005. "Renewable Energy Policies and Markets in the United States," Center for Resource Solutions, San Francisco. http://www.martinot.info/Martinot_et_al_CRS.pdf.

Menz, Fredric, and Stephan Vachon. 2006. The Effectiveness of Different Policy Regimes for Promoting Wind Power. *Energy Policy* 34:1786–1796.

Mez, Lutz. 2009. Renewables in Electricity Generation. In *Governing the Energy Challenge*, ed. Burkhard Eberlein and Bruce Doern, 373–395. Toronto: University of Toronto Press.

Michaelowa, Axel. 2005. The German Wind Energy Lobby. *European Environment* 15:192–199.

National Renewable Energy Laboratory (NREL). 2010. "Estimates of Windy Land Area and Wind Energy Potential by State," February 4, at www.windpoweringamerica.gov/pdfs/wind_maps/wind_potential_80m_30percent.pdf.

Nemet, Gregory, and Daniel Kammen. 2007. U.S. Energy Research and Development. *Energy Policy* 35:746–755.

Neumayer, Eric. 2003. Are Left-Wing Party Strength and Corporatism Good for the Environment? *Ecological Economics* 45:203–220.

Papadakis, Elim. 1989. Green Issues and Other Parties. In *The Greens in West Germany*, ed. Eva Kolinsky, 61–86. New York: Berg.

Paterson, Mathew. 1996. *Global Warming and Global Politics*. New York: Routledge.

Peplow, Mark. 2006. Counting the Dead. *Nature* 440:982–983.

Peterson, Thomas, and Adam Rose. 2006. Reducing Conflicts between Climate Policy and Energy Policy in the US. *Energy Policy* 34:619–631.

Pilat, Dirk. 2006. "The Changing Nature of Manufacturing in OECD Economies," STI Working Paper 2006/9, OECD Directorate for Science, Technology and Industry, October 27.

Portman, Michelle, John Duff, Johann Köppel, Jessica Reisert, and Megan Higgins. 2009. Offshore Wind Energy Development in the Exclusive Economic Zone. *Energy Policy* 37:3596–3607.

Prasad, Monica. 2006. *The Politics of Free Markets*. Chicago: University of Chicago Press.

Price, Jeff. 2002. The Production Tax Credit. *Public Utilities Fortnightly* 140 (May 15): 38–41.

Rabe, Barry. 2004. *Statehouse and Greenhouse*. Washington, D.C.: Brookings Institution Press.

Rabe, Barry, 2006. "Race to the Top," prepared for the Pew Center on Global Climate Change, June.

Reiche, Danyel. 2004. *Rahmenbedingungen für eneuerbaren Energien in Deutschland*. New York: Peter Lang.

Reiche, Danyel, and Mischa Bechberger. 2004. Policy Differences in the Promotion of Renewable Energies in the EU Member States. *Energy Policy* 32:843–849.

Rickerson, Wilson, Florian Bennhold, and James Bradbury, 2008. *Feed-in Tariffs and Renewable Energy in the USA: A Policy Update*. Washington, DC: Heinrich Böll Foundation, May.

Righter, Robert. 1996. *Wind Energy in America*. Norman, OK: University of Oklahoma Press.

Sabatier, Paul. 1988. An Advocacy Coalition Framework of Policy Change and the Role of Policy-Oriented Learning Therein. *Policy Sciences* 21:129–168.

Sabatier, Paul. 2007. Fostering the Development of Policy Theory. In *Theories of the Policy Process*. 2nd ed., ed. Paul Sabatier, 321–336. Boulder, CO: Westview Press.

Sabatier, Paul, and Christopher Weible. 2007. The Advocacy Coalition Framework. In *Theories of the Policy Process*. 2nd ed., ed. Paul Sabatier, 189–220. Boulder, CO: Westview Press.

Schmalensee, Richard. 2010. Renewable Electricity Generation in the United States. In *Harnessing Renewable Energy in Electric Power Systems*, ed. Boaz Moselle, Jorge Padilla, and Richard Schmalensee, 209–232. Washington: RFF Press.

Schreurs, Miranda. 1997. "Domestic Institutions and International Environmental Agendas in Japan and Germany." In *The Internationalization of Environmen-*

tal Protection, ed. Miranda Schreurs and Elizabeth Economy, 134–161. New York: Cambridge University Press.

Scruggs, Lyle. 2003. *Sustaining Abundance*. New York: Cambridge University Press.

Smith, Eric. 2002. *Energy, Environment, and Public Opinion*. Lanham, MD: Rowman and Littlefield.

Snyder, Brian, and Mark Kaiser. 2009. A Comparison of Offshore Wind Power Development in Europe and the US. *Applied Energy* 86:1845–1856.

Sovacool, Benjamin. 2008. *The Dirty Energy Dilemma*. Westport,CT.: Praeger.

Sovacool, Benjamin. 2009. Exploring and Contextualizing Public Opposition to Renewable Electricity in the United States. *Sustainability* 1:702–721.

Der Spiegel. 1986. "Die Klima-Katastrophe—Ozon-Loch, Pol-Schmelze, Treibhaus-Effekt: Forscher Warnen," *Der Spiegel* 1986/33 (August 11, 1986).

Stefes, Christoph. 2010. Bypassing Germany's *Reformstau*. *German Politics* 19:148–163.

Stenzel, Till, and Alexander Frenzel. 2008. Regulating Technological Change. *Energy Policy* 36:2645–2657.

Stern, Frank, and Nicole Wobus. 2008. "Renewable Energy Credit Prices," testimony prepared for the New York State Energy Research and Development Authority by Summit Blue Consulting, November 14.

Swisher, Randall, and Kevin Porter. 2006. Renewable Policy Lessons from the US. In *Renewable Energy Policy and Politics*, ed. Karl Mallon, 185–198. Sterling, VA: Earthscan.

Szarka, Joseph. 2007. *Wind Power in Europe*. New York: Palgrave.

Toke, David, Sylvia Breukers, and Maarten Wolsink. 2008. Wind Power Deployment Outcomes. *Renewable & Sustainable Energy Reviews* 12:1129–1147.

US Energy Information Administration (USEIA), 1996. *Short-Term Energy Outlook*, February.

US Energy Information Administration (USEIA), 2010. *Electric Power Annual*, January.

US Energy Information Administration (USEIA), 2013. *Electric Power Monthly*, February.

van Est, Rinie. 1999. *Winds of Change*. Utrecht, Netherlands: International Books.

Vasi, Ion. 2009. Social Movements and Industry Development. *Mobilization: An International Journal* 14:315–336.

Walz, Rainer. 2007. The Role of Regulation for Sustainable Infrastructure Innovations. *International Journal of Public Policy* 2:57–88.

Watanabe, Rie. 2009. *A Comparative Analysis on Climate Policy Change Processes between Germany and Japan*. Ph.D. dissertation, Free University of Berlin.

Watanabe, Rie. 2011. *Climate Policy Changes in Germany and Japan*. New York: Routledge.

Watanabe, Rie, and Lutz Mez. 2004. The Development of Climate Change Policy in Germany. *International Review for Environmental Strategies* 5:109–126.

Weidner, Helmut. 2002. Environmental Policy and Politics in Germany. In *Environmental Politics and Policy in Industrialized Countries*, ed. Uday Desai, 149–202. Cambridge, MA: MIT Press.

Weidner, Helmut, and Burkard Eberlein. 2009. Still Walking the Talk? German Climate Change Policy and Performance. In *Governing the Energy Challenge*, ed. Burkard Eberlein and Bruce Doern, 314–343. Toronto: University of Toronto Press.

Willey, Joseph. 1998. Institutional Arrangements and the Success of New Parties in Old Democracies. *Political Studies* 66:651–668.

Wiser, Ryan, 2007. "Wind Power and the Production Tax Credit," testimony prepared for a hearing on "Clean Energy: From the Margins to the Mainstream," U.S. Senate Finance Committee, March 29.

Wiser, Ryan, and Mark Bolinger. 2008. *Annual Report on U.S. Wind Power Installation, Cost, and Performance Trends, 2007.* Washington, DC: U.S. Department of Energy, May.

Wüstenhagen, Rolf, and Michael Bilharz. 2006. Green Energy Market Development in Germany. *Energy Policy* 34:1681–1696.

Zahariadis, Nikolaos. 2007. The Multiple Streams Framework. In *Theories of the Policy Process.* 2nd ed., ed. Paul Sabatier, 65–92. Boulder, CO: Westview Press.

II

Environmental Governance and Citizenship from a Comparative Perspective

6

Early Bird or Copycat, Leader or Laggard? A Comparison of Cross-National Patterns of Environmental Policy Change

Thomas Sommerer

On the occasion of international environmental negotiations, like the UN Climate Summit held in Durban, South Africa, in 2011, media commentators regularly divide participating countries into leaders and laggards.[1] In their periodic reviews, the Organisation of Economic Cooperation and Development (OECD) evaluates and compares differences in the environmental performance of governments in the industrialized world. These are only two of numerous illustrations of how, in times of globalization and expanding international cooperation, national governments still play a dominant role in environmental policymaking (see chapter 1, "Introduction: The Comparative Study of Environmental Governance"). Some countries implement ambitious targets and act as pioneers and innovators, while others lag (Jänicke 2005). But these roles often vary across subfields of environmental regulation, and they also might change over time (for more, see chapter 4, "The Three Worlds of Environmental Politics," and chapter 5, "Wind-Power Development in Germany and the United States: Structural Factors, Multiple-Stream Convergence, and Turning Points"). Former latecomers catch up with green leaders, and governments learn from each other's experience or simply emulate policies developed elsewhere. Several studies in environmental politics have confirmed an increasing trend toward convergence and diffusion of domestic policy repertoires (Tews et al. 2003; Prakash and Potoski 2006; Holzinger et al. 2008a, b).

The aim of this chapter is to map dominant patterns in the regulation of environmental protection measures of advanced industrialized nations. For this purpose, I will analyze empirically six dimensions of regulatory change of environmental standards in twenty-four countries from 1970 to 2005. Because this chapter provides a more comprehensive and systematic overview than previous studies, the analytical framework and the empirical results have the potential to contribute to a better understanding of the regulatory performance of national governments, which is one core element of a broader measure of environmental performance, as suggested in chapter 2, "Comparing Environmental Performance."

Comparing Environmental Policy Changes

The analysis in this chapter will be based on the *comparative* study of environmental policy *change*. A comparative perspective reveals similarities and differences in how countries respond to environmental degradation, and a focus on change allows for the display of regulatory dynamics over time. In the following section, I will introduce a theory-informed typology of policy change that draws on a review of specifications of the expected type of change in different theories of policy change and distinguishes six dimensions: the timing, frequency, and direction of change, the level of regulation, and the sequence and the similarity of policies. It represents an attempt to develop an overarching analytical framework for the systematic distinction of different types of change, which is still missing in the research on environmental regulation and elsewhere.

The data set for the statistical analysis will be introduced later in this chapter in the section "Data on Environmental Standards in Twenty-four Countries." Similar to the analysis in chapter 3, "Explaining Environmental Policy Adoption: A Comparative Analysis of Policy Developments in Twenty-four OECD Countries," it builds on policy output data from the ENVIPOLCON database. The data allows for a broad comparison across twenty-four industrialized countries. Time series data from 1970 to 2005 provides a historical perspective from the beginning of environmental policy until recent developments, and it indicates the exact timing of regulatory change. The focus is not on the adoption of policies as in chapter 3, but on changes in metrical standards, such as limit values or recycling targets, that indicate the strength of environmental policy and are thus important for the assessment of the regulatory performance of national governments.

The next section presents the results of the empirical analysis. It reveals major trends in environmental regulation across twenty-four countries and over more than three decades. The data can be interpreted in two different ways. First, it is possible to rank the countries in the sample on six dimensions of change, which gives a highly differentiated picture on leaders and laggards. Whereas some countries rank high (or low) in all dimensions, the analysis also shows that, for example, pioneer countries do not necessarily have the most demanding standards, and that processes of catching up and convergence lead to changes in the ranking. Second, the analysis provides evidence for the empirical relevance of specific types of environmental policy change, like upward and downward shifts, convergence and divergence, first-movers, and parallel change. Because these types of change are derived from expectations about the consequences of different causal mechanisms, the observed pattern gives an indication of their empirical relevance. If an expected outcome is not confirmed—such as a "race to the bottom"—doubts about the explanatory potential of this theory can be raised even without further multivariate analysis.

A Typology of Policy Change

Since the pioneering work by Heclo (1974) and others, the study of policy change has become a popular area of comparative political science. However, this increasingly prominent label is often a misleading one. Strictly speaking, a comprehensive description of policy change figures less prominently in the agenda of many of these studies than an explanatory approach. In quantitative studies, there is a strong tendency to simplify the operationalization of policy change. Large-N research designs often use a binary representation of the adoption of a policy instrument, and sometimes change is not directly operationalized at all. Qualitative case studies often provide a more detailed description of the policy process. Unfortunately, this tends to go hand in hand with a less systematic approach to capturing change in a comparable way that would facilitate the detection of broader patterns.

The empirical disregard of a systematic description of change contrasts sharply with the existence of implicit and explicit assumptions on patterns of policy change at the theoretical level. For example, differences in the direction and timing of change are predicted by theories of international policy diffusion, regulatory competition, or veto players, but these differences hardly translate to empirical research designs.

In the absence of a coherent analytical framework, I introduce a new typology of theoretically relevant constellations of policy change building on six dimensions of policy change, which are derived from the literature of comparative politics, with a specific focus on environmental politics. I begin with an analysis of change-related expectations in nine theories of policy change and conclude this section by presenting the resulting framework.

Several scholars have summarized the multitude of theories that aim at explaining policy change. Kemmerling and Bruttel (2006) identify a "holy trinity": regulatory competition, policy learning, and the effects of domestic institutions. The latter can be divided into three subcategories: theories on path dependence, veto players, and the role of the administrative capacity to develop and revise policies. Jones (1994; cf. Wilson 2000, 250) offers two additional domestic explanations. The first one builds on political demand generated by societal processes, like the mobilization of environmental movements; the other refers to the influence of specific events. Finally, competition and learning as international drivers of change are complemented by theories of imposition and international harmonization (Holzinger and Knill 2005).

The theory of *regulatory competition* ranks prominently among the various concepts to explain environmental policy change (Holzinger and Sommerer 2011). To gain a competitive advantage for domestic industries, a government can decide to make its policy more distinct from its main competitors (Lazer 2001; Holzinger 2003). The widely used metaphor of a downward spiral illustrates an underlying assumption on the direction of policy change. In its classical formulation, the theory of regulatory competition predicts a downward trend of standards; e.g., on environmental protection or labor rights, which will end up "stuck at the bottom?" (Wheeler 2000). Under specific circumstances, a trend of a "race to the top" (i.e., toward stricter regulations) also could occur (Vogel 1995; Scharpf 1997; Holzinger 2003).

A second expectation on the type of regulatory change refers to an increase in the similarity of domestic policies. It is commonly argued that economic competition leads to the leveling of difference across economies, and thus to cross-national convergence (Drezner 2001; Hoberg 2001; Holzinger and Knill 2005). Convergence can be followed up by divergence—because the convergence by one competitor might create an incentive for others to create a new competitive advantage (Lazer 2001, 476; Braun and Gilardi 2006), and again, by convergence, to close an eventual gap (Holzinger 2003).

Theories of *policy learning*, often referred to as "transnational communication" or "voluntary diffusion," are another influential concept in the explanation of regulatory change and have been broadly applied to the study of environmental politics. According to this mechanism, states observe and assess the policy of other countries and transfer adequate solutions to their own polity (Simmons and Elkins 2004; Meseguer 2006; Volden 2006; Holzinger et al. 2008a). While the type of change is rarely modeled explicitly, a number of implicit assumptions can be found. A first expectation refers to policy similarity across countries. A process of learning is expected to lead to a gradual approximation or even a full imitation of foreign models (Holzinger et al. 2008b). Contrary to regulatory competition, divergence is less likely to occur. However, as Strang and Macy (2001, 173) argue, deviation from a model that had been imitated widely could be caused by frustration about immature emulation and failure.

A second dimension of change that is relevant in theories of diffusion and learning refers to the sequence of regulatory shifts. Policy learning is commonly defined in a behaviorist way, meaning that it is based on the observation of existing policy models (Levy 1994). Learning presupposes the pioneering activities of other countries and a time lag between the innovation and an imitation by a foreign government. Third, the same mechanism also has implications for the direction of change. Among multiple different mechanisms, rational learning by the imitation of successful models is an important variation. Learning from success leads to an upward shift in the regulatory level (Volden 2006; Meseguer 2006). Herding as another type of learning is often associated with an inefficient or inadequate outcome, which could be equivalent to a downward shift (Levi-Faur 2002).

Theories of *international harmonization* are also widely used to explain policy change (Dolowitz and Marsh 1996; Martin and Simmons 1998; Holzinger and Knill 2005). It is expected that the legally binding character of agreements at the international or supranational level will influence the domestic policy choice. Because all contracting parties to an international agreement have to implement the same policy, convergence can be the expected outcome (Drezner 2001; Holzinger and Knill 2005). For example, many studies revealed the influence of the European Union (EU) on increasing similarity of the environmental policies of its member states (Héritier et al. 1996; Knill and Lenschow 2000; Jordan 2002; Börzel 2003).

In addition, a specific sequence of policy change is expected. Ideally, international harmonization should cause a synchronous policy response among member states of an international institution. However, evidence from the compliance literature shows that countries often lag in the implementation of international standards, for strategic reasons or because of a lack of administrative capacities (Börzel 2003).

It also is possible to derive a directional expectation from theories of harmonization. In the literature on the European Union, an empirically observed upward shift has been explained by a regulatory contest among member states in influencing common policies (Héritier et al. 1996). Even if it is assumed that the final agreement lies in the middle between high-regulating and low-regulating countries, there is a high likelihood that domestic regulatory levels become stricter in some countries (Holzinger, Knill, and Sommerer 2011, 24). Furthermore, in the case of minimum harmonization, governments that want stricter regulations are given the possibility to opt out and enact standards above the lowest common denominator.

A fourth group of theories on international driving factors of change focuses on *imposition*, where a policy transfer is enforced by a more powerful or influential actor (Holzinger and Knill 2005). A typical example of imposition or conditionality is the EU enlargement process (Schimmelfennig and Sedelmeier 2004). Powerful states can use "carrots and sticks" to convince other countries to follow them, perhaps through trade sanctions (Underdal 1994; Desombre 1995). This mechanism includes two expectations on the type of change. Like international harmonization, imposition should lead to policy convergence. A powerful actor is expected to impose its policy on less powerful ones, leading to growing similarity of policy repertoires (Meseguer 2004, 314; Brune et al. 2004). In addition, the logic of this mechanism necessitates a time lag and the existence of a policy model that could be imposed.

The first of five theories on the influence of the domestic institutional context refers to the power of a *veto player*— both institutional and partisan—to present an obstacle to changing a policy. Formalized by Tsebelis (1995, 2002), this theory has been tested widely across issue areas, but only rarely in the field of environmental politics (Knill, Debus, and Heichel 2010). The veto player theory makes an explicit statement on the frequency of change as an additional dimension. It is assumed that change will occur less frequently in a domestic context with powerful veto players. Friction should be greater in presidential systems than in parliamentary systems because policy change in the former depends on approval from several different bodies (Jones and Baumgartner 2005). However, nothing is said on the direction, the level, or the similarity, apart from an implicit assumption that veto power also can delay political reforms and thus will not lead to pioneering behavior and early change.

Theories of *path dependence* are based on the assumption that sociocultural and socioeconomic factors, together with unique historical constellations, have created different institutions in different countries (Pierson 2004; see Kemmerling and Bruttel 2006). These institutions are assumed to build a continuous framework for regulatory traditions. Two expectations on the type of change can be derived from this theory. First, path dependence will not lead to convergence or divergence, but to the coexistence of different policies, and eventually the occurrence of parallel changes. Second, institutional stability might decrease the frequency of policy change.

A third theoretical approach in the domestic context privileges an explanation of policy change by the *administrative capacity*. It is assumed that organizational resources influence a government's ability to develop new policy and reform existing ones. For example, previous research has shown that the existence of an independent environmental ministry affects the environmental performance of a country (Jänicke 2005, 130). Three expectations on the nature of policy change can be derived from existing research in this field.

First, governments with high administrative and financial capacities have the opportunity to react early (Heclo 1974). The existence of a positive institutional opportunity structure enables a country to become an environmental pioneer (Jänicke 2005; Weidner 2002). Administrative capacities might be the basis for unilateral leadership, an effective national innovation system, and good regulatory effectiveness (Urpelainen 2011; Beise 2004; Furman, Porter, and Stern 2002). Second, ambitious regulations should be the consequence of high administrative capacity (Jänicke 2005). Finally, a higher frequency of change is implicit to the assumption that administrative capacity makes governments more effective.

A fourth category of domestic explanations of policy innovations and reforms can be summarized under the label of *domestic demand*. It is widely agreed that policy change could be affected by domestic political preferences, electoral competition, and the success of issue-specific parties and the pressure of interest groups (Vogel 2000, 267). The societal mobilization of green interest groups is commonly associated with the spread of environmental policies (Jänicke 1990; Jahn 1998, Binder and Neumayer 2005). Vogel (2000) argues that countries with a strong demand for green policies and influential green pressure groups have stronger and better-enforced regulations. Environmental policies also tend to be strengthened during periods of economic prosperity. A high level of pollution, economic wealth, and a widespread post-materialist orientation represent the domestic demand for ambitious environmental policies (Jänicke 1990; Scruggs 2003; Weidner 2002). In addition to the expectations about the direction of change and the regulatory level, domestic demand might have consequences concerning the sequence of change and the similarity of policy responses. Governments under strong domestic pressure will react earlier.

Finally, policy change can also be driven by important *events* (Jones 1994). Transboundary events like the Chernobyl catastrophe should lead

	Timing	Frequency	Direction	Level	Sequence	Similarity
	Early	High	Upward	High	Pioneer	Convergence
					Model	Parallel
	Late	Low	Downward	Low	Synchronous	Divergence
Regulatory Competition			•	•	•	•
Learning and Diffusion			•		•	•
Intern. Harmonization			•	•		
Imposition			•		•	•
Veto Player	•	•				
Path Dependence						•
Innovative Capacity	•	•		•		
Domestic Demand	•			•		
Events					•	•

Figure 6.1

Typology of six dimensions of policy change

to a simultaneous regulatory reaction among national governments affected by such events (Jones 1994; Wilson 2000). According to the literature on policy diffusion, events also could cause similar or even identical responses (True and Mintrom 2001; Brinks and Coppedge 2006).

Figure 6.1 presents the results of this literature review in a systematic way. Overall, six dimensions of policy changes could have been derived from existing theories. The graph includes the main alternatives for all six dimensions and also indicates the relevant theories.

First, some theoretical expectations involve the timing of change. According to theories on the domestic institutional context and the political demand, there will be early birds and latecomers. Second, predictions on the type of change also vary with regard to the frequency of regulatory action. It will be high if the administrative capacity of a government allows for effective regulation, and low if veto players make policy reforms cumbersome. A third dimension refers to the direction of change. A strengthening or a relaxation of policies can be driven by regulatory competition, learning, international harmonization, or imposition. It is important to note that assumptions on the direction of change might be distinct from expectations on the level of regulations; the direction refers to relative changes that do not necessarily lead to policies that are strict (or lax) in absolute terms. Domestic political demand and administrative capacity will lead to ambitious policies, whereas a low level of regulation would be expected to be the end point of a regulatory "race to the bottom."

The fifth dimension of policy change is about the sequence of regulatory action in a comparative perspective. The difference between the dimension of timing and sequencing is minor, but nonetheless important. For the sequence, theoretical expectations refer to the relative position of one country vis-à-vis other countries, not to the earliness or lateness in absolute terms. Pioneer change—change in the absence of a model—can be distinguished from model change and synchronous change. The latter would be the ideal consequence of an event, a common shock, or joint efforts to harmonize a policy at the international level. Model change is expected to be the outcome of learning from the experience of others, either from competition or from imposition. Pioneer change should be the outcome of an attempt to realize a competitive advantage or the result of high administrative capacity at the level of national governments.

The sixth dimension indicates if change affects policy similarity across countries. Convergence is predicted by theories of learning, competition, harmonization, and imposition, but it also could be the consequence of an event that affects more than one country. Divergence could be the result of competition alone. The concept of path dependence features a third alternative: the development on different but parallel tracks.

Data on Environmental Standards in Twenty-Four Countries

These dimensions of policy change will be analyzed empirically in the field of environmental regulation. Therefore, a new data set will be used. This data set comprises information on seventeen metrical environmental standards in twenty-four countries, within an observation period from 1970 to 2005. Originally, it is based on the ENVIPOLCON data set (Holzinger et al. 2008b).[2] In a follow-up to this project, the ENVIPOLCON data with four cross sections for 1970, 1980, 1990, and 2000 (see, e.g., chapter 3) has been transformed into a time-series data set.[3] Together with an update until 2005, the new data set captures each instance of policy change on a yearly basis for the thirty-five-year period. The sample of twenty-four countries includes fourteen member states of the former EU-15 (except Luxembourg), five countries from Central and Eastern Europe that accessed the European Union between 2004 and 2007 (Hungary, Poland, Slovakia, Bulgaria, and Romania), and Norway and Switzerland, members of the European Free Trade Association (EFTA). The sample is completed by three non-European OECD members: the United States, Mexico, and Japan. Thus, major environmental pioneer countries are represented, as well as a large regional cluster. The share of EU member states is considerable, yet not too high. In the first twenty-five years of the observation period, only eleven of twenty-four countries have been part of the European Union.

The selection of environmental regulations in this data set covers different aspects of this policy field (see the list in the appendix at the end of this chapter). There are measures against air pollution (i.e., sulfur content of gas oil; lead content of gasoline; car emissions of CO, HC, and NO_x; and SO_2 and NO_x emissions for combustion plants), policies concerning water quality (zinc, lead, copper, chromium, and BOD discharges into surface water), noise standards for trucks and around highways, and finally, recycling targets for glass and paper.[4] The main advantage of having a broad range of policies is that the empirical analysis does not depend on the specific characteristics of a single policy.

All relevant information and regulatory standards refer to policy output data. They represent the regulatory capacity of the political system—the first dimension in Meadowcraft's ideal measure of environmental performance (see chapter 2), but they do not indicate if these measures are implemented, or if they are effective to solve the underlying problem. However, a preference for output data is justified by the fact that outcome data is influenced by variables outside the political system, and thereby outside the theoretical framework that is of interest for this research area (Heichel and Sommerer 2007).

The data set includes seventeen metrical environmental standards, such as limit values for emissions or waste-recycling targets. These limit values can be strict, or they can be less stringent; they can be identical, or they can differ gradually across countries. The data thus goes beyond the common binary measure of the existence of policy instruments, which is typical in the diffusion literature (see chapter 3). It provides a more comprehensive approach to the study of change because it includes the direction, regulatory level, and gradual similarity of environmental policies.

A Multidimensional Analysis of Environmental Policy Change

The empirical assessment of environmental standards on six dimensions of policy change specified in the section "A Typology of Policy Change," earlier in this chapter, will involve a set of different measures. In the statistical analysis of the timing, frequency, direction, and level of regulatory change, countries are the unit of analysis. For the remaining dimensions, the measurement of change is based on country. With such a dyadic approach, the comparative aspect that is inherent in the dimension of sequence and similarity can be operationalized directly.

The Timing of Environmental Policy Change

In a first step, the timing of environmental policy change in the twenty-four countries of the sample analyzed is analyzed. From the theoretical perspective, the timing could be influence by the administrative capacity, political demand at the domestic level, or by veto power. To assess the timing of change, the time period since the establishment of the seventeen metrical standards is calculated for each country and for each standard. If a limit value for lead in gasoline has been introduced in Finland in 1983, the score for this country is (2005 − 1982 = 23).[5] The average maturity of all seventeen environmental standards is displayed in table 6.1. A high value suggests that a country acted early against environmental pollution, and a low score identifies laggards.

The average maturity across all countries is 19.7 years (table 6.1, column 2). Overall, Japan reacted earlier than other countries, with a policy repertoire that is on average 30.2 years old. While this confirms the image of Japan as an environmental pioneer, Hungary, with the second oldest repertoire (27.1), is normally not among the usual suspects when it comes to green pioneers. Hungary is followed by Belgium (26.1) and Germany (25.4). In general, longstanding EU member states introduced their standards earlier than others, with the notable exceptions of the Netherlands (18.0), the United Kingdom (17.5), and Ireland (13.9). The slowest country was Greece, with an average of only 9.5 years for the repertoire of seventeen standards. Among other countries that reacted late were Romania (11.8), Mexico (14.4), Norway (13.9), and the United States (17.2). Although the United States has been a pioneer in some fields, this is not reflected in the averages here.

To allow for understanding the overall trend better, table 6.1 also includes the results of the timing in four subcategories of regulatory standards, according to different environmental media (columns 3–6). High values, and thus an early adoption of air pollution standards, can be found not only for Belgium and Germany, but also for Sweden (34.9) and the United Kingdom (30.9). By contrast, central and eastern European countries reacted much later in this particular field. The higher average scores for Slovakia, Romania, Poland, and Bulgaria mostly come from

Table 6.1

Policy change in seventeen countries' environmental standards, 1970–2005: Dimensions 1–4

	Time Since Adoption (in years)					Frequency	Direction		Regulatory Level (rank)				
	All	Air	Water	Noise	Waste	N	Upward (in %)	Downward (in %)	1970	1980	1990	2000	2005
Austria	22.0	26.8	15.0	28.5	14.0	41	97.6	2.4	12	16	7	3	6
Belgium	26.1	29.8	30.0	18.0	10.0	56	98.2	1.8	8	9	8	6	11
Bulgaria	17.4	14.4	34.8	0.5	3.0	26	80.8	19.2	13	17	20	19	21
Denmark	21.3	23.9	19.2	27.0	10.5	34	91.2	8.8	17	6	10	7	10
Finland	18.1	29.5	0.0	25.0	11.0	35	94.3	5.7	6	14	15	10	4
France	22.5	27.5	16.8	29.0	10.0	53	98.1	1.9	2	1	4	4	5
Germany	25.5	29.9	22.6	26.0	15.0	57	91.2	8.8	5	3	5	2	2
Greece	9.7	12.9	5.0	13.0	5.0	26	100.0	0.0	17	23	22	21	20
Hungary	27.1	30.3	28.8	26.0	11.0	40	92.5	7.5	18	15	16	15	13
Ireland	13.9	23.9	0.0	14.0	9.0	39	92.3	7.7	24	18	18	16	14
Italy	25.2	25.1	30.0	25.5	13.5	45	97.8	2.2	10	5	3	5	8
Japan	30.2	29.3	36.0	35.0	15.0	38	100.0	0.0	1	2	1	9	7

Table 6.1
(continued)

	Time Since Adoption (in years)					Frequency	Direction		Regulatory Level (rank)				
	All	Air	Water	Noise	Waste	N	Upward (in %)	Downward (in %)	1970	1980	1990	2000	2005
Mexico	14.4	13.5	18.0	23.5	0.0	27	81.5	18.5	24	24	21	24	24
Netherlands	18.0	27.1	0.0	29.5	15.0	47	95.7	4.3	10	7	2	1	1
Norway	13.9	22.5	0.0	17.0	11.0	35	91.4	8.6	24	21	9	13	15
Poland	18.1	12.0	36.0	11.0	5.0	35	88.6	11.4	17	20	19	22	19
Portugal	16.7	18.9	16.0	18.5	8.0	34	97.1	2.9	24	22	23	18	18
Romania	11.8	9.3	19.8	9.5	4.0	23	91.3	8.7	17	19	24	23	22
Slovakia	21.8	18.4	31.0	31.0	3.0	41	85.4	14.6	20	13	17	20	17
Spain	19.5	22.5	20.0	16.5	9.0	40	97.5	2.5	19	11	14	14	16
Sweden	20.5	34.9	0.0	21.5	13.0	29	96.6	3.4	3	4	11	8	3
Switzerland	25.3	26.5	31.0	28.5	3.0	43	97.7	2.3	7	12	6	11	12
United Kingdom	17.5	30.4	0.0	18.0	9.0	48	95.8	4.2	4	10	13	12	9
United States	17.2	18.8	24.8	9.0	0.0	26	100.0	0.0	11	8	12	17	23
Mean	19.7	23.3	18.1	20.9	8.6	38.3	93.9	6.1					

their longstanding regulations on water quality, where their repertoire is considerably older than the average. In turn, it is evident that some countries in the sample do not have water pollution standards at the national level, although they might exist at the regional or local level. This is one aspect that turns Scandinavian countries, the Netherlands, and the United Kingdom into laggards. For noise standards, the results show that a small group of countries—Japan, Slovakia, the Netherlands, France, Switzerland, and Austria—reacted very early. For the latter two, this corresponds to the well-known pressure problem, as both are highly affected by transit traffic in Central Europe. Finally, policies on waste management emerged later than others. In this field, Germany, Japan, and the Netherlands (all with a score of 15.0) have been front-runners for the two recycling quotas in the sample.

The Frequency of Environmental Policy Change

The frequency of regulatory reforms has been identified as a second dimension of change. A low number of changes would be expected as a consequence of strong veto players that slow down decision-making procedures, whereas a country with high organizational resources might generate more change. The seventh column of table 6.1 gives the number of policy shifts for all seventeen standards per country. Overall, the data set contains 918 changes.

Interestingly enough, Germany (57 changes) and Belgium (56) show the highest figures of all twenty-four countries. While it is not surprising that a long regulatory history correlates with more change, the result contradicts some theoretical expectations. The political systems of both countries are notorious for the influence that institutional and partisan veto players exert. Because other countries with a high frequency of regulatory change are long-serving EU member states like France, United Kingdom, Italy, and the Netherlands, a high regulatory activity at the level of EU environmental policy seems to influence frequent changes at the domestic level. Figures for Romania (23), Greece (26) and, to a certain degree, Bulgaria (26), the United States (26), and Mexico (27) suggest a correlation between late reaction and less frequent regulatory changes. However, important deviations from this pattern can be found. Japan has been mentioned as a pioneer, but it shows only a relative small number of subsequent changes (38). The United Kingdom (48) has more than twice the number of changes as Greece but is clearly below the average score for an early timing.

The Direction of Environmental Policy Change

Table 6.1 also displays information on the direction of changes in environmental standards. Theories of international harmonization, imposition, and policy learning predict upward change, whereas regulatory competition among countries could lead to both a strengthening and relaxation of environmental standards. To assess these patterns empirically, I categorize all instances of change according to a numerical increase or decrease in the limit values over time. Percentage figures indicate the relative frequency of upward and downward shifts (columns 8–9).

First of all, an impressive figure of 93.9 percent shows that almost all changes in the data set involved a strengthening of the regulatory level. There are three countries without a single downward shift: Greece, Japan, and the United States. The highest share of downward changes can be observed for Bulgaria (19.2 percent), Mexico (18.5 percent), Slovakia (14.6 percent), and Poland (11.4 percent). This phenomenon can be linked to the adaptation of strict, but not implemented, water quality standards in those countries that in the 1990s have been replaced by more realistic ones (Knill, Tosun, and Heichel 2008).

While it is remarkable that even an environmental pioneer like Germany shows a number of downward changes, the overall trend points toward stricter regulations. In absolute figures, downward movement is rare, with the highest number for Slovakia (6). Thinking in terms of potential explanations, the presence of a downward spiral predicted by the theory of regulatory competition is not confirmed, whereas the observed upward trend could be caused by multiple mechanisms (Holzinger and Sommerer 2011).

The Regulatory Level of Environmental Standards

In the typology of policy change, the regulatory level has been distinguished from the direction of change. Expectations on the absolute level do not necessarily refer to a relative increase or decrease—and vice versa. Theoretically, the existence of ambitious or less ambitious policies could be driven by competition, harmonization, and administrative capacity, or by domestic demand. To study the level of environmental policy, country rankings based on the strictness of seventeen domestic environmental standards are compared over time. The right part of table 6.1 shows country rankings for 1970, 1980, 1990, 2000, and 2005. Because the comparability of policy standards is hindered by different units of measurements and scales, I begin by ranking the countries for each of the seventeen standards separately, with the highest rank given to the country with the most ambitious policy; and ties are allowed. In the second step, the average is calculated for all policies. This score is weighted by the number of standards, so that a country with only few standards, but strict ones, is not overrated. For a better illustration, the average scores are finally transformed to a ranking from 1 to 24.

The data reveals a considerable degree of mobility across countries. Former forerunners like the United States and Japan do not keep pace with the spread of environmental standards elsewhere, and they fall behind after 1990. Other countries, such as the Netherlands, Finland, and Austria, have changed from laggards in the 1970s to high-regulating green countries in 2005. However, several countries have not seemed to alter their role. Germany, France, and (with the exception of the 1980s) Sweden maintain their leading role in strict environmental regulation over time. Mexico, Romania, Bulgaria, Poland, and Greece are among the countries with the least ambitious standards at both the beginning and the end of the observation period. This fact, however, does not imply that these countries did not improve their standards over time.

Although France, Germany, and the Netherlands rank high on both the direction of change and the regulatory level, the results are quite different for many other countries. Thus, they provide an empirical justification for the distinction of these two dimensions. Finland, Sweden, and Japan have high standards but only a few upward changes. On the other hand, Hungary, Spain, and the United Kingdom rank lower on the absolute regulatory level than their number of upward shifts would suggest.

It is also interesting in the broader sense that highly industrialized countries like Japan, the United States, the United Kingdom, and Sweden had the strictest environmental policies the early decades, whereas EU member states took the lead from 1990 on. This pattern could be interpreted as a shift from domestic demand to international harmonization as the major determinant of stringent environmental regulations.

The Sequence of Environmental Policy Change

Beyond the literature on policy convergence and diffusion, the two remaining dimension of policy change are commonly neglected, although expectations about similarity and sequence of regulatory change are immanent to several influential theories. One explanation for the empirical disregard is that these dimensions are difficult to operationalize at the country level. In this section, a dyadic approach is applied to solve this problem. The data set is transformed by a paired comparison, and the change in country A of a pair AB is compared to the existence and level of a "model" policy in the reference country B. Such a pairwise approach has been widely used in the study of international conflicts (Bremer 1992; Kinsella and Russett 2002). In recent years, it has been transferred to the literature on diffusion (Volden 2006; Elkins et al. 2006; Holzinger et al. 2008a).

For the analysis of the sequence of change in environmental standards, I apply such a dyadic approach and operationalize all three constellations of pioneer change, model change, and synchronous change (figure 6.1). According to Volden (2006), model change occurs when a country shifts its regulation toward the policy of another country (such as a neighbor). A country A of a pair AB gets a score of 1 if there was a change in A at time t, and an environmental standard present in a reference country B at time t − 1, and a score of 0 in all other cases. Second, for pioneer change, score 1 is assigned to a constellation where there was change in country A at time t and no model present in country B at t − 1. Finally, synchronous change is defined by a change in countries A and B of a pair AB between t − 1 and t.[6]

In table 6.2, the aggregate figures for the dyadic change of seventeen environmental standards are displayed. In this model, 918 changes at the country level correspond to a total of 21,114 dyadic changes.[7] It is important to note that for the interpretation of change at the level of country pairs, the absolute number of dyadic changes is a hypothetical and inflated figure because it is the result of a multiplication of changes at the country level. However, it allows an assessment of the relative importance of strategic constellations; that is, of eventual imitations, pioneer moves, or common shocks for country A with regard to country B.[8] To facilitate the comparability with the results of the other dimensions, table 6.2 translates the dyadic data for country pairs to figures at the country level.

Overall, the results in table 6.2 illustrate that model change—change in the presence of existing environmental standards abroad—is by far the most common type of change. At the dyadic level, 13,466 model changes can be observed, compared to 3,163 changes without any model in a reference country (pioneer change), and only 1,802 instances of synchronous change.[9]

At the country level, most of the pioneer cases can be found in Germany (272; table 6.2, column 2), France (242), and Belgium (212); these countries were also early regulators in absolute terms (table 6.1). However, the empirical analysis corroborates the theoretical distinction between the timing and the sequence of policy change. There is a group of

Table 6.2
Policy change in seventeen countries' environmental standards, 1970–2005: Dimensions 5-6

| | Sequence of Change | | Model Change | | Similarity | | | | | | Synchronous Change | |
| | Pioneer Change | | All | | Convergence | | Divergence | | Parallel | | | |
	N	%	N	%	N	%	N	%	N	%	N	%
Austria	162	17.2	55,0	58.4	290	30.8	213	22.6	47	5.0	67	7.1
Belgium	212	16.5	70,6	57.3	373	28.9	246	19.2	87	9.2	89	6.9
Bulgaria	119	19.9	38,0	59.6	282	47.2	33	5.5	65	6.9	12	2.0
Denmark	119	15.2	50,5	63.4	236	30.2	211	27.0	58	6.2	94	12.0
Finland	76	9.4	59,9	73.5	326	40.5	222	27.6	51	5.4	99	12.3
France	242	19.9	77,2	64.6	401	32.9	317	26.0	54	5.7	137	11.2
Germany	272	20.8	75,8	59.6	315	24.0	385	29.4	58	6.2	114	8.7
Greece	19	3.2	49,6	81.1	420	70.2	46	7.7	30	3.2	22	3.5
Hungary	199	21.6	53,3	57.8	421	45.7	60	6.6	52	5.5	65	7.1
Ireland	120	13.4	65,5	72.8	305	34.0	302	33.7	48	5.1	139	15.5
Italy	147	14.2	63,4	61.8	329	31.8	263	25.5	42	4.5	138	13.3
Japan	102	11.7	34,8	39.6	162	18.5	164	18.8	22	2.3	26	3.0
Mexico	72	11.6	44,2	68.3	295	47.5	95	15.3	52	5.5	13	2.1

Table 6.2
(continued)

| | Sequence of Change | | | | Similarity | | | | | | Synchronous Change | |
| | Pioneer Change | | Model Change — All | | Convergence | | Divergence | | Parallel | | | |
	N	%	N	%	N	%	N	%	N	%	N	%
Netherlands	235	21.7	62,5	58.8	207	19.2	347	32.1	71	7.5	132	12.2
Norway	113	14.0	57,9	71.2	372	46.2	166	20.6	41	4.4	61	7.6
Poland	64	8.0	56,6	68.5	393	48.8	74	9.2	99	10.5	24	3.0
Portugal	48	6.1	58,9	74.5	394	50.4	158	20.2	37	3.9	81	10.0
Romania	23	4.4	40,2	73.5	353	66.7	19	3.6	30	3.2	11	2.1
Slovakia	107	11.4	63,8	67.7	464	49.2	131	13.9	43	4.6	37	3.9
Spain	88	9.6	64,4	69.8	380	41.3	211	22.9	53	5.6	92	10.0
Sweden	92	13.8	44,9	66.1	250	37.5	170	25.5	29	3.1	86	12.9
Switzerland	173	17.5	55,2	55.9	305	30.8	207	20.9	40	4.2	72	7.3
United Kingdom	206	18.7	76,7	70.5	333	30.2	371	33.6	63	6.7	136	12.3
United States	153	25.6	27,7	45.5	142	23.8	121	20.2	14	1.5	55	9.2
Mean		14.4		64.2		38.6		20.3		5.2		8.1

countries that reacted early, but shows fewer pioneer changes than others. Japan (102), Sweden (92), Spain (88), and Finland (76) introduced environmental standards early, but only in areas where at least a few other countrieswere active as well. On the other hand, countries like the Netherlands (235), the United Kingdom (206), and the United States (153) did not have a broad repertoire of environmental standards in the 1970s and 1980s, but where they introduced regulatory standards, they often have been pioneers. With 25.6 percent of all changes, the United States has the highest share of pioneer changes, followed by the Netherlands (21.7 percent) and Hungary (21.6 percent).

Model change is most common for France (772), the United Kingdom (767), and Germany (758). This is not surprising given the high frequency of absolute changes (see table 6.1). In return, countries with a low absolute number of model changes are the United States and Japan. Regarding the relative frequency of model changes, not only do Greece (81.1 percent), Portugal (74.5 percent), Romania (73.5 percent), and Finland (73.5 percent) bear the mark of latecomers in absolute terms (table 6.1), they also typically emulate environmental standards of other countries .

Synchronous change as the third theoretically relevant constellation of sequence in the change of a policy would be the expected outcome of international harmonization or events that affected more than one country. It already has been said that this type of change is less common. It is most likely to affect EU member states. The highest numbers can be observed for Ireland (139), Italy (138), France (137), and the United Kingdom (136). The results for the relative frequency of synchronous change are similar. At the other end of the scale, synchronous change is least common in Bulgaria (2.0 percent of all changes) and Romania (2.1 percent), both of which are only recent EU members with lower administrative capacity, and two non-European countries Mexico (2.1 percent) and Japan (3.0 percent).

Changes in the Similarity of Environmental Standards

The theoretical expectations on policy similarity, the sixth dimension of change, vary between convergence, divergence, and parallel developments. In this section, changes in the similarity of environmental standards will be assessed on the basis of the dyadic approach as the previous dimension of change. Because similarity and dissimilarity can be studied only when a policy is available in both countries of a pair, in a further extension of the approach by Volden (2006), convergent, divergent, and parallel shifts are distinguished for all categories of model change (see figure 6.1). The similarity of two standards in the countries that make up country pair AB is assessed on a similarity score between 0 and 100.[10] In the second step, scores are compared over time to control for an increase or decrease and then are transformed into a dichotomous variable for either convergence, divergence, or no change in similarity. In a third step, this information is combined with information on model change. For example, a score of 1 for divergence means that change in country A occurs at time t, reference country B had a model at least since t − 1, and this change led to a decrease in similarity between A and B.

In table 6.2 (columns 6–11), dyadic information on the absolute and relative frequency of convergence, divergence, and parallel policy shifts is transformed to the level of countries. Overall, convergent change is by far the most common type of change in environmental standards (7,748). Changes that lead to divergence (4,532) and parallel changes (1,186) occur less often in the sample of twenty-four countries. In absolute figures, the highest frequency of convergence can be found for Slovakia (464), Hungary (421), Greece (420), France (401), and Portugal (394). It has been shown that except for France and Hungary, these countries lag on other dimensions of change. However, they have undergone a process of catching up, partially driven by their entry into the European Union. Like model change, convergence plays a minor role for the United States (142) and Japan (162), but also for two EU member states: the Netherlands (207) and Denmark (236) also converge less often toward other countries in the sample.

The general pattern holds for the relative frequency of convergent changes. Some countries can be identified clearly as copycats. In Greece, more than two of three changes of an environmental standard involved the gradual imitation of foreign models (70.2 percent) and Romania (66.7 percent), and a high share of convergence could be observed for Portugal (50.2 percent). It is less surprising that the relative frequency of convergence is low for the Netherlands (19.2 percent) and Germany (24.0 percent), although Germany shows a high absolute number of convergent changes (315). In general, the pattern of convergence across countries suggests that there was an influence of European harmonization, the imposition of policies in accession talks, or learning from countries with a high reputation for ambitious environmental policies.

The picture is reversed for policy divergence. Countries with a high proportion of pioneer change showed more divergent moves with regard to the existing environmental policies of other countries. The countries with the highest scores are Germany (385), the United Kingdom (371), the Netherlands (347), and France (317). For these countries, divergence is important in relative terms, as is the case for Ireland (33.7 percent), Finland (27.6 percent), and Denmark (27.0 percent). At the end of the scale, it can be observed that for some countries with high scores for pioneer change, divergence from the policies of other countries plays only a minor role. This is the case for Hungary (6.6 percent), the United States (20.2 percent), and Switzerland (20.9 percent), indicating that in these countries, early regulation has not been followed up by subsequent leadership.

Finally, the number of parallel shifts is considerably low. In addition, cross-country variation is less pronounced. This type of change is most common in Poland (10.5 percent), Belgium (9.2 percent), and the Netherlands (7.5 percent) in absolute as well as in relative terms. For all other countries, the scores for the relative frequency of parallel change do not exceed 5 percent.

Conclusion

Over the last four decades, the field of environmental regulation has undergone fundamental changes. The statistical analysis in this chapter attempts to shed light on cross-national patterns in the development and stringency of environmental standards in twenty-four countries from 1970 to 2005. A new approach to the measurement of environmental policy change was introduced that allowed an assessment of the timing, frequency, and direction of policy change but also considered the regulatory level, the sequence of policy shifts, and the similarity of policy repertoires. This technique provides a comprehensive view that goes beyond existing research on environmental politics. It reveals multidimensional similarities and differences of environmental policymaking at the level of nation-states and thus can be seen as one building block for the development of an advance measure of environmental performance of political systems (see chapter 2).

On the basis of the empirical analysis in this chapter, some conclusions can be drawn on major patterns of environmental governance at the level of the nation-state. First, it gives insights into the overall dynamics of the field of environmental politics. For the sample of twenty-four countries, an enormous spread of regulations, and an impressive upward shift in the level of environmental standards have been observed, accompanied by a strong tendency toward convergence (see also Holzinger et al. 2008a; Holzinger and Sommerer 2011). Overall, 918 changes have been captured for a total of seventeen metrical environmental standards, ranging from the field of air pollution to water protection, noise regulation, and waste management. This is equivalent to a mean of 54 changes for each environmental standard, or more than 38 changes per country.

Second, the empirical analysis reveals an interesting pattern of cross-country differences in environmental politics. Some states are "green leaders": they score high on all six dimensions of policy change. In this sample, France, Germany, and, to a lesser degree, Belgium established environmental standards early, were active regulators, and showed a large number of upward shifts and pioneer changes at an ambitious regulatory level. By contrast, other countries seem to be typical environmental laggards. For Bulgaria, Romania, and Mexico, but even for the longstanding EU member states Portugal and Greece, it took longer to establish a broad repertoire of environmental standards. These countries were only rarely pioneers, showed less regulatory activity, and have less ambitious standards in place. However, from the mid-1990s onward, some laggards showed clear signs of convergence toward other countries in the sample, indicating that a process of catching up is under way.

Third, the statistical analysis reveals that the assessment of the regulatory performance of many countries varies with the dimensions of policy change. Three examples illustrate that the inclusion of more than one dimension improves the understanding how advanced a country's environmental politics actually is. Japan is a pioneer, with a long-established repertoire of environmental standards and not a single downward change. On the other hand, the frequency of change there was far below average, and over time, the country lost its top position with regard to the strictness of environmental standards. The Netherlands has been a latecomer, but since then, it has reached a high level of regulation, combined with a large number of strengthened regulations. Similarly, the United Kingdom is an example of a country with late-arriving regulation at a modestly ambitious level, but this impression is contrasted with many upward changes and pioneer regulations in areas of environmental policy where many countries lagged.

Fourth, the empirical analysis in this chapter is also meant to enhance the understanding of the causal mechanisms that drive policy change. All six dimensions of change in the theoretical framework have been deduced from differences in the expected consequences of different theoretical explanations of change. Future research could directly refer to a particular type of policy change for a well-tailored operationalization of the dependent variable. Such an approach allows a better differentiation between competing explanations, so long as these explanations involve different expectations on the type of change. If a theory predicts a particular pattern of change, the indicator for this causal mechanism should be correlated only with this type of change, not with others.

Some observations regarding the explanatory potential of different theories already can be made on the basis of the descriptive analysis in this chapter. Given the upward trend, it is less plausible that regulatory competition, in terms of a race to the bottom, plays an important role in explaining policy change. Furthermore, the empirical pattern suggests a greater influence of domestic factors at the beginning of the observation period, where highly developed and democratic countries tended to have more and stronger environmental regulations. Since then, a process of convergence and catching up undergone by latecomers indicates an increasing influence of international harmonization, learning from foreign policy models, or even imposition.

Appendix

List of Seventeen Environmental Standards

Sulfur content in gas oil

Lead in gasoline

Passenger cars—CO emissions

Passenger cars—NO_x emissions

Passenger cars—HC emissions

Large combustion plants—SO_2 emissions

Large combustion plants—NO_x emissions

Large combustion plants—Dust emissions

Industrial discharges in surface water—Lead

Industrial discharges in surface water—Copper

Industrial discharges in surface water—Zinc

Industrial discharges in surface water—Chromium

Industrial discharges in surface water—BOD

Noise emissions standard from trucks

Motorway noise emissions

Glass reuse/recycling target

Paper reuse/recycling target

Notes

1. See, for example, "Beyond Durban," *The New York Times,* December 17, 2011; "Is Durban Canada's tipping point?" *The Toronto Star,* December 8, 2011; "Agreement to curb HFCs not reached," *The Washington Post,* November 26, 2011.

2. The ENVIPOLCON data set is based on the research project "Environmental Governance in Europe," funded by the European Commission in the Fifth Framework Programme between 2003 and 2006. The data is based on information in official governmental documents and has been collected with the assistance of environmental policy experts in all twenty-four countries.

3. The updated data set is based on the research project "Factors of Policy Change," funded by the German Science Foundation between 2006 and 2009.

4. For more details on the data and the data collection, see Holzinger et al. (2008 a, b) and Sommerer (2010).

5. Similar to this measure for timing the introduction of metrical policy standards, chapter 3 operationalizes the "adoption resistance" for environmental policy instruments.

6. For a technical description of model change, see Volden (2006). Pioneer and synchronous change is applied analogously and was introduced by Sommerer (2010: 123ff).

7. A sample of twenty-four countries leads to 552 directed dyads for each year and 19,872 cases over time.

8. To obtain meaningful results, it was critical that a theory-informed composition of the dyadic sample included major environmental pioneer countries, a network of trade partners, and a regional cluster of neighboring countries. In the literature on conflict studies, this is known as the concept of "politically relevant dyads"—that is, only those country pairs should be included that could be involved in a direct conflict, at least theoretically.

9. The figures for the three constellations do not add up to 21,114. There are many cases where standards are not comparable across countries (e.g., due to different test cycles). In these cases, no value could be assigned.

10. See Holzinger, Knill, and Sommerer (2008) for details on this transformation.

References

Beise, Marian. 2004. Lead Markets: Country-Specific Drivers of the Global Diffusion of Innovations. *Research Policy* 33(6–7): 997–1018.

Bennett, Colin. 1991. Review: What Is Policy Convergence and What Causes It? *British Journal of Political Science* 21(2): 215–233.

Binder, Seth, and Eric Neumayer. 2005. Environmental Pressure Group Strength and Air Pollution: An Empirical Analysis. *Ecological Economics* 55(4): 527–538.

Börzel, Tanja. 2003. *Environmental Leaders and Laggards in Europe: Why There Is (not) a Southern Problem.* Aldershot, UK: Ashgate Publishing, Ltd.

Braun, Dietmar, and Fabrizio Gilardi. 2006. Taking "Galton's Problem" Seriously: Towards a Theory of Policy Diffusion. *Journal of Theoretical Politics* 18(3): 298–322.

Bremer, Stuart. 1992. Dangerous Dyads: Conditions Affecting the Likelihood of Interstate War, 1816–1965. *Journal of Conflict Resolution* 36(2): 309–341.

Brinks, Daniel, and Michael Coppedge. 2006. Diffusion Is No Illusion: Neighbour Emulation in the Third Wave of Democracy. *Comparative Political Studies* 39(4): 463–489.

Brune, Nancy, Geoffrey Garrett, and Bruce Kogut. 2004. The International Monetary Fund and the Global Spread of Privatization. *IMF Staff Papers* 51(2): 195–220.

Desombre, Elizabeth. 1995. Baptists and Bootleggers for the Environment: The Origins of United States Unilateral Sanctions. *Journal of Environment & Development* 4(1): 53–75.

Dolowitz, David, and David Marsh. 1996. Who Learns What from Whom? A Review of the Policy Transfer Literature. *Political Studies* 44(2): 343–357.

Drezner, Daniel. 2001. Globalization and Policy Convergence. *International Studies Review* 3(1): 53–78.

Elkins, Zachary, Andrew Guzman, and Beth Simmons. 2006. Competing for Capital: The Diffusion of Bilateral Investment Treaties, 1960–2000. *International Organization* 60(4): 811–846.

Furman, Jeffrey, Michael Porter, and Scott Stern. 2002. The Determinants of National Innovative Capacity. *Research Policy* 31(6): 899–933.

Heclo, Hugh. 1974. *Modern Social Politics in Britain and Sweden.* New Haven: Yale University Press.

Heichel, Stephan, and Thomas Sommerer. 2007. Unterschiedliche Pfade, ein Ziel? Spezifikation im Forschungsdesign und Vergleichbarkeit der Ergebnisse bei der Suche nach der Konvergenz Nationalstaatlicher Politiken. In *Transfer, Diffusion und Konvergenz von Politiken,* ed. Katharina Holzinger, Helge Jörgens, and Christoph Knill, 107–130. Wiesbaden: vs Verlag für Sozialwissenschaften.

Héritier, Adrienne, Christoph Knill, and Susanne Mingers. 1996. *Ringing the Changes in Europe: Regulatory Competition and Redefinition of the State. Britain, France, Germany.* Berlin: De Gruyter.

Hoberg, George. 2001. Globalization and Policy Convergence: Symposium Overview. *Journal of Comparative Policy Analysis* 3(2): 127–132.

Holzinger, Katharina. 2003. Common Goods, Matrix Games, and Institutional Solutions. *European Journal of International Relations* 9(2): 173–212.

Holzinger, Katharina, and Christoph Knill. 2005. Causes and Conditions of Cross-National Policy Convergence. *Journal of European Public Policy* 12(5): 775–796.

Holzinger, Katharina, and Thomas Sommerer. 2011. Race to the Bottom or Race to Brussels? Environmental Competition in Europe. *Journal of Common Market Studies* 49(1): 315–339.

Holzinger, Katharina, Christoph Knill, and Thomas Sommerer. 2008a. Environmental Policy Convergence: The Impact of International Harmonization, Transnational Communication, and Regulatory Competition. *International Organization* 62(4): 553–587.

Holzinger, Katharina, Christoph Knill, and Bas Arts, eds. 2008b. *Environmental Policy Convergence in Europe: The Impact of International Institutions and Trade*. Cambridge: Cambridge University Press.

Holzinger, Katharina, Christoph Knill, and Thomas Sommerer. 2011. Is There Convergence of National Environmental Policies? An Analysis of Policy Outputs in 24 OECD Countries. *Environmental Politics* 20(1): 20–41.

Jahn, Detlef. 1998. Environmental Performance and Policy Regimes: Explaining Variation in 18 OECD-Countries. *Policy Sciences* 31(2): 107–131.

Jänicke, Martin. 1990. Erfolgsbedingungen von Umweltpolitik im internationalen Vergleich. *Umweltpolitik und Umweltrecht* 13(3): 213–232.

Jänicke, Martin. 2005. Trend-Setters in Environmental Policy: The Character and Role of Pioneer Countries. *European Environment* 15(2): 129–142.

Jones, Bryan. 1994. *Reconceiving Decision-Making in Democratic Politics: Attention, Choice, and Public Policy*. Chicago: University of Chicago Press.

Jones, Bryan, and Frank Baumgartner. 2005. A Model of Choice for Public Policy. *Journal of Public Administration: Research and Theory* 15(3): 325–351.

Jordan, Andrew. 2002. *The Europeanization of British Environmental Policy*. Houndmills, UK: Palgrave Macmillan.

Kemmerling, Achim, and Oliver Bruttel. 2006. "New Politics" in German Labour Market Policy? The Implications of the Recent Hartz Reforms for the German Welfare State. *West European Politics* 29(1): 90–112.

Kinsella, David, and Bruce Russett. 2002. Conflict Emergence and Escalation in Interactive International Dyads. *Journal of Politics* 64(4): 1045–1068.

Knill, Christoph, and Andrea Lenschow. 2000. *Implementing EU Environmental Policy: New Directions and Old Problems*. Manchester, UK: Manchester University Press.

Knill, Christoph, Marc Debus, and Stephan Heichel. 2010. Do Parties Matter in Internationalized Policy Areas? The Impact of Political Parties on Environmental Policy Outputs in 18 OECD countries, 1970–2000. *European Journal of Political Research* 49(3): 301–336.

Knill, Christoph, Jale Tosun, and Stephan Heichel. 2008. Balancing Competitiveness and Conditionality: Environmental Policy-Making in Low-Regulating Countries. *Journal of European Public Policy* 15(7): 1019–1040.

Lazer, David. 2001. Regulatory Interdependence and International Governance. *Journal of European Public Policy* 8(3): 474–492.

Levi-Faur, David. 2002. Herding Towards a New Convention: On Herds, Shepherds, and Lost Sheep in the Liberalization. Politics Papers W6–2002. Nuffield College.

Levy, Jack. 1994. Learning and Foreign Policy: Sweeping a Conceptual Minefield. *International Organization* 48(2): 279–312.

Martin, Lisa, and Beth Simmons.1998. Theories and Empirical Studies of International Institutions. *International Organization* 52(4): 729–757.

Meseguer, Covadonga. 2004. What Role for Learning? The Diffusion of Privatisation in OECD and Latin American Countries. *Journal of Public Policy,* 24(3): 299–325.

Meseguer, Covadonga. 2006. Learning and Economic Policy Choices. *European Journal of Political Economy* 22(1): 156–178.

Pierson, Paul. 2004. *Politics in Time: History, Institutions, and Social Analysis.* Princeton, NJ: Princeton University Press.

Prakash, Aseem, and Matthew Potoski. 2006. Racing to the Bottom? Trade, Environmental Governance, and ISO 14001. *American Journal of Political Science* 50(2): 350–364.

Scharpf, Fritz. 1997. Introduction: The Problem-Solving Capacity of Multi-Level Governance. *Journal of European Public Policy* 4(4): 520–538.

Schimmelfennig, Frank, and Ulrich Sedelmeier. 2004. Governance by Conditionality: EU Rule Transfer to the Candidate Countries of Central and Eastern Europe. *Journal of European Public Policy* 11(4): 661–679.

Scruggs, Lyle. 2003. *Sustaining Abundance: Environmental Performance in Industrial Democracies.* Cambridge, MA: Cambridge University Press.

Simmons, Beth, and Zachary Elkins. 2004. The Globalization of Liberalization: Policy Diffusion in the International Political Economy. *American Political Science Review* 98(1): 171–189.

Sommerer, Thomas. 2010. *Transnationales Lernen als Faktor von Policy-Wandel: Eine Vergleichende Analyse der Umweltpolitik in 24 OECD-Staaten.* Wiesbaden: vs Verlag für Sozialwissenschaften.

Strang, David, and Michael Macy. 2001. In Search of Excellence: Fads, Success Stories, and Adaptive Emulation. *American Journal of Sociology* 107(1): 147–182.

Tews, Kerstin, Per-Olov Busch, and Helge Jörgens. 2003. The Diffusion of New Environmental Policy Instruments. *European Journal of Political Research* 42(4): 569–600.

True, Jacqui, and Michael Mintrom. 2001. Transnational Networks and Policy Diffusion: The Case of Gender Mainstreaming. *International Studies Quarterly* 45(1): 27–57.

Tsebelis, George. 1995. Decision Making in Political Systems: Veto Players in Presidentialism, Parliamentarism, Multicameralism, and Multipartyism. *British Journal of Political Science* 25(3): 289–325.

Tsebelis, George. 2002. *Veto Players: How Political Institutions Work*. Princeton, NJ: Princeton University Press.

Underdal, Ariel. 1994. Leadership Theory: Rediscovering the Art of Management. In *International Multilateral Negotiation: Approaches to the Management of Complexity*, ed. William Zartman. San Francisco: Jossey-Bass.

Urpelainen, Johannes. 2011. Can Unilateral Leadership Promote International Environmental Cooperation? *International Interactions* 37(3): 320–339.

Vogel, David. 1995. *Trading Up: Consumer and Environmental Regulation in a Global Economy*. Cambridge, MA: Harvard University Press.

Vogel, David. 2000. Environmental Regulation and Economic Integration. *Journal of International Economic Law* 3(2): 265–279.

Volden, Craig. 2006. States as Policy Laboratories: Emulating Success in the Children's Health Insurance Program. *American Journal of Political Science* 50(2): 294–312.

Weidner, Helmut. 2002. Capacity Building for Ecological Modernization: Lessons from Cross-National Research. *American Behavioral Scientist* 45(9): 1340–1368.

Wheeler, David. 2000. Racing to the Bottom? Foreign Investment and Air Quality in Developing Countries. Unpublished working paper, the World Bank: 1–25.

Wilson, Carter. 2000. Policy Regimes and Policy Change. *Journal of Public Policy* 20(3): 247–274.

7

The Role of the State in the Governance of Sustainable Development: Subnational Practices in European States

Susan Baker and Katarina Eckerberg

This chapter investigates the role of the state in the governance of sustainable development at the subnational, regional, and local levels in Europe. It is a well-established fact that the nation-state, with its hierarchical formal bureaucracy, is no longer the sovereign power in environmental policymaking (if indeed it ever was). However, we also know that the state, despite having lost its privileged position, is far from absent in most areas of policymaking. In fact, while the state has been severely criticized by scholars of environmental politics and green theory (as discussed in chapter 1, "Introduction: The Comparative Study of Environmental Governance"; see also Barry and Eckersley 2005), there is mounting evidence that the state nonetheless remains a key actor in environmental governance. This chapter seeks to explore the area between these extremes by analyzing sustainable development policymaking in local and regional scales in a comparative perspective.

Specifically, the chapter tries to determine two things: (1) the *scope* of state involvement and (2) the *function* of the state in sustainable development policymaking. By "scope," we mean the extent that the state is an influential (or even essential) actor in the sustainable development policy areas on regional and local scales. By "function," we mean the repertoire of functions, responsibilities, and tasks carried out by the state in sustainable development policymaking. These two analytical dimensions, in combination with data from a large set of cross-national case studies of sustainable development policymaking, will enable us to paint a detailed and nuanced picture of the contemporary role of the state.

To assess these two aspects of state involvement, the investigation is structured around four key themes, which are identified by the governance literature as central aspects of the transformation of public policymaking: (1) multi-level governance, (2) networks and public/private partnerships, (3) participation, and (4) market-based instruments. We use these four themes to interrogate systematically a range of empirical findings to identify patterns of responses and to draw some insights into contemporary sustainable development policymaking in a comparative context. The logic behind the selection of these policy types and instruments is that they are all paradigmatic examples of new forms of governance in which the state is assumed to play a very limited role. Any observations of a more comprehensive state involvement in these "hard cases" of governance, therefore, should be relevant to the generic discussion of the contemporary role of the state. By doing so, the chapter also contributes to one of the key themes of this entire book: identifying cross-national patterns of environmental governance and processes of policymaking, particularly at the regional and local levels.

Study Design

Empirically, we draw upon our research, undertaken with a dozen European colleagues, that examined the processes of planning, funding, and implementing sustainable development at the subnational level in European Union (EU) member states and in Norway (Baker and Eckerberg 2008a). As such, the comparison focuses on policy processes and institutions, operating in different structural and cultural contexts and how these determine the outcomes of governance efforts (see the introduction to this volume).

The data used in this chapter is drawn from a number of sources. It includes comparative research from the REGIONET project, which examined regional models of sustainable development across Europe (Lafferty and Narodoslawsky 2004), with particular emphasis on National Sustainable Development Strategies, in the United Kingdom, France, Ireland, and Austria; central funding programs for sustainable development in Sweden, Germany, the Netherlands, and Denmark (Eckerberg et al. 2005); and processes of urban governance for sustainable development under the auspices of the EU-funded DISCUS project, including Denmark, Sweden, Norway, Finland, the United Kingdom, Belgium, the Netherlands, Germany, France, Spain, Portugal, Italy, Greece, Estonia, Russia, Lithuania, Slovakia, Poland, Bulgaria, Romania, and Croatia (Evans et al. 2004). In addition, it draws from case studies of sector policies, examining institutional capacities in Spanish local government (Hanf and Morata 2008), policy diffusion in the German *Länder* (state) (Kern 2008); sustainable development and waste management policy in Ireland (Connaughton et al. 2008), rural sustainable development in Norway (Hovik

2008), and local governance for sustainable development in the Netherlands (Coenen 2008). Taken together, these studies offer a rich empirical data set for analyzing the role of the state in environmental governance. It should be mentioned, however, that the empirical research was somewhat uneven in its treatment of variables, particularly in that some of the research projects paid less attention to issues of market governance, and although they dealt with networks, they did not necessarily address patterns of citizen participation.

Definitions and Key Concepts

Before we proceed further, a brief clarification of key terms used in this chapter is needed. These key concepts and terms are, without exception, rather difficult to encapsulate in a simple and straightforward definition. We employ the classic Brundtland definition of sustainable development, a form of development to which the European Union has made declaratory and legal commitment, while setting aside more complex discussions as to how this definition is operationalized in practice. For the purposes of this chapter, we use the term "state" broadly, to embrace all institutions of government in an organized political community, including the system of public administration. "Subnational government" refers to public, administrative entities operating at regional and local levels, which may have legislative powers devolved to them. The term "governance" has become somewhat notorious within academic studies because it is used in multiple ways, including as a descriptive label, as a theoretical concept used to explain empirical patterns of governing, and as a normative prescription, seeking to formulate rules and features of "good governance" practices (Jordan 2008). This confusion has many causes, not least that the earlier literature on governance tended to be both fragmented and noncumulative (see Baker 2012 for a further discussion of this topic). Our use of the term "governance" is discussed in more detail below, where we depict current concerns and developments within the literature on governance.

Structure of the Chapter

We begin this discussion with a short history of the research on governance and the fact that, in the initial period, it tended to underestimate the importance of the central state as a policy actor, but more recently has pointed to the coexistence of hierarchical steering alongside the operations of networks and market governance. We then turn to a more in-depth elaboration of the role of the state in governance, using the four key themes drawn from the literature on governance as our analytical framework for analyzing the empirical evidence from European states. Finally, we draw some conclusions for understanding the interrelationship between the state, on the one hand, and market and participatory and network forms of governance, on the other.

Governance and the State

It is a well-established finding that most industrialized and democratic states have undergone a move from government to governance since the 1970s (Pierre and Peters 2000). The extent, timing, and consequences of this transformation remain the subject of some debate among scholars, but most would agree that the processes of steering and policymaking have undergone a comprehensive reorganization in recent decades. Moreover, the governance literature has now redirected its attention, shifting from an earlier preoccupation with so-called new governance patterns that tended to claim that the central state has been displaced as a policy actor, to reinvestigations of the interactions that exist within and between the different agents and actors involved in governance processes.

At its most general level, the concept of governance is used to capture modes of coordinating, managing, steering, and guiding action in the realm of public affairs (Baker and Eckerberg 2008a). Governance, it should be pointed out, is something that governments have always done. Beyond this broad understanding, specific governance styles can be identified, which are typically grouped into three ideal types: hierarchies, markets, and networks. Hierarchies involve traditional forms of top-down control and regulation by the state, through legislation and regulation, fiscal and monetary measures (macroeconomic policy), strategic planning, and political brokerage. Governance can occur through market-based forms of resource allocation, and market governance (not to be confused with economic markets) is a public governance style. Finally, governance can occur through organized networks, including economic and business interest associations and involving various forms of public-private collaboration (Pierre and Peters 2000). Participation of actors drawn from civil society is also seen as an important element, although the relationship between network governance and participatory practices is not entirely clear from the literature. There is a lack of consensus as to whether "network governance" refers only to the involvement of organized economic interests or also encompasses practices designed to facilitate the engagement of civil society in policymaking. Within these ideal types, there are differences in the roles of institutions of public administration, the functioning of policymaking processes and systems, the types of policy tools and instruments in use and the involvement of private actors (see Baker 2012 for fuller elaboration).

Historically, a different weight has been given to the various styles of governance in societal steering. Analysis of these mixes has focused on the extent to which the state has political and institutional *capacity* to steer and how the state relates to other influential actors (Jänicke and Weidner 1997; see also chapter 9, "Decentralization and Deforestation: Comparing Local Forest Governance Regimes in Latin America"). This brings attention to the interactions that exist between the different styles at different points in time and across multiple levels of governance. In the postwar period, governments relied more heavily on hierarchical steering, especially regulation and sanctions (Peters and Pierre 2003). This pattern of governance came under pressure as early as the 1970s as a result of the increasing complex, dynamic, and interdependent nature of contemporary policymaking, which in turn makes it very difficult for the state to have the knowledge and resource capacity to tackle problems unilaterally (Kooiman 1993). Globalization and Europeanization added further pressures (Kohler-Koch and Eising 1998), resulting in actors, agencies, and processes operating at the supranational levels having greater influence on domestic policy. Europeanization also has enhanced the importance of the subnational, regional, and local levels (Keating and Loughlin 1997). The fiscal crisis experienced by many western states toward the end of the twentieth century added further pressures, particularly in states with highly developed welfare regimes. Under the influence of neoliberal ideologies, which pointed to the inherent inefficiencies and ineffectiveness of what they referred to as "big government," many Western states placed greater reliance on the other governance styles (markets and networks). This led to the introduction of new strategies for public service production and delivery (Kooiman 2000, 150), albeit with differences between the various welfare state regimes (Mydske et al. 2007).

As a result of these various and complex influences, contemporary processes of governance make greater use of input from organized interests, involve participation of civil society actors, and rely upon a wider range of policy tools and instruments, especially for policy implementation. We should note, however, that several of the features associated with the contemporary shift in governance patterns are actually well-established models of exchange between public and private actors (Peters and Pierre

2003, 3; Painter and Pierre 2005, 2). Schmitter's work on corporatism drew attention to the institutionalized system of exchange between the public and private sectors in western European states (Schmitter 1989). In addition, as shown in chapter 6, "Early Bird or Copycat, Leader or Laggard? A Comparison of Cross-National Patterns of Environmental Policy Change," there has been an upward expansion of regulatory governance in the environmental arena in the past thirty-five years, a development that has strengthened the role of the state in structuring society's relationship with nature. Having said this, however, contemporary developments do raise issues concerning the nature of the relationship between hierarchical actors, drawn from across multiple levels of governance, and organized interests, drawn from both the economic sector and from civil society, especially as they relate to the environmental role of the state (Meadowcroft 2005).

At first, the study of governance under these changing conditions gave rise to claims that the displacement of policymaking and the related institutional capacity downward in the political system, outward to agencies and nongovernmental organizations (NGOs), and upward to transnational institutional systems such as the European Union resulted in the "hollowing out" of the state (Rhodes 1997). However, it is now accepted that claims about the diminution of the state have overstated the case and that hierarchical governance continues to play a role in public policy and in societal steering. While governing mechanisms no longer rest on the authority and sanctions of the state alone, this does not necessarily mean that the state is less viable, or indeed less important (Smith 2003; Pierre 2000).

Governance and Sustainable Development

Our work in this chapter is informed by previous research on the governance of sustainable development (see Jordan 2008 for a summary of this literature). These studies focus on governance practices aimed at the promotion of sustainable development, as well as the *specific* governance requirements that ought to be introduced if society is to move along a more sustainable development trajectory. It highlights the distinctive governance challenges, given the scale of social transformation required and the character of the steering logic involved in the promotion of sustainable development, including the need to steer in the context of uncertainty, to take account of the long term, and to promote both horizontal and vertical policy coordination (see Nilsson and Eckerberg 2007; Meadowcroft

2007). Attention also has turned to the need to introduce more antici-patory and reflexive governance approaches (see Voß, Bauknecht, and Kemp et al. 2006). As such, our investigation of governance in environ-mental policy practices forms part of the need to "re-invent states, rather than to reject or circumvent them" (Eckersley 2004, 3).

Analyzing Sustainable Development Policies in European States

We now turn to the empirical analysis of the role of the state in sus-tainable development governance. As mentioned above, we have struc-tured this chapter around four key themes: (1) multi-level governance, (2) networks and public/private partnerships, (3) participation, and (4) policy instruments, in this case the use of market-based instruments. The enhancement of the role of multi-level authorities, the increased use of networks and partnerships in public policy, the facilitation of participa-tory processes, and the use of a greater array of policy tools are seen as quintessential characteristics of contemporary governance (Jordan 2008). While using these four themes, we are aware that empirical data does not always map neatly to such categorizations. For example, participatory processes and networks governance can overlap, and multilevel gover-nance processes can operate both down a vertical axis (running from the European Union to the national and the subnational levels) and across a horizontal axis (for example, between similar levels of public authorities across the member states). We first offer a detailed description of our four analytical themes and then move on to an analysis of the role and func-tion of the state in these policy areas.

Multilevel Governance

Within the European Union's system of governance, sovereignty is ex-ercised in common at the Community level. Member state functions (also at subnational levels) and Community functions are interwoven, and institutions depend on one another to form a functioning whole. The European Union also promotes the use of policy networks in the for-mulation and implementation of its policies. To capture this complexity, the European Union is now typically characterized as a system of multi-level governance (Hooghe and Marks 2001). Multi-level governance can be understood as the exercise of policymaking across the various levels of EU and member-state governments and outward to policy networks and civil society. Beyond this definitional understanding, the concept is used to emphasize the increased interdependencies between these levels.

Policymakers, for example, make increased use of the expertise of environmental NGOs not only to inform policy, but also to implement it, an input that lends legitimacy to the EU integration process (so-called input legitimacy). This conceptualization has led to new exploration of the power sharing and resource dependencies between the levels (Bache and Flinders 2004), including in relation to policy capacity; that is, the ability of different governance levels to act in pursuit of public policy goals. Capacity is here linked not just to formal competences, but also to institutional capacity, understood as administrative resources, finance, and expertise.

It is generally recognized that multi-level strategies to promote sustainable development need to improve capacity, as stressed in Agenda 21, the action plan of the United Nations to implement sustainable development, particularly those that require enhanced engagement of subnational authorities (Lafferty and Eckerberg 1998). There are different ways of looking at capacity. We are interested in "policy capacity," which is the ability to marshal the necessary resources to make intelligent collective choices about, and set strategic directions for, the allocations of scarce resources to public ends (Painter and Pierre 2005, 2). Capacity-building instruments are widely used instruments of public policy and play a role in the enhancement of governance processes (Painter and Pierre 2005). They can include provision of technical assistance through giving grants for skills training. This increases accountability by developing the management skills that facilitate compliance with national government requirements (Radin 2003, 608). Policy capacity is particularly relevant for the governance of sustainable development, not least because network styles of governance have the potential to affect negatively state capacity to steer collective action toward more sustainable forms of consumption and production. The concept of capacity also can consider the capacity that rests outside the state, especially within civil society. Here, the role of sustainable citizenship in the promotion of sustainable development, especially in states where there is high trust in state institutions, takes on a new importance (see chapter 8, "Sustainable Citizenship: The Role of Citizens and Consumers as Agents of the Environmental State").

The interdependencies that exist within the system of multi-level governance shape efforts to promote sustainable development (Radin 2003, 608). Here, framework policies and objectives are set at the European Union and then further specified at national levels, while roles and responsibilities must be distributed across the different levels of governance, consistent with each nation's resources and capacity (Meadowcroft 2002).

EU member states directly confront this challenge when, having begun to formulate sustainable development strategies in the 1990s, they had to address the capacity and resource shortfalls of their subnational authorities, which were charged with implementation. This led many central states to adopt new programs and plans, including financial packages, that were aimed at enhancing the capacity of regional and local authorities to act in pursuit of sustainable development; the empirical evidence for this change is presented in this chapter.

Networks and Public/Private Partnerships

The concept of governance also captures the dispersal of policymaking and policy delivery among a variety of private and public actors (Rosamund 2004, 121). As discussed above, contemporary governance patterns can be distinguished from a more hierarchical approach by the enhanced role of policy networks, and to a lesser extent, of participatory processes involving civil society actors, in policymaking. Indeed, so central is the role now played by networks that Rhodes's understanding of contemporary governance sees it as synonymous with network governance (Rhodes 2000; 1997).

The development of network governance focuses attention on the exercise of power and the play of politics as public and private actors, at various levels, negotiate over policy. Some evidence is presented below as to how the state manages these networks, and with what consequences, to pursue sustainable development. Attention is also given to the institutional dimension, especially both formal and informal arrangements put in place to facilitate the engagement of policy networks in governance processes.

Participation

Participation occurs through institutional settings that bring together actors at some stage in the policymaking process. They are best seen as located along a continuum, ranging from allowing only a minor, consultative role for non-state actors to more deliberative processes in which actors have a major say in shaping policy goals through dialogue and social learning (Baker 2012). There is a range of well-developed arguments in support of the claim that the promotion of sustainable development is best undertaken through the enhancement of practices of participatory democracy (Dryzek 2005). Indeed, the mobilization of social actors, especially through participatory practices, is seen as essentially for the purpose of sound governance in pursuit of sustainable development. This is because not only does it provide a critical voice and scrutiny of public policy, but also it helps to support both the means and ends of its achievement (Eckersley 2004, 246). These include functional arguments centered on the claim that they lead to better policy decisions and improved implementation prospects (lending so-called output legitimacy to policy) to more ideologically based claims that they help to improve democratic practice (lending input legitimacy).

Despite these claims, however, arguments abound that participatory practices are often weak in terms of traditional political accountability and representation. The capacity for participatory processes to provide a meaningful forum for deliberations also depends heavily on the type of formal access given to policymaking; the stage in the policymaking process at which participation is allowed; the opportunity structures that exist within the policy process to influence policymaking; and the institutional constraints that are placed on them and that are present more generally (Hallstrom 2004). The empirical material presented in this chapter addresses several of these issues.

Market-Based Instruments

The use of so-called market-based instruments has grown steadily since the early 1970s, as has their range, which now extends from subsidies through to emission charges and tradable permits (OECD 1998). Their use within the environmental policy arena has been extensively studied (Lenschow 1999; Jordan et al. 2003; Holzinger et al. 2006). The development of less bureaucratic, more flexible, "soft" policy instruments is seen as a key feature of contemporary governance (Lenschow 1999). They emphasize the closer cooperation that exists between public and private actors in the formulation and implementation of public policy.

The use of market instruments creates positive incentives for actors to cooperate voluntarily in environmental governance. This contrasts with the more adversarial relationship between government and economic actors, which was characteristic of "command and control", regulatory approaches to environmental management, particularly between the 1970s and 1980s in Europe. Despite their growing importance, we need to be mindful of the fact that the use of market instruments does not replace the regulatory approach, but rather that cooperative arrangements and legally non-binding agreements between public and private actors are combined with hierarchical interventions.

It is often argued that harnessing market forces for environmental governance helps a state to cope with the growing complexity of environmental policy against the background of its limited institutional and administrative capacity. The question of whether and to what extent our evidence supports this view forms a main theme in our empirical presentation. However, before we present our findings, it may be useful to give a brief summary of the issues raised so far in relation to the state's engagement in the governance of sustainable development.

We have argued that contemporary governance, while continuing to use hierarchical interventions, increasingly involves the use of "softer" steering instruments, of multilevel governance processes, and of policy networks and civil society participation in policymaking and delivery (Lafferty 2004). This new mix of governance styles, it is claimed, has resulted in major shifts in the role of the state, such that the state now operates in interdependent ways with other actors and can no longer impose its policy, but must negotiate both policy and implementation with partners in the public, private, and voluntary sectors (Stoker 2000, 98)—actors that in turn operate across multiple levels of government. Our attention now turns to whether and to what extent this results in any changes in the steering role of the state as it relates to the promotion of sustainable development.

Evidence from European States

The four themes outlined above are used in this section to structure the presentation of the empirical evidence, collected from the studies of governance in pursuit of sustainable development at the subnational level across a large number of European states.

Multi-level Governance

The influence of the European Union on the governance of sustainable development can be discerned across all the countries studied, specifically through its role as regulator and as formulation of strategic plans (for example, the EU Sustainable Development Strategy, outlined in the 2001 publication "A Sustainable Europe for a Better World"). Berger and Steurer (2008) show how a National Sustainable Development Strategy (NSDS), devised to comply with EU requirements, functions as a coordination tool for subsequent subnational policy developments. Coordination mechanisms that they found were used by the central state include guidance on local action plans, financial support, and training programs. Nevertheless, they found that the engagement of the subnational level should not be taken for granted, in that stimulating such engagement presented several challenges. Central governments face the difficult task of securing commitment from, and enhancing the competence of, the subnational levels, especially in relation to policy objectives and measures, without which *national* sustainable development trajectories could be blocked (Berger and Steurer 2008). Furthermore, while discussions of multi-level governance often assume that lower levels of government are executing policies within national frameworks, evidence from Spain suggests that the central government still has not taken any significant action to encourage, stimulate, or guide regional and local action in relation to the pursuit of sustainable development, and many subnational levels of government have taken on this task alone (Hanf and Morata 2008, 117). Similarly, in Germany, the *Länder* often become most active in areas when the federal government has failed to act (Kern 2008).

In addition, due to different administrative and policy traditions, the nature of the relationship between the multiple levels of governance varies greatly across the member states. In some countries (the United Kingdom, France), there is considerable decentralization of powers, while in others (Ireland), centralized policymaking still prevails, and those with federal government systems (Austria, Germany) exercise great policy autonomy and financial independence at the regional level. The greater the degree of autonomy, the easier it is both for locally specific needs and interests to be taken into account, and for subnational authorities to operate channels for communication and policy transfer directly with the EU level. In this context, Evans et al. (2008) found that both participatory governance processes and more substantive achievements in relation to sustainable development are chiefly found among local governments that have a high level of fiscal, legal, and political autonomy.

The German case provided the most obvious example of the need to take account of constitutional arrangements (Kern 2008). In Germany, the role of local authorities can vary, including from program to program, as does the extent of involvement of private-sector actors (Baker and Eckerberg 2008b). In contrast, the Irish case is illustrative of a highly centralized administrative system that continues to obstruct action at the subnational level. However, even in the case of Ireland, long recognized as one of the most centralized states in the European Union, there has been a growing emphasis on the role of regional authorities in promoting partnerships and in identifying their own regional sustainable development priorities (Connaughton et al. 2008). There was also evidence that, despite the strong engagement at the local level in implementing nationally defined sustainable development strategies, the local level is increasingly working as an independent actor in promoting locally relevant strategies, as in the Dutch case (Baker and Eckerberg 2008b).

These differences across the European states draws our attention to the distinction between multi-level governance, which refers to steering and public management, and multi-level politics; that is, to the distribution of power across the different levels of government (Smith 2003). In other words, the study of policymaking for sustainable development has to consider how the power relations between actors shape these processes. Indeed, as Hovik (2008) showed in her study of rural sustainable development networks in Norway, this play of power is a key determinant shaping the outcomes of governance processes.

Hanf and Morata (2008) also paid attention to the role of institutions and found evidence of "path dependency," in that sustainable development becomes interpreted through existing structures, procedures, and patterns of interactions. The need to be aware of the institutional framework in which policies are prepared, developed, and implemented also was pointed out by Baker and Eckerberg, particularly in their Swedish study (Baker and Eckerberg 2007, 2008b). Furthermore, while the European Union has helped to increase capacities at the regional level (Keating and Loughlin 1997), this capacity enhancement must not be seen as leading necessarily to enhance the pursuit of sustainable development. Berger and Steurer (2008) found that other strategies, such as national development or spatial plans, often exercise a dominant influence when compared to NSDSs. They also revealed tensions between different EU programs. The EU Structural Funds, for example, were found to be more concerned with traditional economic development and, more recently, social cohesion, rather than sustainable development (Berger and Steurer 2008). In both Ireland and France, for example, the pursuit of sustainable development at regional level proved to be only weakly linked to the NSDS. Instead, it was influenced by the National Development Plan in Ireland and local political interests in France (Berger and Steurer 2008, 40), often to the detriment of the sustainable development component.

Research also revealed that, contrary to the general trend, the European integration process could support hierarchical governance in some cases. In her study of the impact of Local Agenda 21 among the German federal states, Kern (2008) found that *Länder* competencies have been restricted rather than extended by European integration. Connaughton et al. (2008) also confirm that compliance with EU environmental regulation gives central government a strong role (especially through centralized institutions such as the environmental protection agency), and that the need for regulatory compliance can distort the use of market and network governance arrangements at the subnational level.

However, where such networks arrangements are strong, they can support innovative approaches. Towns and cities that are consistently high in achievement in pursuit of sustainable development are those that have worked in European networks (Evans et al. 2008). These networks can become conduits for the transfer of best practices. Although not an EU member, Norway also exploits European networks to share experiences and knowledge on rural sustainable development (Hovik 2008). Similar patterns of policy transfer as they relate to interorganizational policy learning also were exposed in Germany (Kern 2008). This reveals how policy transfer and benchmarking, particularly at the subnational levels, has become important for governance. Hanf and Morata (2008) also see policy transfer as an essential element of contemporary patterns of governance. As such, the role of policy learning and transfer is important for governance, especially given the ascendency of the so-called open method of coordination within the European Union, a method that promotes horizontal policy transfer and learning (Nilsson and Eckerberg 2007). However, within this method, there can be tensions between exploiting past learning to standardize around best practices on the one hand, and maintaining adaptability and avoiding "lock in" to outmoded routines on the other. This problem has arisen in the United Kingdom, where the desire for public/private partnerships to build upon past successes and established forms of best practices often works against the adoption of new initiatives and innovative approaches (Baker and Eckerberg 2008b).

Networks and Public/Private Partnerships

All the countries studied showed evidence of enhanced cooperation between the private and public sectors in support of sustainable development. This cooperation can extend from the engagement of interest groups in policy formulation to construction of public/private partnerships for program delivery. In Norway, this provided new opportunities for local politicians and private stakeholders to take shared responsibility for sustainable rural development (Hovik 2008). However, Hovik also drew attention to Scharpf's argument that these partnerships are being conducted under "the shadow of hierarchy" undermining Rhodes's original claim that we were witnessing the emergence of "governance without government" (Rhodes 1997). The presence of networks does not indicate, in itself, changes in the pattern of governance. The composition of a network, its remit, and the influence exercised by public sector partners remain vital. The activity of networks can be subject to hierarchical steering by government, where the state can initiate networks and shape both their frames of reference and membership. In Norway, for example, only a minor consultation role was given to subnational networks for rural sustainable development, and rarely were such networks involved in exchange and negotiation based on mutual trust (Hovik 2008; see also chapter 10, "Enforcement and Compliance in African Fisheries: The Dynamic Interaction between Ruler and Ruled," and chapter 11, "Causes and Consequences of Stakeholder Participation in Natural Resource Management: Evidence from 143 Biosphere Reserves in Fifty-Five Countries"). Furthermore, government actors used their privileged position to promote their own interests within these networks, enhancing traditional hierarchical patterns of governance (Hovik 2008).

This evidence cautions us to take into account the constraints, whether direct or indirect, that governments can impose on the activities of networks (Stoker 2000). Baker and Eckerberg's research on economic instruments used by central governments to promote sustainable development at subnational level also supports these ideas, showing how market and network governance enhances rather than diminishes the role of the state. Capacity enhancement measures that involve the provision of grants and subsidies were shown to strengthen the hand of central governments in steering the engagement of networks in governance processes, particularly in Sweden, Germany, and the United Kingdom (Baker and Eckerberg 2008b). Similarly, the Irish study indicated that hierarchical steering from central governments remains an integral feature of policy processes that use networks (Connaughton et al. 2008). Evans et al. (2008) also voiced concern that an exclusive focus on network governance can underestimate the role that local governments continue to play in innovating, supporting, and nurturing sustainable development planning and policy processes.

The case of waste management in Ireland further illustrated the difficulties of relating methods and practices of participatory governance to the existing structures and processes of government. In Ireland, growing dependence on the private sector has reduced the strategic control that local authorities exercise over waste management (Connaughton et al. 2008). Tensions also exist between partnership arrangements and traditional practices of representative democratic. Evans et al. (2008) pointed out where new governance arrangements create ambiguity and uncertainty in the eyes of both policymakers and the public about who is responsible and accountable for policy. Given that accountability and legitimacy are key tenets of democratic governance, maintaining a strong steering role for the state, at both the central and the local levels, becomes all the more important.

The Irish case directed attention to the need to explore the underlying rationale for partnership arrangements. In practice, these arrangements are often driven by new public management principles, not necessarily by the pursuit of sustainable development (Connaughton et al. 2008). The neoliberal belief in the power of the market was a particularly important driver in the United Kingdom. The UK Lottery Fund, for example, while acting as a key source of funding for local sustainable development initiatives, nevertheless prioritizes public/private partnerships aimed at efficiency in policy delivery, often to the determent of wider social and ecological aims (Baker and Eckerberg 2008b). Likewise, Coenen (2008) stressed that public/private partnerships are often motivated by efficiency considerations, but that at times, concerns about enhancing the legitimacy of policymaking processes (so-called input legitimacy) are also important. Evidence of the use of network governance here shows that such experimentation is not necessarily stimulated by the principles of sustainable development alone.

In short, our research findings suggest that, while network governance has increased, there is no causal connection between this and a commitment to promote sustainable development. The discourse of sustainable development has helped to open up a new political space within many European governments, legitimized through the vocabulary contained in Agenda 21. However, this new way of working is not necessarily stimulated by, nor is it confined to, the sphere of sustainable development policymaking.

Participation

A key characteristic of new governance is that it affords increased opportunities for citizens or the general public to have more direct input into the making of public policy. In part, this participation aims at strengthening the legitimacy of public-sector institutions (Peters and Pierre 2003, 3). The Dutch policy of political renewal, as discussed by Coenen (2008), provides an excellent example of this trend.

However, our evidence suggests that participatory processes are not without problems. For example, while there are several well-established

Dutch approaches to stakeholder engagement, participation has often occurred too late in the decision-making process to allow groups to exert influence over policy decisions (Coenen 2008). Hovik also pointed out the reluctance of public policymakers to open up the policy processes to extensive civil society participation (Hovik 2008). Similar limited participation of civil society was found by Baker and Eckerberg (2008b), but with the Dutch and Danish cases providing interesting exceptions. Evans et al. (2008) and Hanf and Morata (2008) revealed that the options available to civil society to participate in cooperative policymaking remain limited.

The findings of Evans et al. (2008) suggested that effective (or what they call "dynamic") governing for sustainable development is most likely to occur when governments work closely with civil society agents. This reinforced the idea of "bracing" social capital (Rydin and Holman 2004), where a strong relationship is formed between a limited group of actors with an interest in local sustainable development issues. Nevertheless, while the study of the UK Lottery Fund also revealed that local authorities tend to rely upon a limited number of groups participating in funding schemes, the extent to which these formed an example of bracing social capital that is oriented toward the pursuit of sustainable development remained in doubt (Baker and Eckerberg 2008b). Similarly, Connaughton et al. (2008) cautioned about the relationships that are established between the community and the state. In Ireland, NGOs, especially those operating at the local level, tend to remain outsiders. Organizations that present radical views (such as proposing "zero waste" strategies) remain on the fringes, with fewer opportunities to influence the agenda or policy outcomes. The Norwegian case also exposed similar problems (Hovik 2008). Here, participation practices tended to be driven by emphasis on problem-solving capacity rather than on principles of stakeholder participation—behavior that is at odds with the more radical requirements of sustainable citizenship (as discussed in chapter 8). In addition, local councils tend to put the interests of their local constituency above the willingness to deliberate and negotiate agreements with external actors in a network.

These findings again point to some difficulties in combining the logic of representative democratic government and the logic of network governance. Yet ample evidence points to the reciprocal benefits arising from strong relationships between local government and civil-society organizations, for both social cohesion and the pursuit of sustainable development (Evans et al. 2008).

Market-Based Instruments

Market based instruments were found to be in use in all the countries studied (Baker and Eckerberg 2008a). However, it proved difficult to maintain a sharp distinction between new and old policy tools, and the old tools (for example, financial instruments) were often used not just for the purposes of hierarchical steering, but also to promote network governance (Baker and Eckerberg 2008b). Furthermore, as pointed out above, many of the market-based instruments, such as voluntary agreements and public/private partnerships, required strong state engagement to both kick-start and oversee their progress. The catalyzing effect of economic instruments also was found to work best where government efforts are linked to existing activities, such as Local Agenda 21 (Baker and Eckerberg 2008b).

Evidence further suggested that some tools could lead to unintended consequences for governance processes. For example, governments often use competitive procedures for the allocation of funds, which can lead to self-interested behavior among partnerships, which in turn can threaten to destroy the basis of future sustainable development partnerships. Funding allocation also tends to favor quantifiable measures over more qualitative issues inherent in sustainable development (Baker and Eckerberg 2008b). The Irish case proved particularly revealing in that the use of public/private partnerships for the management of waste has diminished the control of local authorities over the strategic management of waste. In addition, privatization, which often brings increased prices, led in the Irish case to an increase in illegal dumping, while commercialization turned waste into a profitable commodity—a process that does little to encourage waste reduction. These examples show how the use of market-based instruments may not necessarily be good for the promotion of sustainable development.

Conclusion

The empirical evidence presented in this chapter has pointed to the continuing role of the central state in the governance of sustainable development. With regards to the scope of state involvement in governance-type arrangements, a clear finding in our study is that the state has far from abandoned the policy arena. This evidence has a direct bearing on our understanding of the role of the state in the pursuit of sustainable development. The empirical evidence presented here supports recent work on the centrality of the state in governing environmental problems. In practice, neither network nor market governance operates without state engagement nor can the pursuit of sustainable development be effective without the firm hand of the state in steering it. Thus, we hold that new governance practices "cannot secure long-term ecological viability in the absence of active steering by the state" (Meadowcroft 2005, 7).

Thus, the scope of state involvement does not show signs of significant retrenchment, even in the governance-heavy policy area selected for analysis. Quite the opposite, in fact: the state has been shown to be a key player in initiating and coordinating the sustainable development planning processes. Contrary to claims that the nation-state no longer plays a central role in environmental policy (see in particular Spargaaren and Mol 2008 and the introduction to this volume), we find that central government continues its formal exercise of power, which sees it establishing framework legislation, developing strategies, initiating funding mechanisms, and spurring the subnational level to engage in appropriate policies. In short, we find that the governance of sustainable development at the subnational levels is still highly dependent on traditional national and supranational government structures, processes, and policy priorities. The main *function* of the state in the analyzed policy areas is, on the one hand, to coordinate between different interests, and on the other, to stimulate new policy initiatives. National legislation and policy priorities remain key drivers for sustainable development (as evidenced by chapter 6), as are central government steering mechanisms and instruments. In addition, the state has been seen to both initiate and coordinate policy networks and to retain a great deal of power over the nature and functioning of network forms of governance. Evidence also points out how the use of new environmental policy instruments can strengthen the hand of the state by supporting hierarchical governance. Thus, far from being simple, the relationship between market and network governance and the state has been shown to be complex and dense.

In presenting this comparative evidence, this chapter thus supports efforts to infuse discussion of contemporary governance with a discussion of the traditional political themes of power. This, in turn, points to the importance of combining the discussion of governance processes with an acknowledgment of the centrality of formal constitutional arrangements and processes in framing the pursuit of sustainable development. Institutions structure political situations and shape political outcomes, as they influence not just actors' strategies but their goals, and they mediate their relations of cooperation and conflict (Smith 2003) as confirmed by many of the chapters in this book. In other words, political institutions matter to the governance of sustainable development. The empirical material presented in this chapter has been able to shed some light on the role of political institutions in contemporary governance processes, particularly as the earlier literature on governance has been criticized for ignoring this aspect (Pierre and Peters 2000).

As to the larger question of whether contemporary forms of governance help to promote a coherent approach to sustainable development, we are less certain in our verdict because the empirical research presented here points both to competing policy goals in practice and to frequent inconsistency between declaratory level-commitments and actual policy measures for sustainable development. Clearly, we have not yet arrived at the stage in the development of the "green state" where sustainable development has displaced conventional economic priorities and where the pursuit of sustainable development has become so internalized and integrated that it becomes "natural" (Lundqvist 2001, 469). As a result, we have not found it possible to establish any causal connection between the pursuit of sustainable development and the emergence of contemporary patterns of governance.

References

Bache, Ian, and Matthew Flinders, eds. 2004. *Multi-level Governance*. Oxford: Oxford University Press.

Baker, Susan. 2012. The Governance Dimensions of Sustainable Development. In *European Union, Governance, and Sustainability*, ed. Verena Bitzer, Ron Cörvers, Peter Glasbergen, and Inge Niestroy. Den Haag: Open University Press.

Baker, Susan, and Katarina Eckerberg. 2007. New Governance and Sustainable Development in Sweden: The Experience of the Local Investment Programme. *Local Environment: The International Journal of Justice and Sustainability* 12(4): 325–342.

Baker, Susan, and Katarina Eckerberg, eds. 2008a. *In Pursuit of Sustainable Development: New Governance Practices at the Sub-National Level in Europe.* Abingdon: Routledge/ECPR Studies in European Political Science.

Baker, Susan, and Katarina Eckerberg. 2008b. Economic Instruments and the Promotion of Sustainable Development: Governance Experiences in Key European States. In *In Pursuit of Sustainable Development: New Governance Practices at the Sub-national Level in Europe*, ed. Susan Baker and Katarina Eckerberg, 50–73. Abingdon: Routledge/ECPR Studies in European Political Science.

Barry, John, and Robyn Eckersley, eds. 2005. *The State and the Global Ecological Crisis*. Cambridge, MA: MIT Press.

Rosamund, Ben. 2004. New Theories of European Integration. In *European Union Politics*, ed. Michelle Cini, 109–127. Oxford: Oxford University Press.

Berger, Gerhard, and Richard Steurer. 2008. National Sustainable Development Strategies in EU Member States: The Regional Dimension. In *In Pursuit of Sustainable Development: New Governance Practices at the Sub-national Level in Europe*, ed. Susan Baker and Katarina Eckerberg, 29–49. Abingdon: Routledge/ECPR Studies in European Political Science.

Coenen, Frans. 2008. New Interpretations of Local Governance for Sustainable Development in the Netherlands. In *In Pursuit of Sustainable Development: New Governance Practices at the Sub-national Level in Europe*, ed. Susan Baker and Katarina Eckerberg, 190–207. Abingdon: Routledge/ECPR Studies in European Political Science.

Connaughton, Bernadette, Brid Quinn, and Nicholas Rees. 2008. Rhetoric or Reality: Responding to the Challenge of Sustainable Development and New Governance Patterns in Ireland. In *In Pursuit of Sustainable Development: New Governance Practices at the Sub-national Level in Europe*, ed. Susan Baker and Katarina Eckerberg, 145–168. Abingdon: Routledge/ECPR Studies in European Political Science.

Dryzek, John. 2005. *The Politics of the Earth*. Oxford: Oxford University Press.

Eckerberg, Katarina, Susan Baker, Agneta Marell, Katrin Dahlgren, Adrian Morley, and Niklas Wahlström. 2005. *Understanding LIP in Context: Central Government, Business and Comparative Perspectives*. Report 5454. Stockholm: Swedish Environmental Protection Agency.

Eckersley, Robyn. 2004. *The Green State: Rethinking Democracy and Sovereignty*. Cambridge, MA: MIT Press.

Evans, Bob, Marko Joas, Susan Sundback, and Kate Theobald, eds. 2004. *Governing Sustainable Cities*. London: Earthscan.

Evans, Bob, Marko Joas, Susan Sundback, and Kate Theobald. 2008. Institutional and Social Capacity Enhancement for Local Sustainable Development: Lessons from European Urban Settings. In *In Pursuit of Sustainable Development: New Governance Practices at the Sub-national Level in Europe*, ed. Susan Baker and Katarina Eckerberg, 74–95. Abingdon: Routledge/ECPR Studies in European Political Science.

Hallstrom, Lars K. 2004. Eurocratising Enlargement? EU Elites and NGO Participation in European Environmental Policy. *Environmental Politics* 13(1): 175–193.

Hanf, Ken, and Francesc Morata. 2008. Institutional Capacities for Sustainable Development: Experiences with Local Agenda 21 in Spain. In *In Pursuit of Sustainable Development: New Governance Practices at the Sub-National Level in Europe*, ed. Susan Baker and Katarina Eckerberg, 99–121. Abingdon: Routledge/ECPR Studies in European Political Science.

Holzinger, Katharina, Christoph Knill, and Ansgar Schäfer. 2006. Rhetoric or Reality? New Governance in EU Environmental Policy. *European Law Journal* 12(3): 403–420.

Hooghe, Lisbeth, and Gary Marks. 2001. *Multi-Level Governance*. New York: Rowan and Littlefield.

Hovik, Sissel. 2008. Governance Networks Promoting Rural Sustainable Development in Norway. In *In Pursuit of Sustainable Development: New Governance Practices at the Sub-national Level in Europe*, ed. Susan Baker and Katarina Eckerberg, 169–189. Abingdon, UK: Routledge/ECPR Studies in European Political Science.

Jänicke, Martin, and Helmut Weidner, eds. 1997. *National Environmental Policies: A Comparative Study of Capacity-Building*. Berlin: Springer Verlag.

Jordan, Andrew. 2008. The Governance of Sustainable Development: Taking Stock and Looking Forwards. *Environment and Planning. C Government & Policy* 26(1): 17–23.

Jordan, Andrew, Rudiger K. Wurzel, and Anthony R. Zito, eds. 2003. *New Instruments of Environmental Governance? National Experiences and Prospects*. London: Frank Cass.

Keating, Michael, and John Loughlin, eds. 1997. *The Political Economy of Regionalism*. London: Frank Cass.

Kern, Kristine. 2008. Sub-national Sustainable Development Initiatives in Federal States in Germany. In *Pursuit of Sustainable Development: New Governance Practices at the Sub-National Level in Europe*, ed. Susan Baker and Katarina Eckerberg, 122–144. Abingdon: Routledge/ECPR Studies in European Political Science.

Kohler-Koch, Beate, and Rainer Eising, eds. 1998. *The Transformation of Governance in the European Union*. London: Routledge.

Kooiman, Jan, ed. 1993. *Modern Governance: Government-Society Interaction*. London: Sage.

Kooiman, Jan. 2000. Societal Governance: Levels, Modes, and Orders of Social-Political Interaction. In *Debating Governance*, ed. Jon Pierre, 138–166. Oxford: Oxford University Press.

Lafferty, William M., ed. 2004. *Governance for Sustainable Development: The Challenge of Adapting Form to Function*. London: Edward Elgar.

Lafferty, William M., and Katarina Eckerberg, eds. 1998. *From the Earth Summit to Local Agenda 21: Working Towards Sustainable Development*. London: Earthscan.

Lafferty, William M., and Michael Narodoslawsky, eds. 2004. *Regional Sustainable Development in Europe—The Challenge of Multi-Level, Cross-Sectoral, Cooperative Governance*. Oslo: ProSus.

Lenschow, Andrea. 1999. Transformation in European Environmental Governance. In *The Transformation of Governance in the European Union*, ed. Beate Kohler-Koch and Rainer Eising, 39–60. London: Routledge.

Lundqvist, Lennart J. 2001. A Green Fist in a Velvet Glove: The Ecological State and Sustainable Development. *Environmental Values* 10(4): 455–472.

Meadowcroft, James. 2002. Politics and Scale: Some Implications for Environmental Governance. *Landscape and Urban Planning* 61(2–4): 169–179.

Meadowcroft, James. 2005. From Welfare State to Ecostate. In *The State and the Global Ecological Crisis*, ed. John Barry and Robyn Eckersley, 3–24. Cambridge, MA: MIT Press.

Meadowcroft, James. 2007. Who Is in Charge Here? Governance for Sustainable Development in a Complex World. *Environmental Policy and Planning* 9(3–4): 193–212.

Mydske, Per Kristen, Dag Harald Claes, and Amund Lie. 2007. *Nyliberalisme—Ideer og Politisk Virkelighet*. Oslo: Universitetsforlaget.

Nilsson, Måns, and Katarina Eckerberg, eds. 2007. *Environmental Policy Integration in Practice: Shaping Institutions for Learning*. London: Earthscan.

OECD. 1998. *Evaluating Economic Instruments*. Paris: OECD.

Painter, Martin, and Jon Pierre. 2005. Unpacking Policy Capacity: Issues and Themes. In *Challenges to State Policy Capacity: Global Trends and Comparative Perspectives*, ed. Martin Painter and Jon Pierre, 1–18. Basingstoke, UK: Palgrave/Macmillan.

Peters, B. Guy, and Jon Pierre. 2003. Introduction: The Role of Public Administration in Governing. In *Handbook of Public Administration*, ed. Guy Peters and Jon Pierre, 1–10. London: Sage.

Pierre, Jon. 2000. Introduction: Understanding Governance. In *Debating Governance*, ed. Jon Pierre, 1–12. Oxford: Oxford University Press.

Pierre, Jon, and Guy B. Peters. 2000. *Governance, Politics, and the State*. London: Macmillan.

Radin, Beryl A. 2003. The Instruments of Intergovernmental Management. In *Handbook of Public Administration*, ed. Guy Peters and Jon Pierre, 607–618. London: Sage.

Rhodes, Roderick A. W. 1997. *Understanding Governance: Policy Networks, Governance, Reflexivity and Accountability*. Buckingham, UK: Open University Press.

Rhodes, Roderick A. W. 2000. Governance and Public Administration. In *Debating Governance*, ed. Jon Pierre, 54–90. Oxford: Oxford University Press.

Rydin, Yvonne, and Nancy Holman. 2004. Re-evaluating the Contribution to Social Capital in Achieving Sustainable Development. *Local Environment* 9(2): 117–133.

Scharpf, Fritz. 1997. *Games Real Actors Play: Actor-Centered Institutionalism in Policy Research*. Oxford: Westview Point.

Schmitter, Philippe C. 1989. Corporatism Is Dead, Long Live Corporatism. *Government and Opposition* 24(1): 54–73.

Smith, Andy. 2003. Multi-Level Governance: What It Is and How It Can Be Studied. In *Handbook of Public Administration*, ed. Guy Peters and Jon Pierre, 619–628. London: Sage.

Spargaaren, Gert, and Arthur Mol. 2008. Greening Global Consumption: Redefining Politics and Authority. *Global Environmental Change* 20(3): 386–393.

Stoker, Gary. 2000. Urban Political Science and the Challenge of Urban Governance. In *Debating Governance*, ed. Jon Pierre, 91–109. Oxford: Oxford University Press.

Voß, Jan-Peter, Dierk Bauknecht, and René Kemp, eds. 2006. *Reflexive Governance for Sustainable Development*. Cheltenham, UK: Edward Elgar.

8

Sustainable Citizenship: The Role of Citizens and Consumers as Agents of the Environmental State

Michele Micheletti, Dietlind Stolle, and Daniel Berlin

"There is no notion more central in politics than citizenship, and none more variable in history, or contested in theory" is how political theorist Judith Shklar (1991) explains citizenship's importance as a political configuration.[1] Other scholars call it a "momentum concept" (Lister 2007, 49, quoting Hoffman 2004, 138), developing in response to societal problems. How citizenship is conceptualized and practiced, therefore, has a central relevance to reaching political goals. Thus, citizenship as a notion on how individuals relate to the state, interact in civil society, and assume responsibility for the collective good should play a key role in the emergence of the environmental state. An increasingly important citizenship discourse discussed in this book (see chapter 12, "Conclusion: An Emerging Environmental State?") focuses on how citizens, both individually and collectively and in different settings, can participate in sustainable development. Even governments and nongovernmental organizations (NGOs) declare that citizen awareness and involvement are essential for solving the complex problems of climate change, global human rights violations, world poverty, deforestation, overuse of the oceans, global trafficking, and unequal distributions of power between the Global North and South.

The Significance of Citizen Awareness

Citizen awareness is important in different ways. This includes how citizens articulate interests to politicians and government and engage in civil society as well how they engage in consumption. Governments, NGOs, policymakers, and researchers increasingly focus on unsustainable consumption (particularly meat eating and car driving) as important factors in accelerating climate change (UNEP 2011). They argue that private consumption must be integrated into actions for the common good, including state efforts at solving environmental problems. Environmental scholars have studied private consumption's role in creating environmental problems (Princen, Maniates, and Conca 2002; Maniates & Meyer 2010; Valor 2008). While some maintain that including choices involved in private life must be part of governance of the common good (Ostrom 1990; Spaargaren 1997; Spaargaren and Mol 2012; Stern et al. 1997; Warde and Southerton 2012), others argue that this is neoliberalism and illustrates the shift of responsibilities from government to market mechanisms and from citizens to consumers (Dean 2010; Clarke et al. 2007; Maniates 2001), a viewpoint that we question here.

This chapter departs from the discourse on sustainable citizenship that includes an argument for embracing private lifestyle and choices in a fuller view of the responsibilities entailed in citizenship. It develops the concept of sustainable citizenship, systematizes its normative responsibility claim on all societal actors and institutions, and investigates it empirically. The first two sections introduce sustainable citizenship and its responsibility claim, briefly compare the model with other citizenship models, show how it develops a different notion of collective action, and discuss how it can be operationalized empirically. Then we explore its potential measurements in the European (and particularly Swedish) context and demonstrate that several behavioral and attitudinal dimensions presently found in surveys are useful for studying it, but also point to the need for further measures to capture its complexity and breadth. Thus, we directly address the debate about consumption's role in political problem solving. The next sections focus on the real-life presence of sustainable citizenship among individuals. The empirical analysis examines some of the sociodemographic and attitudinal attributes characterizing sustainable citizens. Importantly, we try to understand whether sustainable citizens differ from other citizens in terms of age, gender, education, political placement, socioeconomic resources, and views of the responsibilities associated with governmental and other institutions. We explore if sustainable citizens hold different general expectations about "good citizenship," if their level of trust in governmental and other institutions differs from other citizens, and whether or not they believe that individuals can be politically efficacious and important agents of change in the field of sustainability. The final section draws conclusions from the empirical studies, identifies plausible reasons for some surprising discrepant findings, and offers ideas about the future study of sustainable citizenship.

Defining Sustainable Citizenship

The term "sustainable citizenship" originated in the discourse on the United Nations (UN) report *Our Common Future*, which recognized the need to check continued and accelerated environmental destruction and its negative effects on economic and social development globally. The report included common pool resource use, sociopolitical (social equity) challenges, and economic growth that "meets the needs of the present without compromising the ability of future generations to meet their own needs" in its definition of sustainable development (WCED 1987, chapter 2, paragraph 1). It was followed by innovative multipronged efforts by government and civil society involving education and participation that reveal both similarities and differences from the state-led corporatism discussed elsewhere in this book. Better practice in the field of production and consumer lifestyles was particularly called for to "meet the basic needs of the poor, and reducing wastage and the use of finite resources in the production process" (UN 1992, paragraph 5). The idea was to give government, civil society, citizens, consumers, and business additional roles and responsibilities so that these goals of safeguarding the global common good could be met worldwide in all societal sectors. The notion of sustainable citizenship came into the discussion through these efforts. As discussed below, it integrates and transcends traditional citizenship models; involves a different conceptualization of the common good; integrates additional values, attitudes, and practices intimately tying private life with the common good; and entails additional forms of collective and individual action identified in this chapter. Thus, it offers a broader and deeper configuration of citizenship (cf. Lister 2007; Barry 2005).

Other models, particularly environmental citizenship (Dobson 2003, 2007), also define citizenship in a more comprehensive fashion. They expand the liberal and civil republican citizenship models in four important ways. First, they view citizenship as not just concerning relationships within the nation-state but as involving cosmopolitan (Delanty 2000) or nonterritorial elements and practices going beyond the present citizen obligations and rights toward government and civil society. The argument is that liberal and civic republican citizenship models—configured with the nation-state in mind and developed successfully in industrial society before environmental problems became complex global threats—cannot sufficiently cover the responsibilities and practices necessary to safeguard the global common good today (Barry 2005; Lister 2007; Dobson and Sáiz 2005; van Steenbergen 1994). Second, the more recent reconfigurations adopt nonreciprocality as their key element and give citizens responsibilities regardless of whether they, their immediate community, or their nation-state benefit from them. Third, while the older models define the common good by simply aggregating the stakes of people into a collective interest, the more recent ones define it as global sustainability (Barry 2005, 26; Dryzek 1987; Pennington 2001), thus implying a collective commitment across territorial boundaries and additional forms of collective action. Fourth, following feminist study, they maintain that good values, attitudes, and practices in the private sphere of the family and the economic sphere of business and consumer society must be integral parts of citizenship. This implies that consumers and corporate actors must begin to take action based on their responsibilities for sustainable development. Thus, these more recent versions view the key determinant of whether an action constitutes citizenship as not only if government and civil society are involved, but also "*what* a person does and with what public consequences, rather than *where* they do it" (Lister 2007, 57). Citizenship is, therefore, a more total relationship of being (identities, practices, and governance), involving a series of responsibilities (not just obligations and rights) of a vertical (individual-government) and horizontal (individual-individual) nature.

Sustainable citizenship has emerged as a broader and more inclusive citizenship notion remedying theoretical weaknesses identified in environmental citizenship—its emphasis on "environmental sustainability" and "environmental space" (cf. Dobson 2003, 2007, 2010; Barry 2005, 23–24), and citizens' responsibility for their "ecological footprint" is defined largely as produced by the "material relationship of individual people with their environment" (Dobson 2003, 106). This is said to restrict sustainable development to the two pillars of economic growth and environmental protection, thereby neglecting or deemphasizing the third pillar of sociopolitical equity. Scholars therefore began to theorize sustainable citizenship to "encompass economic, social, political, and cultural spheres" upon which to build collective solutions to the severe, complex challenges facing politics globally (Barry 2005, 24; cf. Bullen and Whitehead 2005, 504, 507; Lister 2007; van Poeck et al. 2009). Sustainable citizenship addresses even nonreciprocal relations directly toward humans and animals even if there is not an environmental impact. Noteworthy is that some challenges in these nonreciprocal relations have not been sufficiently addressed in governmental policy measures, particularly in the field of working conditions in the global textile industry and unsustainable eating habits.

We identify as a second major problem the environmental model's limited temporal view of citizenship as involving exclusively "[t]he obligations of the ecological citizen . . . towards generations yet to be born" (Dobson 2003, 106). Given *Our Common Future*, this is understandable, but it does not explicitly highlight the ramification of historical instances of common pool plundering and legacies of colonialism on how citizens today should assess their (nonreciprocal) relationships of responsibility. Neither does it consider how these relationships often create path dependencies and habits reproducing unsustainable values and practices within government, corporate institutions, and the individual exercise of free choice. Thus, the criticism is that environmental citizenship is ahistorical and characterized by what post-colonial theorists call "First World," "ecological imperialist" views (Huggan 2004).

Reconfiguring citizenship as sustainable citizenship avoids these problems by incorporating normative components from earlier citizenship models and including more responsibility-taking on the part of different societal actors. It can be defined as nonreciprocal responsibility for the three relationships involved in sustainable development. Its normative claim is that good citizenship requires that people individually and collectively assume (1) "temporal responsibility" for destructive historical practices and for future generations, (2) "material responsibility" for how consumption affects nonhumans (animals, nature), and (3) broader "spatial responsibility" than just within the nation-state (Bullen and Whitehead 2005; van Poeck et al. 2009; Lister 2007; cf. Young 2006). Its scholars even claim that it goes further than environmental citizenship in challenging underlying structural causes for unsustainable trade-offs and choices, an aspect that is not always the focus of government (Barry 2005, 23–24) and implying that a greater number of private choices are involved with governance for the common good (O'Riordan 2002). Its key characteristic is citizen engagement of a broad, bold, and radical nature; its proponents claim it takes place in "every waking minute of everyday" (Bullen and Whitehead 2005, 513). Sustainable citizenship infuses more practices and spheres of living in political responsibility, thus waking them from their apolitical state. Obviously, it might open up opportunities for new forms of governance regulation benefiting the environmental state's development.

Empirical research on sustainable citizenship can take different trajectories, including closely examining governmental, NGO, and corporate policy. Given its normative claims, it should even focus on the role of ordinary everyday life in citizenship responsibilities (e.g., studying how individual citizens seek to inform themselves and engage with government, civil society, and corporations) and the role of private consumption, the topic of this chapter. There are additional compelling reasons for this chapter's consumption focus. First, as noted in the public discourses following *Our Common Future*, institutional and individual consumer choice can seriously affect the balance between economic growth, global sociopolitical equity, and common pool resource use, as highlighted in the discussion of the meat consumption's importance in the climate change debate (UNEP 2011). Second, there is general agreement that private life has assumed more meaning in contemporary modern democracies (Soper and Trentmann 2007), thus explaining many NGOs' increased focus on shopping in their mission statements and mobilizing campaigns (e.g., Peattie 2010; Spaargaren and Oosterveer 2010; Micheletti and Stolle 2012; Stolle and Micheleti 2013). Third, governments focus increasingly on implementing incentives for citizens and corporations to practice and promote "leaner" consumption (e.g., energy saving in fuel use, heating technology, and sustainable architecture), which expands responsibilities for sustainability beyond governments on all levels (King and King 2005; Lemos and Agrawal 2006; Diamond 2009). So the argument is that the sheer reality of the significance of private choice and their mostly negative sustainability effects make it important—if not crucial—to include a scrutiny of private life and lifestyle in an assessment of good citizenship.

Operationalizing Sustainable Citizenship

Citizens practicing sustainable citizenship integrate the three pillars of sustainable development into everyday practices. For this to occur, they must accept the sustainability view of the common good and apply it in their different societal roles as voters, organizational members, consumers, and so on. They must have awareness and knowledge about the complex relationships between economic growth, sociopolitical equity, and environmental protection and the trade-offs commonly made when exercising free choice in public and private settings. The most common survey measures focus mostly on citizenship practice in the electoral and civil society realm along with their related values and knowledge. These measures stress important and powerful channels for political expression and mechanisms of influence; few of them include an individual's direct responsibility for engaging in complex problem solving. They also tend to view citizen involvement as more reciprocal than nonreciprocal in orientation and do not investigate fully how sustainability and the common good can be promoted on various societal levels and spheres.

Thus, not only the theoretical notions but also the empirical measures to capture sustainable citizenship practice need to be expanded and the present ones used differently. We suggest that it is analytically important to distinguish attitudinal and behavioral aspects of sustainable citizenship. Sustainable citizens can be recognized only because they have a basic *interest* in sustainability's three pillars, which prompts them to be aware of the extra-territorial spatial, political, environmental, and social consequences of their actions and those of others. We also believe that sustainable citizens can be distinguished from other citizens because they not only have an interest in these issues but are *concerned* and care about them by, for instance, putting sustainability's concern for the environment and the developing world high on their own and the political agenda. The interest is, therefore, one aspect of attitudinal sustainability citizenship.

Sustainable citizenship studies should also include the attitudinal orientations and values tapping the readiness of individuals to take more nonreciprocal responsibility for spatial, temporal, and material relationships. Of course, this preparedness also can involve more traditional citizenship practices (e.g., voting or donating to causes). Sustainable citizenship might vary on a scale from individual self-sacrifice for the common good to a more modest position about living in a way that at least aims at minimizing harm to the global environment and to others' general societal well-being. For example, this could be measured by attitudes about the willingness to change one's lifestyle, personal economic well-being, and the amount of effort that individuals put into contributing to more sustainable living.

Mental or attitudinal readiness for at least some self-sacrifice of material conditions is only one aspect of sustainable citizenship. Behaving according to such values is something else, and it demands a bit more. Such behavior is extremely difficult to measure in surveys; it is subject to the tendency of respondents to follow social norms and reply in a socially desirable or "politically correct" manner. Other statistical or qualitative measures of a more objective nature ideally should be added to take full account of the existence of sustainable citizenship. Here, we focus on self-reported attitudes about consumption of products reflecting sustainable values. More traditional citizenship conceptions include behavioral survey items concerning voting and campaign-related political activities; environmental citizenship taps green behavior (e.g., recycling, energy use, or transportation patterns). Sustainable citizenship's measures ideally should account for a broader range of behaviors involving values about climate change, deforestation, overfishing, animal welfare, and human rights. One exemplary survey measure concerns shopping decisions because citizens who shop based on environmental, political, ethical, and economic considerations practice some aspects of sustainable citizenship and even voluntarily engage in market-based collective action that has developed as part of sustainability's problem solving. Interestingly, several such initiatives (e.g., labeling schemes) receive government support. However, to probe more fully an individual's degree of sustainable citizenship, behavioral measures ideally should even account for more specific motivations (for example, shopping decisions related to a host of aspects of sustainable living) and also tap the regularity with which the motivations are used daily.

Thus, sustainable citizenship must be operationalized as a multidimensional phenomenon involving a series of measures tapping attitudes, values, and practices concerning nonreciprocal responsibility in both public and private life. It ideally should measure the interdependences of the three pillars of sustainable development and formulate new measures tapping nonterritorial responsibilities to make the world a better place to live (promoting the common good) and items directly probing attitudes about societal diversity (the social pillar of sustainable development) and biodiversity (the environmental pillar). Particularly as more recent citizenship notions require shifts in attitudes at a deeper level (Dobson and Sáiz. 2005, 158), this readiness for sacrifice needs to be tapped as well.

Other political attitudes and values believed to be important for realizing sustainable citizenship, citizenship expectations, political trust, and political efficacy, are included in this study (see also chapter 6, "Early Bird or Copycat, Leader or Laggard? A Comparison of Cross-National Patterns of Environmental Policy Change"). While not directly measuring immediate reasons for practicing sustainable citizenship, they help identify the attitudinal profile of sustainable citizens. It is also important to assess citizenship expectations, defined as a shared set of beliefs about the citizen's role in politics, because they shape citizens' orientations on what is expected of them and what they expect of themselves, can mold citizens' political behavior, and promote a healthy, well-functioning political community. Thus, clear expectations about one's responsibility for the common good, as the one formulated by sustainable citizenship, potentially can reinforce new styles of political and collective action (Dalton 2008b, 78) and the pursuit of sustainable democracy (see chapter 7, "The

Role of the State in the Governance of Sustainable Development: Sub-national Practices in European States"). The strength of how individuals view different citizenship expectations, therefore, might influence choices and practices that affect the present and the future. How citizens relate to them can be informed through education, opinion formation, and even past events. Obviously, governmental and other institutions play a key role here.

Recent empirical citizenship research has focused on capturing two main dimensions of citizenship expectations: citizen duty and engaged citizenship (Dalton 2008ab). "Duty citizenship" is measured by items formulating expectations about the significance of voting, obeying laws, performing jury and military service, reporting crime, and not cheating on social benefits, all of which are vital for legitimate and well-ordered political systems and mostly nation-state-oriented in focus. "Engaged citizenship," another important part of democratic governance, is taped by survey items that ask about helping those who are worse off than oneself, holding independent opinions, understanding other's opinions, volunteering, joining civic groups and political parties, and generally being active and interested in politics (McBeth et al. 2010). Closely related, "active citizenship" has two distinguishable dimensions: one relating to protest and activities oriented toward societal and political change, including political actions to hold governments accountable; the other concerning community-oriented activities (Hoskins and Mascherini 2009). Both engaged and active citizenship underscore the public sphere's importance and largely also have a nation-state focus. What is missing are fuller measures tapping more effectively citizenship expectations specifically on global sustainable development (the past and future temporal relationships of citizenship) and a more comprehensive view of the spatial relationship. Neither is the important claim about taking responsibility in political problem solving through private lifestyle changes sufficiently measured.

The cross-national surveys on citizenship, involvement, and democracy contribute via measures of ethnic and gender diversity, the role of welfare services in citizenship, and social solidarity (e.g., putting others' interests before your own, not treating immigrants worse than native citizens, and helping people in the rest of the world worse off than oneself) (Dalton 2008b; van Deth et al. 2007; Petersson et al. 1989). A few surveys even include the aspects of taking individualized responsibility in political problem solving and acting on one's own initiative, a necessary component for practicing sustainable citizenship. Some surveys, including those discussed above, contain a question about citizenship's material relationships, the primary one being the expectation about choosing products for political, ethical, or environmental reasons, even if they cost a bit more. This expectation is a measure of the inclusion of the private and economic sphere (material relationships of citizenship). Ideally, other items for measuring a more expanded view of citizenship should be included.

Comparative Studies

Sustainable citizenship has not yet been measured comparatively. Studies have assessed comparative rates of environmental behavior (Oreg and Katz-Gerro 2006) and tapped citizenship's material relationships by measuring organic consumption levels (Thogersen 2010). Few have examined the complexity of attitudes and behaviors making up sustainable citizenship and, therefore, how they might contribute to growing the environmental state. Some research focuses on political consumerism (Stolle et al. 2005; Stolle and Micheletti 2013), a related phenomenon but without the same normative claim. Political consumerism is a political engagement ranging from consumer choice of green or fair trade goods to those promoting nationalism or the exclusive support of selected religious or ethnic groups. Sustainable citizenship, in contrast, focuses on the normative concerns of green values and sociopolitical equality that can motivate more sustainable shopping choices and sustainable lifestyles, among other practices; levels of sustainable citizenship can range from high to low. However, political consumer acts have been measured and are some of the few available cross-national measures to tap sustainable citizenship practice. Political consumerism is defined as the evaluation and choice of producers and products with the aim of changing ethically or politically objectionable institutional or market practices (Micheletti et al. 2006, xiv-xv). Certain political consumer practices are even increasingly promoted by governments when they initiate and support ecolabeling and other labeling or rely on them in their public procurement policies.

The limited data for comparative political consumerism survey research indicate that it is an increasingly important form for citizens' political engagement in Western democracies (Inglehart 1997, 313; Norris 2002; Petersson et al. 1998; Boström et al. 2005; Dalton 2008b). Its rise is based on postmaterialist values (e.g., protecting the environment) (Inglehart 1997). Available World Values Survey data clearly shows boycotting (deciding not to purchase goods or patronize stores for political, ethical, and environmental reasons) has risen in the countries where valid comparisons can be made. Although other forms of activism often involving environmental issues (e.g., petition signing, demonstrations) are also very widespread at the end of the twentieth century, boycotting has risen most over time. It is more than four times as likely in 1999 compared to 1974 (Stolle et al. 2005).

Table 8.1 presents comparative measures of boycotts and "buycotts" (deliberately purchasing certain goods for environmental, ethical, or political reasons) across twenty-one democracies in Europe and the United States in 2003. Noteworthy is that in all but two countries (Greece and Italy), buycotting is more common than boycotting—in many cases twice as common or even more. Different reasons might account for this pattern. Boycotting might depend on public campaigns and mobilization, although the practice is no longer aggressively promoted by most NGOs focusing on sustainability because of its potential adverse effects on people and nature. Buycotting is supported by the availability of labeling and monitoring schemes that allow citizens to make responsible shopping choices; this "choice architecture" differs across countries and is much more prevalent in the European North than in southern and eastern Europe. When trying to distinguish countries with high and low levels of political consumerism (the percentage of individuals who buycott, boycott, or both), it becomes apparent that countries with similar geographic, economic, and political characteristics cluster together. The top five political consumer countries are Sweden, Switzerland, Denmark, Finland, and Germany; the five lowest are Portugal, Italy, Poland, Hungary, and Greece. Interestingly, a country's level of political consumerism also seems to correspond to the relative strictness of environmental standards (see chapter 6). Finland, Germany, and Sweden are top-ranked in terms of both political consumerism and regulatory strictness, while Poland, Portugal, and Greece are at the bottom of both these rankings (See chapter 4, "The Three Worlds of Environmental Politics," for corresponding rankings of countries' environmental performance). Other important factors include a country's social capital, level of democracy, and development of its service sector (Stolle and Micheletti 2013).

Data on political consumerism cannot easily distinguish actions of a sustainable citizenship nature from those that are not. Some forms of buycotting, however, are more closely related (e.g., fairtrade shopping, a consumer choice increasingly promoted by governments and NGOs alike). Fairtrade involves more nonreciprocal relationships than eco-friendly shopping and, therefore, reflects a more expanded spatial concept of citizen responsibility. It is also the best available comparative measure to explore how shopping consciously can be used as individualized collective

Table 8.1
The cross-national spread of political consumerism

Country	% Not Active at All	% Boycotting (%)	% Buycotting (%)	Boycotting and/or Buycotting (%)	Ranking of Political Consumerism	Fairtrade Coffee: Average per capita Consumption in grams	Fairtrade Coffee: Average Market share (%)	Ranking by Market Share
All Countries	69.2	16.8	26.5	30.8	N/A	N/A	N/A	N/A
Austria	65.5	21.5	29.7	34.5	9	58.55	0.85	8
Belgium	69.2	12.8	27.0	30.8	10	73.51	0.99	6
Canada	N/A	N/A	N/A	N/A	N/A	26.69	0.41	13
Czech Rep.	73.7	10.8	22.6	26.3	14	0	0	17
Denmark	52.2	22.9	43.8	47.8	3	120.34	1.34	5
Finland	52.9	26.8	41.8	47.1	4	26.00	0.23	14
France	63.5	26.6	28.0	36.5	8	46.00	0.86	7
Germany	54.7	26.1	39.2	45.3	5	38.43	0.56	10
Greece	88.0	8.5	6.6	12.0	16	0	0	17
Hungary	88.3	4.8	10.5	11.7	17	0	0	17
Ireland	71.5	13.6	24.6	28.5	11	30.52	1.36	4
Italy	89.9	7.6	6.5	10.1	19	5.09	0.09	15
Japan	N/A	N/A	N/A	N/A	N/A	0.42	0.01	16
Netherlands	71.6	10.4	25.7	28.4	12	186.54	2.88	2

Table 8.1
(continued)

Country	% Not Active at All	% Boycotting (%)	% Buycotting (%)	Boycotting and/or Buycotting (%)	Ranking of Political Consumerism	Fairtrade Coffee: Average per capita Consumption in grams	Fairtrade Coffee: Average Market share (%)	Ranking by Market Share
Norway	58.6	20.3	36.7	41.4	6	66.17	0.72	9
Poland	89.2	3.6	9.8	10.8	18	0	0	17
Portugal	91.8	3.4	6.9	8.2	20	0	0	17
Slovenia	88.3	5.1	9.6	11.7	17	0	0	17
Spain	86.5	8.0	11.6	13.5	15	0	0	17
Sweden	39.7	32.5	55.1	60.3	1	46.13	0.56	10
Switzerland	49.6	31.4	44.6	50.4	2	194.19	2.94	1
United Kingdom	59.8	26.1	32.3	40.2	7	52.06	2.27	3
United States	71.9	18.2	22.9	28.1	13	23.29	0.56	10

Entries in columns 1–4 represent percentages of political consumerism practiced per country. Data source: European Social Survey 2002/2003 (Jowell et al. 2004). Column 5 contains the rankings of column 4 (political consumerism). Columns 7 and 8 represent domestic sales of fairtrade coffee in grams, divided by the total population of each country and the percentage of total national coffee sales that is fairtrade, respectively (for 2000–2006). The measures capture individual and some institutional purchases of fairtrade coffee.

Data sources: Coffee data 2000–2003: FLO 2005; coffee data 2004–2006: FLO 2007; population data: World Bank 2007. Total coffee consumption per capita 2000–2004, all countries except Canada: International Coffee Organization (ICO) 2006. Total coffee consumption per capita 2000–2004, Canada: ICO 2005. Column 9 has the rankings of fair-trade coffee market shares.

action (Micheletti 2010) to help counteract colonial legacies and promote sociopolitical equity in the developing world (temporal relationship); it even includes some environmental criteria. Fairtrade International (FLO) records comprehensive data for sales of fairtrade goods. We examine the cross-national variation in fairtrade coffee data because it is by far the highest-selling fairtrade good and has been the fairtrade movement's emblematic campaign and mobilizing commodity (data from Stolle and Micheletti 2013, 99). Of course, these results only represent one instance of the practice of sustainable citizenship and do not show how much responsibility for citizenship's material relationship "really occurs"–only how much this particular activity occurs.

Table 8.1 presents data on both political consumerism and fairtrade coffee purchases. In comparative terms, Switzerland and the Netherlands purchase the most fairtrade coffee, with Denmark close behind, and Italy and Japan bringing up the rear; some countries (mostly in eastern and southern Europe) did not have any fairtrade coffee purchases.[2] Because coffee consumption levels vary, it is useful to look at market share data, indicating the portion of the total national coffee market captured by fairtrade. This measure changes the rankings a bit: the top two remain unchanged, but the United Kingdom rises to third, just ahead of Ireland and Denmark. Fairtrade coffee consumption has spiked sharply in the United Kingdom (data not shown), surpassing Switzerland in 2005 as the top global consumer of this product in terms of market share. This share reached an estimated 4.51 percent in 2006. By both measures, Japan and Italy rank lowest.[3]

How well do rates of reported acts of individual political consumerism match up with the actually observed market shares of fairtrade coffee? Overall, self-reported political consumer activity relate fairly strongly,[4] thus substantiating the survey data, but with two caveats. First, the actual percentage of people claiming to participate in political consumerism is far higher than the actual market share or sale of this fairtrade product. As discussed earlier, political consumerism captures a variety of shopping decisions beyond those of sustainable development. The survey data also does not depict the regularity of making ethical or political shopping decisions; the sales data does not distinguish between individual political consumers or larger institutional ones (e.g., governmental agencies, hospitals, and schools). These considerations might explain some mismatch between survey and market/sales data.

The other caveat is that there are a number of countries in which the two measures collide. For example, Germany and Finland rank relatively low for both fairtrade coffee measures and high on reported political consumerism; the United Kingdom ranks high on fairtrade coffee, which its survey results would not have led us to expect. In particular, Sweden, the country that has ranked the highest in political consumerism in previous comparative surveys and ranked fourth in the world as a coffee-consuming nation (Global Market Information Database 2002), is not highest in fairtrade coffee purchases. It may be that in some of the highest-ranking countries in self-reported political consumerism, citizens prefer activities that focus more on taking material responsibility for "environmental sustainability" through buying eco-labeled or organic goods rather than for "sociopolitical sustainability" by buying fairtrade coffee. Alternatively, it could be that these goods are not widely available, citizens lack awareness for why they should purchase them, and governmental and other institutions do not promote them.

Multidimensional Sustainable Citizenship

To create a more robust account of the multidimensionality of sustainable citizenship at the individual level of consumption and private life, attention must shift to another data source. We focus on various attitudinal and behavioral dimensions of sustainable citizenship, which to our knowledge is only available in the Swedish national representative *Consumption and Societal Issues Survey* (2009), with 1,053 respondents. Sweden also might be considered a critical case of widespread behavioral sustainable citizenship practice because of its high political consumerism level, the crucial role played by government in promoting environmentalism, its position as one of the top-ranked countries with strict environmental standards, and governmental and civil society promotion of individual taking of responsibility (Swedish Government 2005, see chapters 6 and 7). The data set draws out sustainable citizenship's various dimensions and helps in analyzing the sociodemographic factors explaining its practice and its attitudinal correlates. Sustainable citizenship is operationalized by a combination of whether people are interested in and concerned about sustainability problems (*awareness*), whether they are mentally prepared to support policies for sustainable development that might affect economic growth and their own material well-being negatively (*attitudes*), and whether they take sustainability concerns into consideration in, for example, their daily shopping decisions (*behavior*). Our focus is on how consumer practice and private life is politicized and activated as an area for thinking about sustainability and acting it out at the individual level.

Two different measures are used for the awareness dimension: *concern for sustainability threats* (an additive index formed by respondents' worries for the state of the environment, poverty in the Third World, and human rights globally) and *interest for sustainability issues* (measured by interest in environmental issues, human rights, and developing countries/ the Third World). For the *attitudinal* dimension, five survey items form an index, probing respondents' willingness to promote a greener society and better living conditions for people in other parts of the world, even if this would mean a reduction in economic growth and material well-being. The *behavioral dimension* involves a direct measure of the material relationship of citizenship: whether people buy food because the products are "environmentally friendly," "manufactured under good working conditions," and to "support animal husbandry." The awareness, attitude, and behavior indexes were combined into a scale of *sustainable citizenship* probing important values within sustainable citizenship on public well-being and the common good. The scale ranges between 0 and 40, with a mean of 24 (Std.D = 6.7; see the appendix on p. 343).

Portrait of Sustainable Citizens: Supporting Sustainability or Fleeing Politics?

Sustainable citizenship is demanding. It requires citizens to put their private life in a more political context by relating it to processes of sustainable development at home and abroad. Agenda 21, , the voluntary action plan evolving from the 1992 UN Conference on Environment and Development (UNCED) held in Rio de Janeiro, Brazil, and other related processes have shown that information about and understanding of one's own responsibilities and, particularly, education are important, and therefore they should contribute to the practice of sustainable citizenship (Swedish Government 2004; UNESCO 2012). Sustainable citizenship also might depend on finding stores that stock more sustainable products, and even on one's economic circumstances (Valor 2008). It might be correct, therefore, to assume that sustainable citizenship is practiced more easily in urban areas, where more stores stock fairtrade and other sustainability goods, environmental problems are prevalent, and family and personal income is generally higher than in rural areas. Awareness of and interest in sustainability issues, thus, might be more dominant in urban areas and for individual citizens in higher income categories. In addition, and as shown elsewhere in this book, even public policy and the political debate might matter.

Although all Swedish political parties address sustainability in some way, the Green Party and Left Party stand out as the most eager promoters of lifestyle change and material sacrifices for sustainable development; e.g., the Green Party's support for policies giving bonuses to drivers of energy-efficient cars and charging a fee for inefficient vehicles, and the Left Party's goal of introducing a special deduction for the energy conversion of existing single-family homes. Previous research finds that their supporters are more interested in, concerned with, and willing to prioritize sustainability matters (Jagers 2009) and are overrepresented among political consumers (Micheletti et al. 2012). Therefore, we assume that ideological self-placement on the left-right scale is relevant for practicing sustainable citizenship.

Finally, gender and age should matter. Earlier studies show that women are at least equally if not more active in political and societal involvements bridging the public-private divide (Stolle and Hooghe 2011; Ferrer-Fons 2004; Petersson et al. 1989, 1998; Stolle and Hooghe 2006; Stolle and Micheletti 2013). We also expect that young generations—socialized in an era of value change that stressed, at least in part, other-oriented, postmaterial considerations more than old-school industrial ones and who learned about sustainability in school, at home, or through media outlets—might be more supportive of less hierarchical and more individualized[5] forms of citizenship responsibility and collective action (UNESCO 2011; Tilbury et al. 2002; van Poeck et al. 2009; Inglehart and Welzel 2005).

Although focusing on the four dimensions of citizenship cannot put an exact percentage on the practice of sustainable citizenship, the survey results indicate that about 19 percent of the respondents score over 30 out of 40 points on the sustainable citizenship scale, and thus might be characterized as sustainable citizens. On the individual 10-point dimensions, 51 percent score over 7.5 points on "concern" and 28 percent over 7.5 points on "interest." The corresponding scores for attitudes and behavior are somewhat lower in general (19 percent and 16 percent respectively; results not shown).

Table 8.2 reports three regression models on the basic characteristics of Swedish sustainable citizens. As expected, more women than men practice it; gender is a distinguishing factor in all regression models. Further, involvement in sustainable citizenship increases somewhat with age. The fact that older generations usually have more economic resources available for sustainable citizenship's material relationship, particularly when it comes to the behavioral dimension of choosing premium priced green and fair trade food, probably plays a role here (cf. Stolle and Micheletti

Table 8.2
The socioeconomic and attitudinal correlates of sustainable citizenship in Sweden

	Model 1		Model 2		Model 3	
	Socioeconomic		Citizenship Expectations		Trust and Efficacy	
	N = 975		N = 975		N = 921	
	B	SE	B	SE	B	SE
Socioeconomic Position and Ideology						
Gender (woman = 1)	3.0**	0.41	1.9**	.33	2.1**	0.33
Age	0.04**	0.01	0.03*	0.01	0.04**	0.01
Education (university = 1)	1.3**	0.44	.032	0.35	-0.11	0.35
Children	-0.075	0.55	-0.38	0.43	-0.49	0.43
Economic Situation	0.02	0.09	-0.06	0.07	-0.09	0.07
Area of residence (large city = 1)	0.81	0.46	0.41	0.35	0.42	0.35
Ideology (left = 0)	-0.64**	0.09	-0.25**	0.07	-0.19**	0.08
Citizenship Expectations						
Solidarity citizenship			2.2**	0.11	2.1**	0.11
Duty citizenship			0.07	0.11	.06	0.11
Information-seeking citizenship			0.01	0.14	.08	0.14

Table 8.2
(continued)

	Model 1		Model 2		Model 3	
	Socioeconomic		Citizenship Expectations		Trust and Efficacy	
	N = 975		N = 975		N = 921	
Trust						
Representative institutions					−.03	0.09
International institutions					−.08	0.10
Sustainability institutions					0.50**	0.12
Efficacy						
Institutions					−0.13	0.09
People and corporations					0.39**	0.09
Constant	22.9**	0.9	7.4**	1.13	5.1**	1.28
Model fit						
Adjusted R^2	0.12		0.47		0.50	
Standard error of regression	6.3		4.9		4.8	

Notes: Cell entries are unstandardized b-values (B), standard errors (SE), and model performance statistics. * $p < 0.05$; ** $p < 0.01$

2013). However, even when controlling for the perceptions of one's economic situation the generational differences in sustainable citizenship practice persist.[6] Further analyses reveal no age differences on the attitudinal dimension (mental preparedness) or concern for sustainability threats (e.g., climate change, environmental degradation, world poverty, and human rights violations; analysis not shown), indicating that young people are as concerned about sustainability issues as older ones. The respondents' subjective perceptions of their own economic situation are uncorrelated with sustainable citizenship.

Moreover, as expected, education is important because awareness of products, producers, and social and environmental conditions is crucial. The highly educated are somewhat more likely to practice sustainable citizenship, but this effect does not hold in the regression models that include citizenship expectations, trust, and efficacy perceptions. Also, people on the political left are more likely to engage in sustainable citizenship, which most likely reflects the kinds of issues (such as climate change, human rights violations, and world poverty) that they believe are complex political problems requiring various forms of responsible action on the part of political and corporate institutions, civil society, and individuals. This result accords with comparative political consumerism research (Stolle and Micheletti 2013) and holds in all the analyses. The socioeconomic variables, economic circumstances, whether one has children, and area of residence have no effect, against some expectations, but this confirms that economic elements are not necessarily important factors. Apparently in Sweden, at least the economically better and worse off practice it to a similar degree. Worthy of note is the fact that if car driving (another environmentally significant behavior) is included in our measure, people with a relatively good economy and those having children score lower. This might be an effect of striking a sustainability trade-off so that family life does not become burdensome, or perhaps even because the local state fails to provide good public transportation in more rural (and car-dependent) areas. Where one lives also becomes a significant factor, with urban citizens more sustainable than others. Obviously, these results show that governmental effort is necessary to create opportunities for individuals to develop better sustainable citizenship practices.

Belief System of Sustainable Citizens

Citizenship expectations are important for political and societal development and therefore, they should be important for greening democracy, sustainable development, and growing the environmental state. Typically, voting, obeying laws, and putting the collective interest before one's self-interest are examples of good citizenship. These concerns are also important for sustainable citizenship, but additional expectations are now necessary. One way of studying this is by investigating whether the individuals characterized as sustainable citizens reveal different views of citizenship expectations.

The 2009 survey includes several citizenship expectations. Three different dimensions were revealed in a factor analysis, which seem to reflect both traditional and more recent notions of good citizenship. The *duty expectation* contains beliefs about obeying laws and the welfare state's codes of conduct; the *solidarity expectation* involves showing consideration and concern for others inside and outside one's own country, and the *information-seeking expectation* probes citizen views on being prepared to take more active and voluntary individualized responsibility for political and societal developments (see the appendix on p. 343). Sustainable citizens generally believe that all expectations on all three dimensions are more important, indicating that they are basically more politically and societally conscientious. Like other citizens, they believe that duty citizenship is most important. However, the solidarity expectation shows the strongest difference, indicating that sustainable citizenship is underpinned by a strong logic of political and societal solidarity. This is further corroborated by model 2 in table 8.2, which accounts for sociodemographic characteristics; those scoring higher on the solidarity expectation also practice more sustainable citizenship, while the duty and information-seeking expectations do not relate to practicing sustainable citizenship more than other kinds of citizenship. As the solidarity expectation involves more nonreciprocal notions of responsibility (see the Appendix), sustainable citizens seem to understand the spatial and material relationships of citizenship differently than others. Thus, they believe that citizenship involves a broader and deeper context than simply the duty relationship to the state and exercise of democratic freedoms for one's own benefit. This particular result shows that it is important to include more aspects of sustainable development into theorizing citizenship, its empirical modeling, and even how it is taught and encouraged by governments, NGOs, and others.

Another set of potentially influential values on the propensity to engage in sustainable citizenship involve an assessment of a sense of personal and institutional efficacy about the sources of societal change. How people perceive government's problem-solving capacity is one important gauge of which societal actors can resolve sustainability problems best. If citizens believe that government either does not recognize or cannot control new societal uncertainties and risks, thus indicating some assessment of state failure in sustainable problem solving, they may decide to try additional arenas and methods to help address these issues (Beck 1997; Inglehart 1997; Ostrom 1990; Spaargaren and Mol 2008; McFarland 2010). For this to occur, they must believe that their voluntary efforts matter. Following collective action scholarship (Finkel and Muller 1998; Teorell et al. 2007), the decision to get more involved has been found to depend on whether one believes in the ability of people similar to oneself to influence societal developments. Thus, mobilization to collective action not only depends on good public and governmental campaigning, but also assessment of effectiveness.

If this is correct, sustainable citizens should have more positive assessments than others about the efficacy or ability of individuals, groups of individuals, and others to influence politics, which can help develop policy capacity through networks and participation (see chapter 7). Therefore, we investigate the perceptions that distinguish the efficacy of *political institutions* and of *people* and *corporations* in their ability to change society, all of which are important in theorizing sustainable citizenship. The efficacy of political institutions includes questions about politicians, public authorities, and international organizations (the European Union, United Nations, and World Trade Organization), thus also tapping the spatial relationship included in sustainable citizenship. The efficacy of nongovernmental actors includes oneself, people more generally, and corporations (see the appendix on p. 343) and follows sustainable citizenship's conception of collective action. What stands out in table 8.2 is that sustainable citizens believe more strongly in the capacity of nongovernmental actors to initiate societal change and transform current practices and institutional orders, even when other sociodemographic characteristics are taken into account. Institutional efficacy, however, does not help to distinguish sustainable citizens from others (see model 3 in table 8.2), implying that sustainable citizens do not believe more or less than others that society can be changed through institutions.

Some might even argue that sustainable citizens' belief in people's and corporations' efficacy is so high because it compensates for perceptions of low-performing political institutions. This claim requires an analysis of institutional trust, or the confidence in how different organizations and institutions manage their work. To tap properly into institutional trust's effects on sustainable citizenship, trust is divided into *trust in international organizations* (the European Union, United Nations, and World Trade Organization), *in national representative institutions* (the government and the Parliament) and *in "sustainability institutions"* (major government agencies and NGOs in the field of environment, sustainable development, and consumption); see the appendix (p. 343) for more details.

Table 8.2 (model 3) shows that sustainable citizens do not trust national and international institutions more or less than others, but they do display significantly more trust in sustainability institutions—environmentally oriented public agencies, the Swedish Consumer Agency, and environmental and consumer organizations—when sociodemographic characteristics are accounted for. Their main task is to mold sustainable values, attitudes, and practices by providing or supporting various sustainable practices, including sustainable shopping through information on why and how these practices can be nurtured (e.g., through labeling schemes). The correlation between sustainable attitudes and behavioral patterns on the one hand, and trust in sustainability institutions on the other, is not surprising.

The combined analysis shows that Swedish sustainable citizens are neither disgruntled nor distrusting, neither unhappy with the institutional environment of their societal context nor obedient and uncritical supporters of political institutions. They view their private lives as part of political problem solving, but this does not mean that they seek to flee more traditional citizen political practices (e.g., voting and joining political groups). They have just as much trust in national and international institutions as others and just as great belief in traditional governmental institutions' capacity to influence societal development.

Importantly, two beliefs are related to high degrees of sustainable citizenship practice: high trust in institutions dealing with issues of sustainable development, and strong faith in the capacity of people and corporations to change society. The explanatory power of these two results is interesting, considering the fact that the highest levels of political consumerism are found in countries where the state plays a relatively prominent role by enforcing stricter environmental regulations (see chapter 6). Rather than being driven by antiestablishment feelings and the desire to avoid politics, sustainable citizens hold a good degree of trust for some political institutions but do not rely completely on them to solve sustainability problems. Therefore, they are critical and efficacious citizens (cf. Norris 1999) who use their own capabilities and lifestyles to pursue societal betterment and who believe that mobilized individuals and corporations can actually help to bring about societal change. They are sustainable developers who want to pitch in and help governments and other powerful institutions improve the global well-being of all. These findings offer important advice to government and civil society on how they can engage citizens in growing the environmental state.

Conclusion

This chapter penetrates the theoretical and empirical study of sustainable citizenship. It uses the available literature to develop a positive theory on the three nonreciprocal relationships involved in the economic, environmental, and social justice pillars of sustainable development. It specifies operational measures from this theory and employs available cross-national comparative data and a specialized Swedish survey to study its value base and actual practice. A new definition of the common good was formulated; additional measures of citizenship attitudes, awareness, and practices were employed. Accordingly, citizenship is more than the social contract between the state and individuals. It is also a moral pact that people must make about the role of the past in the present and the role of the present in the future global common good. No wonder, then, that this more total relationship of multidimensional vertical and horizontal responsibility challenges governance. Responsibility here is difficult to realize and integrate into policymaking on all levels of government and needs the support of other actors, spheres, and practices.

The analysis here shows that studying sustainable citizenship requires additional measures than those typically used in surveying. To begin to explore it, successfully employed measurements of cross-national political consumerism, a closely related phenomenon, were drawn upon. The most important finding is that several countries standing out as high and low on self-reported political consumerism also have similar patterns around actual comparative market-share measures of sustainable products (fairtrade coffee purchases). Sweden and the United Kingdom are noteworthy exceptions. Sweden ranks highest in political consumerism but seventh in fairtrade coffee consumption and ties with the United States as tenth in market share of fairtrade coffee. The United Kingdom has spiraling fair trade coffee consumption but ranks only seventh in political consumerism. Available survey data allow for further analysis of the Swedish case by focusing on public *awareness* about sustainability problems, *attitudes* for sustainable development, and their actual *practice* (behavior). The sociodemographic characteristics explaining Swedish sustainable citizenship practices parallel several of those of political consumerism. Gender, education, political left-leanings, and institutional trust levels stand out as important factors. Thus, better individual sustainable practice depends largely on the character and function of governmental, nongovernmental, and even corporate institutions.

The research also sheds light on the study of citizenship expectations by revealing how previous scholarship has not fully measured citizenship as a more inclusive, multidimensional, and normative practice. Importantly, it finds that sustainable citizens believe more than others that solidarity-oriented expectations are part of good citizenship. Additional analyses from the Swedish survey show that sustainable citizens feel more solidarity with other people in their own country, in Europe, and in less developed parts of the world.[7] Given the growing importance of stakeholding for good governance, this result is important for discussions on growing the environmental state (see, e.g., chapter 7). Finally, sustainable citizens also believe more in the efficacy of ordinary citizens and NGOs to help change society and are more trusting of sustainability institutions. They seem to believe in acting together with others and using their private lives more individually to support the common good for future generations. The degree to which this represents a more nonreciprocal view of one's citizen responsibilities can be debated.

The findings raise questions about sustainable citizenship and green governance in general. Further reflection on the results for Sweden and the United Kingdom suggests that exploring three interpretations more systematically in future research might be a valuable pursuit. First, the strong relative rise in UK fairtrade consumption might be explained by the British government's and NGOs' forceful support for fairtrade labeled goods, particularly coffee. Sweden, in contrast, has lagged behind, even in establishing fair trade towns, an effort requiring local political, governmental, and NGO support (Micheletti and Isenhour 2010; cf. Malpass et al. 2007). The higher Swedish self-reported political consumer levels might be explained further by the fact that Sweden was an "early bird" on environmental policy legislation (see chapter 6), and its sustainability institutions stress that individuals have a role to play in taking responsibility for environmental sustainability. Indeed, twice as many respondents in Sweden report often purchasing food products because they are environmentally friendly compared to sociopolitical reasons such as workers' conditions.[8]

These results can be compared to the environmental laggard, the United Kingdom, which lacks its own national eco-labeling scheme, nor is it a member of the Global Ecolabelling Network. Moreover, Britain has generally emphasized governmental over consumer-driven actions to target business (Jordan et al. 2006). This institutional context might help explain why fewer UK citizens appear to be mobilized to political consumerism. Unfortunately, there is no available UK data to explore these dynamics further.

Second, the role of collective guilt for past wrongs might explain the varying fairtrade coffee results. Here, Britain's colonial legacy and historical experience with slavery has been important public reasons for engaging in fairtrade (including purchasing fairtrade coffee). This contrasts greatly with Sweden, a country with no politicized colonial history working to trigger buying fair trade coffee. Similarly, Sweden's outlier political consumer status, apparently based on green shopping and good knowledge of various eco-labeling schemes, might be explained by the legacy of the Swedish model, a trustful societal/corporatist partnership that helped create a society characterized by higher levels of social capital, organizational membership, and socioeconomic equality, all characteristics correlating with political consumerism and sustainable citizenship (Stolle and Micheletti 2013; Katzenstein 2003; see chapter 7). Therefore, a country's political history, political culture, and past experience with grand political problem solving might explain different aspects of sustainable citizenship, including those of awareness, attitudes, and practice analyzed in this chapter. More generally, collective memory, civil society's vibrancy, cross-class cooperation, and post-industrialism are some of the possible important general indicators for future sustainable citizenship research. Such an analysis would entail focusing more systematically on political and other institutions and if and how governments develop and promote sustainability choice architecture (e.g., incentives and labeling schemes) for nudging individuals into various collective actions promoting sustainable development (John et al. 2011).

Yet to examine deep-seated value change, individual level studies are still necessary. What must be developed are multiple measures to probe how individuals struggle to bridge everyday public and private life and new measures to facilitate studying how individual citizens and others deal with the difficult trade-offs involved with balancing sustainable development's three pillars (e.g., car driving). Formulating a "sustainable footprint" index to evaluate cross-country levels of actual (not just attitudinal or self-reported) sustainable citizenship practice also could enrich this emerging field of study and might even help government gauge its efforts in promoting sustainable development.

Environmental scholars, therefore, are encouraged to expand their focus from primarily concentrating on state and governmental responses to environmental crises (cf. Mol and Buttel 2002; Barry and Eckersley 2005; Meadowcroft 2005) to ones including citizenship expectations and practices. A more complete model of society's responses to contemporary complex global environmental problems also should (and perhaps must) include changes in the expectations placed on citizens. In particular, studies seeking to improve our understanding of how different types of emerging environmental states interact with the sustainable citizenship model appear necessary. The differences between the United Kingdom and Sweden seem to suggest that there are unexplored dynamics warranting more scholarly attention.

Finally, more effort should be put into formulating citizenship expectations to explore these and other dynamics. New expectations involving sustainability should include ideas about actively seeking information on corporate behavior, being prepared to consume less to fight climate change, and even thinking about how personal lifestyle affect the well-being of future generations. These new expectations should address the important political role of powerful spheres and institutions other than government, most prominently the market arena and corporations. Together with the established citizenship expectations discussed in this chapter, they seem necessary for developing a strong and effective value base for growing an environmental state that furthers democracy for all.

Notes

1. This chapter reports results from two Swedish Research Council–funded projects: "Political Consumption: Politics in a New Era and Arena" and "Sustainable Citizenship: Opportunities and Barriers for Citizen Involvement in Sustainable Development." Aina Gallego Dobón, John Gaventa, Fay Lomax Cook, and Barbara Arneil offered valuable comments on a draft of this chapter presented at the "Dynamics of Citizenship in the Post-Political World" conference at Stockholm University in May 2010. Jenny Wisen finalized the bibliographical references.

2. There are parallels with other findings in this volume. The relative market share of fairtrade coffee is correlated with the relative strictness of environmental regulation in a country (chapter 6); the correlation is not as strong as for political consumerism and regulatory strictness (Somer's $D = 0.27$; $p = 0.04$).

3. Both Japan (7) and Italy (8) come in just below the top five countries in Sommerer's ranking of regulatory strictness in 2005.

4. Self-reported political consumerism relates to fairtrade coffee per capita sales at $0.563***$ (N=21), and to fairtrade coffee market share at $0.513**$ (N=21), meaning that the claimed engagement of citizens in our survey data and the actual market share of fairtrade products correlate fairly strongly with each other.

5. Our definition of individualization follows Ulrich Beck (1997), Andrew McFarland (2010), and others. It involves ways in which individuals can act politically in a more creative and scattered fashion to mobilize critical mass for attempting to change distributions of power and influence in society. Individualization processes have the potential to politicize private choice and life by putting them in a larger societal context. This conception differs from, for instance, Michael Maniates (2001), who maintains that individualized effort represents a depoliticization of environmental destruction and green justice and flight from politics.

6. When sustainable citizenship is broken down into its four different elements (concern, interest, attitudes, and behavior), the differences between younger and older people are attributed primarily to the fact that older people are more frequently shoppers of sustainable food. Interest in sustainability issues tends to increase with age if we use the linear age variable, but there are no significant differences when eighteen- to twenty-nine-year-olds are compared with those over thirty.

7. Correlation between sustainable citizenship and solidarity is 0.36 for Africa, 0.31 for Asia, 0.32 for Latin America, and 0.30 for "other parts of the world." For Sweden, Europe, and the United States, the correlations are 0.1, 0.16, and 0.18 respectively. All the coefficients are statistically significant at the 0.001 level. N is above 990 in all instances.

8. A total of 93 percent say that they have purchased an eco-labeled product within the last twelve months, compared to 77 percent who bought fair trade products within the same period. According to these results, 11 percent buy eco-labeled products several times a week, compared to 3 percent for fair trade products. Asking specifically about food purchases, 29 percent say that that they often or very often choose food on the grounds that the product is eco-friendly, and 12 percent choose food on the grounds that the products were manufactured under good working conditions.

References

Barry, John. 2005. Resistance Is Fertile: From Environmental to Sustainability Citizenship. In *Environmental Citizenship: Getting from Here to There*, ed. Andrew Dobson and Derek Bell, 21–48. Cambridge, MA: MIT Press.

Barry, John, and Robyn Eckersley. 2005. An Introduction to Reinstating the State. In *The State and the Global Ecological Crisis*, ed. John Barry and Robyn Eckersley. Cambridge, MA: MIT Press.

Bauman, Zygmunt. 2007. *Consuming Life*. London: Polity Press.

Beck, Ulrich. 1997. *The Reinvention of Politics: Rethinking Modernity in the Global Social Order*. Cambridge, UK: Polity Press.

Boström, Magnus, Andreas Føllesdal, Mikael Klintman, Michele Micheletti, and Mads P. Sørensen, eds. 2005. "Political Consumerism: Its Motivations, Power, and Conditions in the Nordic Countries and Elsewhere." *TemaNord* 2005:517. http://www.norden.org/sv/publikationer/publikationer/2005-517.

Bullen, Anna, and Mark Whitehead. 2005. Negotiating the Networks of Space, Time, and Substance: A Geographical Perspective on the Sustainable Citizen. *Citizenship Studies* 9(5): 499–516.

Clarke, John, Janet E. Newman, Nick Smith, Elizabeth Vidler, and Louise Westmarland. 2007. *Creating Citizen-Consumers: Changing Publics and Changing Public Services*. London: Sage.

Dalton, Russell J. 2008a. Citizenship Norms and the Expansion of Political Participation. *Political Studies* 56(1): 76–98.

Dalton, Russell J. 2008b. *The Good Citizen: How a Younger Generation Is Reshaping American Politics*. Washington, DC: CQ Press.

Dean, Mitchell. 2010. *Governmentality: Power and Rule in Modern Society*. 2nd ed. London: Sage.

Delanty, Gerard. 2000. *Citizenship in a Global Age: Society, Culture, Politics*. Buckingham, UK: Open University Press.

Diamond, David. 2009. The Impact of Government Incentives for Hybrid-Electric Vehicles: Evidence from US States. *Energy Policy* 37(3): 972–983.

Dobson, Andrew. 2003. *Citizenship and the Environment*. Oxford: Oxford University Press.

Dobson, Andrew. 2007. Environmental Citizenship: Toward Sustainable Development. *Sustainable Development* 15:176–185.

Dobson, Andrew, and Ángel Valencia Sáiz. 2005. Introduction. *Environmental Politics* 14(2): 157–162.

Dryzek, John. 1987. *Rational Ecology: Environment and Political Economy*. New York: Basil Blackwell.

Fairtrade International (FLO). (2005a). FLO International: Coffee. http://www.fairtrade.net/sites/products/coffee/markets.html. Accessed February 23, 2005.

Fairtrade International (FLO). (2007). FLO International: Coffee. http://www.fairtrade.net/coffee.html.

Ferrer-Fons, Mariona. 2004. "Cross-National Variation on Political Consumerism in Europe: Exploring the Impact of Micro-level Determinants and its Political Dimension." Paper presented at the ECPR Joint Sessions, Uppsala, Sweden.

Finkel, Steven E., and Edward N. Muller. 1998. Rational Choice and the Dynamics of Collective Political Action: Evaluating Alternative Models with Panel Data. *American Political Science Review* 92(1): 37–49.

Global Market Information Database. 2002. *Coffee Consumption (Most Recent) by Country*. Euromonitor. http://www.nationmaster.com/graph/foo_cof_con-food-coffee-consumption.

Hoffman, John. 2004. *Citizenship beyond the State*. London: Sage.

Hoskins, Bryony L., and Massimiliano Mascherini. 2009. Measuring Active Citizenship through the Development of a Composite Indicator. *Social Indicators Research* 90(3): 459–488.

Huggan, Graham. 2004. "Greening" Post-colonialism: Ecocritical Perspectives. *Modern Fiction Studies* 50(3): 701–733.

Inglehart, Ronald. 1997. *Modernization and Postmodernization*. Princeton, NJ: Princeton University Press.

Inglehart, Ronald, and Christian Welzel. 2005. *Modernization, Cultural Change, and Democracy*. New York: Cambridge University Press.

International Coffee Organization (ICO). (2005). Historical Data. http://www .ico.org/asp/display5.asp.

International Coffee Organization (ICO). (2006). Overview of the Coffee Market. http://dev.ico.org/documents/icc93-5e.pdf.

Jagers, Sverker C. 2009. In Search of the Ecological Citizen. *Environmental Politics* 18(1): 18–36.

John, Peter, Sarah Cotterill, Alice Moseley, Liz Richardson, Graham Smith, Gerry Stoker, and Corinne Wales. 2011. *Nudge, Nudge, Think, Think: Experimenting with Ways to Change Civic Behaviour*. London: Bloomsbury Academic Publishing.

Jordan, Andrew, Rüdiger Wurzel, Anthony Zito, and Lars Brückner. 2006. Consumer Responsibility-Taking and Eco-labelling Schemes in Europe. In *Politics, Products, and Markets: Exploring Political Consumerism*, ed. Michele Micheletti, Dietlind Stolle, and Andreas Follesdal, 161–179. Rutgers, NJ: Transaction Publishers.

Jowell, Roger and the Central Co-ordinating Team, European Social Survey (ESS) 2002/2003. 2004. Technical Report Edition 2. London: Centre for Comparative Social Surveys, City University. http://www.europeansocialsurvey.org.

Katzenstein, Peter J. 2003. Small States and Small States Revisited. *New Political Economy* 8(1): 9–30.

King, Nancy J., and Brian J. King. 2005. Creating Incentives for Sustainable Buildings: A Comparative Law Approach Featuring the United States and the European Union. *Virginia Environmental Law Journal* 23(3): 397–460.

Kymlicka, Will, and Wayne Norman. 1994. Return of the Citizen: A Survey of Recent Work in Citizenship Theory. *Ethics* 104(2): 352–381.

Lamb, Harriet. 2007. The Fair Trade Consumer. In *The Handbook of Organic and Fair Trade Food Marketing*, ed. Simon Wright and Diane McCrea, 54–82. Oxford: Blackwell Publishing.

Lemos, Maria Carmen, and Arun Agrawal. 2006. Environmental Governance. *Annual Review of Environment and Resources* 31:291–325.

Lister, Ruth. 2007. Inclusive Citizenship: Realizing the Potential. *Citizenship Studies* 11(1): 49–61.

Malpass, Alice, Paul Cloke, Clive Barnett, and Nick Clarke. 2007. Fairtrade Urbanism? The Politics of Place beyond Place in the Bristol Fairtrade City Campaign. *International Journal of Urban and Regional Research* 31(3): 633–645.

Maniates, Michael F. 2001. Individualization: Plant a Tree, Buy a Bike, Save the World? *Environmental Politics* 1(3): 31–52.

Maniates, Michael F., and John M. Meyer, eds. 2010. *The Environmental Politics of Sacrifice*. Cambridge, MA: MIT Press.

McBeth, Mark K., Donna L. Lybecker, and Kacee A. Garner. 2010. The Story of Good Citizenship: Framing Public Policy in the Context of Duty-Based Versus Engaged Citizenship. *Politics & Policy* 38(1): 1–23.

McFarland, Andrew S. 2010. Why Creative Participation Today. In *Creative Participation: Responsibility-Taking in the Political World*, ed. Michele Micheletti and Andrew S. McFarland, 15–33. Boulder, CO: Paradigm Press.

Meadowcroft, James. 2005. From Welfare State to Ecostate. In *The State and the Global Ecological Crisis*, ed. John Barry and Robyn Eckersley, 3–24. Cambridge, MA: MIT Press.

Micheletti, Michele. 2010. *Political Virtue and Shopping: Individuals, Consumerism, and Collective Action*. 2nd ed. New York: Palgrave.

Micheletti, Michele, and Cindy Isenhour. 2009. Political Consumerism. In *Consumer Behaviour: A Nordic Perspective*, ed. Karin M. Ekström, 133–152. Lund: Studentlitteratur.

Micheletti, Michele, and Dietlind Stolle. 2012. Sustainable Citizenship and the New Politics of Consumption. *Annals of the American Academy of Political and Social Science* 644(1): 88–120.

Micheletti, Michele, Andreas Føllesdal, and Dietlind Stolle. 2006. *Politics, Products, and Markets: Exploring Political Consumerism Past and Present*. Rutgers, NJ: Transaction Publishers.

Micheletti, Michele, Dietlind Stolle, and Daniel Berlin. 2012. Habits of Sustainable Citizenship: The Example of Political Consumerism. In *The Habits of Consumption*, eds., Alan Warde and Dale Southerton, 141–163. Helsinki: Helsinki Collegium for Advanced Studies. http://www.helsinki.fi/collegium/e-series/volumes/volume_12/index.htm.

Mol, Arthur P.J., and Frederick H. Buttel. 2002. The Environmental State under Pressure: An Introduction. In *The Environmental State under Pressure*, ed. Arthur P.J. Mol and Frederick H. Buttel, 1–11. Cambridge: Emerald Group Publishing Limited.

Norris, Pippa, ed. 1999. *Critical Citizens: Global Support for Democratic Governance*. Oxford: Oxford University Press.

Norris, Pippa. 2002. *Democratic Phoenix: Reinventing Political Activism*. New York: Cambridge University Press.

O'Riordan, Tom. 2002. Green Politics and Sustainability Politics. *Environment* 44(3): 1–2.

Oreg, Shaul, and Tally Katz-Gerro. 2006. Predicting Proenvironmental Behavior Cross-Nationally: Values, the Theory of Planned Behavior, and Value-Belief-Norm Theory. *Environment and Behavior* 38(4): 462–483.

Ostrom, Elinor. 1990. *Governing the Commons. The Evolution of Institutions for Collective Action.* Cambridge: Cambridge University Press.

Peattie, Ken. 2010. Green Consumption: Behavior and Norms. *Annual Review of Environment and Resources* 35:195–228.

Pennington, Mark. 2001. Environmental Markets vs. Environmental Deliberation: A Hayekian Critique of Green Political Economy. *New Political Economy* 6(2): 171–190.

Petersson, Olof, Jörgen Hermansson, Michele Micheletti, Jan Teorell, and Anders Westholm. 1998. *Demokrati och Medborgarskap.* Stockholm: SNS.

Petersson, Olof, Anders Westholm, and Göran Blomberg. 1989. *Medborgarnas Makt.* Stockholm: Carlssons.

Princen, Thomas, Michael Maniates, and Ken Conca. 2002. *Confronting Consumption.* Cambridge, MA: MIT Press.

Shklar, Judith. 1991. *American Citizenship: The Quest for Inclusion.* Cambridge, MA: Harvard University Press.

Soper, Kate, and Frank Trentmann. 2007. Introduction. In *Citizenship and Consumption*, ed. Kate Soper and Frank Trentmann, 1–16 New York: Palgrave.

Spaargaren, Gert. 1997. *The Ecological Modernization of Production and Consumption: Essays in Environmental Sociology.* Ph.D. thesis, Wageningen University.

Spaargaren, Gert, and Arthur P. J. Mol. 2008. Greening Global Consumption: Redefining Politics and Authority. *Global Environmental Change* 18(3): 350–359.

Spaargaren, Gert, and Arthur P. J. Mol. 2012. The Role of Citizen-Consumers in Globalizing Environmental Politics. In *A Handbook Of Globalisation And Environmental Policy.* 2nd ed., ed. Frank Wijen, Kees Zoeteman, and Jan Pieters, 684–720. Cheltenham, UK: Edward Elgar Publishing.

Spaargaren, Gert, and Peter Oosterveer. 2010. Citizen-Consumers as Agents of Change in Globalizing Modernity: The Case of Sustainable Consumption. *Sustainability* 2(7): 1887–1908.

Stern, Paul C., Thomas Dietz, Vernon W. Ruttan, Robert H. Socolow, and James L. Sweeney, eds. 1997. *Environmentally Significant Consumption: Research Directions.* Washington, DC: National Academy Press.

Stolle, Dietlind, and Marc Hooghe. 2006. Consumers as Political Participants? Shifts in Political Action Repertoires in Western Societies. In *Politics, Products, and Markets: Exploring Political Consumerism*, ed. Michele Micheletti, Dietlind Stolle, and Andreas Follesdal, 265–288. Rutgers, NJ: Transaction Publishers.

Stolle, Dietlind, and Marc Hooghe. 2011. Shifting Inequalities. *European Societies* 13(1): 1–24.

Stolle, Dietlind, Marc Hooghe, and Michele Micheletti. 2005. Politics in the Supermarket: A Comparative Study of Political Consumerism as a Form of Political Participation. *International Political Science Review* 26(3): 245–269.

Stolle, Dietlind, and Michele Micheletti. 2013. *Political Consumerism. Global Responsibility in Action.* Cambridge: Cambridge University Press.

Swedish Government. 2004. Att lära för hållbar utveckling Betänkande av Kommittén för utbildning för hållbar utveckling [Teaching Sustainable Development. Report from Official Report of the Swedish Government from the Public Committee on Education for Sustainable Development] (SOU 2004:104). Stockholm.

Swedish Government. 2005. Tänk om!—En handlingsplan för hållbar konsumtion för hushållen (Skr.2005/06:107). ("Rethink! An Action Plan for Sustainable Consumption for the Household"). Stockholm.

Teorell, Jan, Paul Sum, and Mette Tobiasen. 2007. Participation and Political Equality: An Assessment of Large-Scale Democracy. In *Citizenship and Involvement in European Democracies: A Comparative Analysis*, ed. Jan W. Van Deth, José Ramón Montero, and Anders Westholm, 384–414. New York: Routledge.

Thogersen, John. 2010. Country Differences in Sustainable Consumption: The Case of Organic Food. *Journal of Macromarketing* 30(2): 171–185.

Tilbury, Daniella, Robert B. Stevenson, John Fien, and Danie Schreuder, eds. 2002. *Education and Sustainability: Responding to the Global Challenge.* Gland, Switzerland: IUCN Commission on Education and Communication (CEC).

UNCED (United Nations World Commission on Environment and Development). 1987. *Our Common Future.* Oxford: Oxford University Press.

UNEP. 2011. *Towards a Life Cycle Sustainability Assessment: Making Informed Choices on Products.*

UNESCO. 2011. Education for Sustainable Development. Accessed January 17, 2011. http://www.unesco.org/new/en/education/themes/leading-the-international-agenda/education-for-sustainable-development.

UNESCO. 2012. *Shaping the Education of Tomorrow: 2012 Full-Length Report on the UN Decade of Education for Sustainable Development.* Paris: UNESCO.

United Nations. 1992. Chapter 4, Changing Consumption Patterns. In *Agenda 21: Earth Summit—The United Nations Programme of Action from Rio.* http://www.un.org/esa/dsd/agenda21.

Valor, Carmen. 2008. Can Consumers Buy Responsibly? *Journal of Consumer Policy* 31(3): 315–326.

Van Deth, Jan W., José Ramón Montero, and Anders Westholm, eds. 2007. *Citizenship and Involvement in European Democracies: A Comparative Analysis.* New York: Routledge.

Van Poeck, Katrien, Joke Vandenabelle, and Hans Bruyninckx. 2009. Sustainable Citizenship and Education. In *Conference Proceedings, University of Stirling, Scotland, 23–26 June* 2009. http://www.ioe.stir.ac.uk/events/index.htm.

Van Steenbergen, Bart. 1994. Towards a Global Ecological Citizen. In *The Condition of Citizenship*, ed. Bart Van Steenbergen, 141–152. London: Sage.

Warde, Alan, and Dale Southerton, eds. 2012. *The Habits of Consumption, Studies across Disciplines in the Humanities and Social Sciences 12.* Helsinki: Helsinki Collegium for Advanced Studies.

World Bank. (2007). World Development Indicators. http://data.worldbank.org/products/data-books/WDI-2007.

World Commission on Environment and Development (WCED). 1987. *Our Common Future: Report of the World Commission on Environment and Development.*

Young, Iris Marion. 2006. Responsibility and Global Justice: A Social Connection Model. *Social Philosophy & Policy* 23(1): 102–130.

III

Natural Resource Management in a Comparative Perspective

9

Decentralization and Deforestation: Comparing Local Forest Governance Regimes in Latin America

Krister Andersson, Tom Evans, Clark C. Gibson, and Glenn Wright

In the last three decades, governments around the world have taken a creative approach to state building, often moving away from the traditional approach of centralization toward policies of decentralization. Indeed, the governments of most less-developed countries have decided to decentralize at least part of their natural resource governance regimes.[1] The results have been extremely mixed. In addition, despite the length of time that these policy experiments have been in force, there is little scientific evidence on what makes decentralized natural resource governance work.

Exploring Government Responses to Decentralization

This chapter responds to an ongoing public policy debate about the contextual conditions that affect the effectiveness of decentralization policy (Treisman 2007). Here, we address the question of how local governance institutions respond to decentralization and what effect this response has on the stability of the resource base.

To address this question, we employ a four-pronged strategy. First, we frame the issue of decentralization within a new institutionalist perspective and develop a testable hypothesis on the environmental effects of decentralization. As in chapter 10, "Enforcement and Compliance in African Fisheries: The Dynamic Interaction between Ruler and Ruled," and chapter 11, "Causes and Consequences of Stakeholder Participation in Natural Resource Management: Evidence from 143 Biosphere Reserves in Fifty-Five Countries," we view state capacity as an important determinant of the effectiveness of local governments in both formally decentralized and formally centralized settings. We view de jure decentralization as less important than the capacity to carry out policy.

Second, we use a comparative research design that exploits the variation in institutional conditions both within and across national policy regimes. Our unit of analysis is local government, and we study a random sample of 300 local government territories in three different national regimes with varying degrees of decentralization: highly decentralized (Guatemala), highly centralized (Peru), and semi-decentralized (Bolivia). As in chapter 1, "Introduction: The Comparative Study of Environmental Governance," we argue that the comparative approach is among the most fruitful avenues for research in resource management, and our analysis compares across at least two levels of analysis—municipal government and national government.

Third, we use comparable, time-series observations on a wide array of variables related to resource governance and environmental outcomes for all 300 municipal territories. We use remote sensing technology to create time-series observations of forest cover for our sampled municipalities for three different dates. Finally, we employ robust regression techniques to test the hypothesis and interpret the results of those tests.

This chapter has seven sections. The next one provides a background to the evolution of modern forestry policy in the developing world. In the third section, we review the core findings in the decentralization literature related to forest governance. The fourth section outlines our approach to the study of decentralized governance, followed by a description of our data and methods. In the fifth section, we present our results, and we discuss the causes in the sixth section. We offer our conclusions in the final section.

The Evolution of Forest Policy

Until the 1970s, central governments in most countries viewed the governance of forests through the lens of economic development (Arnold 1992; Wunsch and Olowu 1990). While forests could be managed either by the state or private entities, the value of forest protection was determined by the value of their stock and the flow to the market. If the market considered the land beneath the trees to be more valuable than the wood, governments generally did not stand in the way of forest clearing (Richards and Tucker 1988).

The last thirty years have seen significant shifts in ideas about forests and their governance. Decentralization has emerged as a key component to ideas about effective public policy, democracy, and the environment. International donors and multilateral lending agencies now fund scores of projects that incorporate decentralization as part of their goals (United Nations Food and Agriculture Organization 2001; Inter-American

Development Bank 1994; OECD 1997; World Bank 1988; World Bank 1997). By the late 1990s, a World Bank survey found that 85 percent of the world's less-developed countries had launched processes toward increased decentralization of their public sectors (Bahl 1999). Many scholarly works laud the positive effects of decentralized governance, although a growing group of scholars have expressed skepticism toward decentralization as a policy reform panacea (Agrawal 2001; Andersson 2003; Smoke 2003; Platteau 2004).

At the same time, there is significant disagreement over the precise definition of the term "decentralization," with politicians and scholars using the term to apply to a widely divergent array of phenomena, from deconcentration (in which national government bureaucracies grant regulatory authority and decision-making power to regional subbureaucracies within national government hierarchies) to reforms that create subnational governments and grant them significant taxing, spending, and regulatory authority.

Here, we concern ourselves with two types of decentralization. First, we examine the effects of the formal devolution in forest policy, in which central governments grant local governments the authority and mandates to govern forest resources—which we refer to as "de jure decentralization." Second, we examine the effects of "de facto decentralization," which we define as the extent to which governments develop the capacity to govern policy (in this case, forestry policy). We identify the parameters of decentralization in our case countries below.

According to proponents of decentralization, making lower-level officials responsible for the provision of more goods and services should result in more efficient, flexible, equitable, accountable, and participatory government (Bish and Ostrom 1973; Chubb 1985; Crook and Manor 1998; Ferejohn and Weingast 1997). This argument says that unlike national-level agencies, local politicians and officials will design more appropriate policies because they are more familiar with their environments and their citizens' needs.

Policymakers, donors, and scholars long frustrated by the lackluster outcomes of centralized environmental policy in developing countries have sought to make municipal governments responsible for protecting environmental resources: forests are one of the principal targets of their efforts (United Nations Food and Agriculture Organization 1999; World Resources Institute 2003). Latin America, which hosts about two-thirds of the world's remaining rain forest (Singh 2001) is certainly no exception, and within Latin America, no other countries have taken the decentralization reforms of their forestry sectors further than Guatemala and Bolivia have (United Nations Food and Agriculture Organization 1999; United Nations Food and Agriculture Organization 2001).

In Guatemala, where forestry decentralization is extensive, municipal governments can issue and tax logging permits on any type of forestry property (private, public, or municipal) for up to 10 cubic meters of timber per household per year; and the municipality can own forest property, and create and enforce its own management rules within it, so long as these rules comply with the national forestry law. Municipal governments also can rent out part of their forest land to local citizens and charge user fees for services provided. In addition to the revenue raised by the municipalities themselves, municipal governments receive 50 percent of the centrally collected timber harvesting taxes (although most municipalities still receive very small amounts). Guatemala fits the criteria for a highly decentralized case.

In Bolivia, where forestry decentralization is limited, the authority of the municipal governments is largely limited to performing the tasks that the central government asks of them. Bolivia fits the criteria for a semi-decentralized case. According to Ribot (2002), this form of decentralization is now the most common natural resource policy in less-developed countries. In Bolivia's forestry sector, the role of the municipal government is to assist the central government in promoting sustainable forest management among local users, as well as monitoring and enforcing the centrally defined rules. Bolivian municipalities also receive intergovernmental transfers but, like Guatemala, the official sums are not significant to most. The central government owns all forest resources in the country, including trees on individual private property.

In Peru, there is no decentralization of forest governance responsibilities to local governments, and regional governments have been mandated formal authority only recently (and after the data used here was gathered). As such, it fits our criteria for a "pre-decentralization" baseline case in our comparative research design. In contrast to the relatively decentralized forest governance structures of Bolivia and Guatemala, Peru's central government has resisted pressures from both donors and several groups of local governments to decentralize the responsibilities related to forest governance. Although local governments in Peru frequently can be involved in forest policy, local forestry policy often falls outside of what is legally permissible under Peruvian law (Government of Peru 1975; Government of Peru 2001; Soria 2003; International Timber Trade Organization 2004).

In all three countries, regardless of central government mandates, our data suggests that local governments often are involved in forestry policy, frequently spending resources on reforestation, enforcement, education, and other forest management activities in ways similar to conventional forestry policy in the developing world. In addition, scholars have studied the factors that lead local governments to devote more resources to forest policy (Andersson and Gibson 2006; Andersson and Van Laerhoven 2007; Gibson and Lehoucq 2003). Here, we extend the literature by connecting local forest policy with biophysical outcomes (in this case, forest cover).

These experiments in decentralized forest governance policy in Latin America have received substantial attention from policy analysts and scholars. Next, we review this growing body of literature and their core hypotheses.

Previous Research

While theoretical thrusts, geographical areas, and thematic foci vary a great deal across studies in the decentralization literature, many studies arrive at similar findings. These are the core hypotheses of the decentralization literature.

First, most studies agree that positive outcomes in decentralized environmental governance are unlikely without popular participation in local government decision making (Blair 2000; Larson 2002; Singleton 1998). Moreover, such participation has little meaning unless the local government is authorized to make decisions regarding the existing rules for resource use, an attribute that is not always part of the local mandate (Andersson and Van Laerhoven 2007; Agrawal and Ribot 1999; Agrawal and Ostrom 2001).

Second, most researchers now agree that positive outcomes in a decentralized environmental governance framework rely on local governments being downwardly accountable to resource users (Crook and Manor 1998; Ribot 2002; Smoke 2003). While democratic elections of local officials seem to be a necessary condition for this to occur, they are hardly sufficient. Traditional and informal social networks characterized by severe power asymmetries and patronage relationships often trump the formal structures of democratic elections and hamper any real democratic decentralized governance of natural resources (Andersson 2003; Platteau 2004).

Third, successful decentralized governance of natural resources relies on the technical capacity of the local unit to which governance responsibilities have been devolved (Andersson, Gibson, and Lehoucq 2004; Contreras-Hermosilla and Vargas Ríos 2002; Pacheco 2000; World Bank 1988). Even if local governments are downwardly accountable and include users in decision making, such efforts are not likely to lead to positive outcomes unless the governance system is capable of generating appropriate technical responses to the observed problems.

Fourth, essentially all studies now agree that without a secure source of funding, local governments can do little about environmental issues (Fiszbein 1997; de Mello 2000; Pacheco 2000). Despite the widely recognized need for financial resources, most local governments in decentralized regimes in developing countries have a largely underfunded mandate (Andersson 2004; Andersson, Gibson, and Lehoucq 2006; Boone 2003; Gibson 1999).

Regarding the effects of decentralization itself, results have varied widely. Some scholars have found no effect (Treisman, 2007), others have found substantial positive effects (Faguet 2004; Ribot 2002; Ribot 2008; World Bank 2003), and still others have found that results vary substantially depending on a range of other variables (Andersson 2004; Andersson and Gibson 2007; Andersson and Ostrom, 2008; Faguet and Sanchez 2008; Gibson and Lehoucq 2003; Seabright 1993).

A large number of other scholars also have argued that the capacity of governments to carry out policy is an important predictor of many desirable policy outcomes, including government stability, efficiency, public service provision, and economic growth (Acemoglu, Ticchi, and Vindigni 2006; Besley and Persson 2009; Besley and Persson 2010; Chang 1993; Thies and Sobek 2010; Ziblatt 2008). In many of these studies, the ability to generate revenues through taxation often has been used as a proxy for the ability of the state to carry out policy (Besley and Persson 2009; Besley, Persson, and Sturm 2010; Ziblatt 2008).

The two other chapters in this part of the book (chapters 10 and 11) agree on this last point—where government capacity is constrained, effective common pool resource governance is unlikely. Chapter 10 notes, for example, that users will be unwilling to restrain their own individual harvests of resources if they cannot count on the state to sanction the noncompliance of harvest limits. Because enforcement is costly where there is a high rate of noncompliance, human-natural resource systems will tend to settle into one of two choices: a high-enforcement/high-compliance equilibrium, or a low-enforcement/low-compliance equilibrium. We would add that where government capacity is weak (and therefore, where governments are unable to enforce harvest limit agreements), a high-enforcement/high-compliance equilibrium will be out of the question, and a tragedy of the commons–like situation likely will emerge. As chapter 11 argues, we see corruption as one possible cause of low government capacity.

Here, we test the effects of de facto decentralization—essentially, local government capacity—against the effect of the range of institutional incentives described above, as well as the effects of de jure decentralization, defined as the presence or absence of national-level decentralization reforms that provide municipal governments with the authority to regulate forestry. We argue that de facto decentralization—local government capacity—is more important for the success of forestry policy than de jure decentralization.

In addition, we seek to address three important limitations of the extant decentralization literature. First, while many studies consider how local variations in institutional arrangements affect the performance of the local governments (i.e., Agrawal and Ribot 1999; Crook and Manor 1998; Larson 2002; Pacheco 2000), most are case studies that focus on a very small sample of local units within a single country or subnational region. Second, many of the existing studies do not use measures of governance outcomes as their dependent variables. Instead, they use proxy measures—such as participation and public resource allocations—which may, or may not, be linked systematically to environmental outcomes (Blair 2000; Fiszbein 1997; Larson 2002; Pacheco 2000). Finally, as Duit notes in chapter 1, few studies in natural resource management use large-N statistical techniques, and even fewer use longitudinal data to analyze the environmental effects of decentralization.

Our Approach

Policy reforms, such as decentralization, do not translate automatically into environmental outcomes. It is, therefore, crucial to analyze the processes in the middle of a causal chain linking policies with outcomes. We argue that the effects of a policy change depend especially on the role played by local institutional arrangements in incentivizing local governance actors—local politicians in particular—to pursue particular policies.

Our approach builds on the work of the "new institutionalism" school of political economy (Bates 1981; Horn 1995; Knight 1992; North 1990; Ostrom 1990). We emphasize the value of considering institutions at multiple levels, drawing on earlier work that analyzes institutions as "two-level games" (Putnam 1995), "nested action arenas" (Ostrom 2005), or systems of multilevel governance (Hooghe and Marks 2003). We recognize that institutional arrangements are nearly always made up of several layers of social orders—from local micro-interactional orders to international and transnational arrangements—and that the relationships of complementarity and contradiction between these layers are crucial.

We use these insights to build a model for the analysis of decentralized resource governance. Through this approach, we highlight the ways in which decentralization reforms are filtered by institutional arrangements to produce outcomes that are visible on the landscape. The key point of our approach is that the configuration of local institutional arrangements shape the extent to which decentralization affects the environment. We also recognize that to assess the influence of decentralization reforms on policy outcomes, it is necessary to go beyond traditional measures of formal decentralization—measures that simply treat decentralization as a dichotomous variable and assigns a value of 0 if the government of a country has not passed any formal decentralization reforms, and a value of 1 if it has. Here, we measure the extent to which local governance actors have taken actions related to service provision into their own hands, regardless of what the formal mandate says. This is arguably a more accurate and context-sensitive measure of decentralization.

Our general hypothesis is that *the stronger the decentralized local government institutions, the more likely that local governments will be able to promote forest conservation.* Decentralization may make policymakers more accountable to community pressures for sustainable forest governance, but unless local actors actively create institutional arrangements that can handle the challenges of providing and producing public services to the electorate effectively, decentralization has little meaning. This is especially true in natural resource management policy, where a failure in state capacity to enforce rules on harvest will tend to promote a tragedy of the commons–like equilibrium (as discussed in chapter 10). The ability to collect revenue often has been used as a proxy for state capacity in the literature—in effect, if governments are able to extract taxes from residents, they also will be able to carry out most other policy-related tasks such as enforcing rules (Besley and Persson 2009; Thies 2009; Thies and Sobek 2010). Therefore, we use revenue generated from local sources as a proxy for local institutional capacity.

For example, a local government that has an official mandate on paper, but does not translate its mandate into real actions on the ground, is arguably not a decentralized entity in practice. On the other hand, a local government in a centralized governance structure that nevertheless is taking actions into its own hands and organizes local tax revenue collection, is a de facto decentralized entity.

To test this hypothesis in the context of Bolivia, Guatemala, and Peru, we have constructed proxy measures for de facto and de jure decentralization, which are described in tables 9.1 and 9.2 and in the narrative of the next section.

Data and Methods

To test the study's main hypotheses, we rely on a comparative research design and longitudinal data through survey work in 300 selected municipal territories, as well as the interpretation of satellite images of forest cover for each unit of analysis: the municipal territory. Bolivia, Guatemala, and Peru are ideal cases for a comparative study of decentralized forest governance. They share a number of essential biophysical, socioeconomic, historical, and cultural characteristics, but differ on the variable of theoretical interest to this study: decentralization. All three countries are Latin American, midlevel developing countries with large rural and indigenous populations, significant forest cover, frequent land use–related disputes, and locally elected mayors. But the three countries differ a great deal when it comes to the degree of decentralized governance structure in each country's forestry sector. The amount of regulatory power that each national government grants to its local governments fits along a continuum between a great deal of local decision-making autonomy (Guatemala), to moderate amounts (Bolivia), to virtually no local decision-making power in the forestry sector (Peru).

There are three major data sources for this study: (1) surveys of local governance actors (from 2000 and 2007), (2) census/archive data (from 2000 and 2007), and (3) satellite images (1993, 2000, and 2007). In each of the 300 selected municipalities, we interviewed the elected mayor who held office in 2000 and 2007. In addition, we interviewed municipal forestry officials and community leaders in order to triangulate responses in 2007. Each face-to-face interview took approximately one-and-a-half to two hours. The survey instrument (258 questions) was designed to elicit information regarding the interviewee's policy priorities, staff, relationship with central and nongovernmental agencies, and relationship with citizens. It uses a variety of techniques to understand political incentives and behaviors. We checked several of the interview responses with archival data and found the survey instrument to be highly reliable.

In addition, we use government statistics from the three countries and topographic data created using digital elevation models, as well as forest cover data that was generated using remote sensing techniques (Landsat TM imagery). Digital terrain models were used to generate estimates of the percentage of land in each municipality above a 12 percent grade—that is, the slope above which mechanized agriculture is not feasible—to control for the suitability of local land for mechanized agriculture.

The main independent variables of interest are the proxies for de facto and de jure decentralization. We test the effects of these two variables on measures of forest conditions. De jure decentralization is a dummy variable that identifies whether the municipality was located in a formally decentralized regime; therefore, this variable is coded 1 for Bolivia and Guatemala and 0 for Peru. De facto decentralization is the importance of local sources of income to the municipal government—a proxy for the degree to which the municipal government is independent of the central government. This variable is derived from the following survey question: "How important is the revenue collected by your local government relative to other sources of income?" Responses to this question were coded from a 0 (no local income) to 5 (much more important) than other sources of income.

To test the effects of de facto and de jure decentralization on forest conditions, we use extradispersed poisson regression. This estimation technique is appropriate for our data because our dependent variable is a proportion that cannot be transformed to normal because of a large number of zeros, and the mean and variance are not identical (Hoffman 2004; Rabe-Hesketh and Skrondal 2008). Finally, to address the autocorrelation problems inherent in the use of panel data, we use a Generalized Estimating Equation (GEE) approach, which allows us to control for possible within-unit correlation (Frees 2004; Rabe-Hesketh and Skrondal 2008).

We also test the effects of de facto and de jure decentralization on the assessment of mayors, community leaders, and forestry officials of the importance of forestry as a policy priority compared to other policy areas. Because this variable follows a normal distribution, we use a GEE cross-sectional, time-series linear regression with unstructured intra-unit correlation assumptions, where intra-unit correlations on the dependent variable are derived from the data and standard errors are adjusted accordingly. We also test these two models with several other estimation techniques.

We use two different dependent variables to illustrate different parts of the causal chain between policymakers' choices and forest change. First, we seek to explain the effects of decentralization on forest outcomes (here, total forest cover). Therefore, our first dependent variable is the percentage of the municipal surface area covered in forest. This data was generated using remote sensing (aerial photography and satellite imagery) and coded using geographic information system (GIS) software. We use data for 2000 and 2006 in Peru, 2001 and 2006 in Guatemala, and 2000 and 2007 in Bolivia.

In our second model, we seek to identify the effect of decentralization on forestry policy. Therefore, our second dependent variable is a subjective question asked of mayors in our forestry policy survey. Mayors were asked to describe the relative importance of forestry policy during their term in office, from "much less important" to "much more important" than other policy areas. Because mayors generally are hesitant to describe any policy areas as "much less important" than others, we normalized these responses by taking the mean of all nonforestry policy areas (i.e., public health, transportation, security, education, agriculture, roads and streets, electrification, potable water, sewage, and trash collection) and subtracting the resulting value from the value assigned to forestry. The result is a normally distributed variable with the mean at –0.69.

We also use controls for several alternative explanations for forest conservation and forestry policy, including community organization pressure, central government supervision, nongovernmental organization (NGO) pressures, elite education (i.e., the mean education level of the mayor, municipal forestry official, and the interviewed community leader), and municipal budget size. These controls should address any potential concerns about the confounding effect of these alternative explanations. NGO and community organization pressure is coded from 1 to 5 ("much less frequent" to "much more frequent" than demands for other services), and central government supervision frequency also ranges from 1 to 5 ("never" to "very frequently"). Elite education is the average years of schooling of all interviewed municipal officials, and municipal budget size is the size of the municipal budget in millions of US dollars. The road density data that we use here as a control was available for only one time period in all three countries. Therefore, the 2001 road density data was used as a proxy for road density in 2007 as well. In all three countries, this is the kilometers of road per square kilometer of surface area (logged).

In addition to the results that we report here, we also performed a number of robustness checks on our models. First, we tested for the effects of outliers using deviance, Pearson, and Anscombe residuals. The exclusion of outliers only made our results more significant. We also tested the normality of our residuals (we observed no problems with residuals deviating from normality) and retested our models with heteroskedasticity-robust negative binomial regression and observed no substantive differences between estimation techniques. In addition, we tested the sensitivity of our models to the inclusion and exclusion of control variables; although the biophysical controls are important for the significance of our independent variables of interest, our results are relatively insensitive to the exclusion or inclusion of controls. Finally, we tested our GEE models with different intra-unit correlation assumptions, observing no differences in the direction or significance of our variables of interest.

Results

The multivariate regression results presented in table 9.1 provide support for the hypothesis that the strength of local institutions for taxation is positively associated with forest conservation. Municipalities that have more financial autonomy experience less forest loss and invest relatively more in forestry activities.

Another key finding is that de jure decentralization seems to have a very different effect on forest outcomes compared to de facto decentralization: de jure decentralization appears to have an inconsistent effect on outcomes (and the significance of the de jure variable is also less robust). De jure decentralization is positively associated with forest cover but negatively associated with the relative importance of forestry policy. Municipalities in de jure more decentralized national regimes actually appear to consider forestry *less* important than municipalities in less decentralized regimes.

The model results are consistent with previous research findings that road density, NGO pressure, the local financial importance of forestry, and topography affect forest change. By including variables for road density and slope, we control for each municipal territory's propensity for being targets for profitable conversions from forests to agriculture, a major driver of deforestation in tropical countries. The slope variable, which measures the proportion of land that is not feasible for industrial agriculture because it is steeper than the conventional 12 percent threshold, is

Table 9.1

The relationships between decentralization and forest cover and policy

Dependent Variable	Forest Cover (pct.)	Relative Importance of Forestry Policy
Model	Extradispersed poisson regression (square root of dispersion adjustment)	Population-averaged panel regression with unstructured correlation assumption
De jure decentralization	1.104 (0.002)**	−0.645 (0.034)*
De facto decentralization	0.166 (0.000)**	0.125 (0.005)**
Community pressure	0.040 (0.376)	0.224 (0.000)**
Frequency of central government supervision	0.046 (0.162)	0.044 (0.200)
NGO pressure	0.090 (0.015)*	0.093 (0.016)*
Bolivia	−0.872 (0.000)**	−0.309 (0.221)
Elite education	0.010 (0.431)	0.003 (0.796)
Municipal budget (thousands of $US)	−0.000 (0.420)	−0.000 (0.877)
Road density (logged)	−0.194 (0.000)**	−0.026 (0.565)
Slope above 12% (pct.)	−0.002 (0.122)	0.003 (0.045)*
Forest cover (lagged, pct.)	0.004 (0.022)*	
Forest cover (pct.)		0.010 (0.000)**
Constant	1.557 (0.000)**	−1.629 (0.000)**
Observations	375	376
Municipalities	217	218

p values in parentheses

+ significant at 10%; * significant at 5%; ** significant at 1%

positively correlated with forestry priority ($p < 0.05$), but has no statistically significant effect on forest cover (model 1).

The results from these multivariate regression models provides evidence that de facto decentralization matters for forest conservation in these three countries. Our de facto decentralization variable is a measure of how local governance actors respond to the opportunities offered to them through decentralization reforms. A municipality that organizes itself to raise its own revenue depends less on external sources and thus enjoys more autonomy. We see this as an important proxy for local institutional capacity.

According to the regression results, the de facto decentralization variable has a positive and statistically significant effect on forest cover ($p < 0.001$) and on forest governance investments ($p < 0.01$). Taken together, these results suggest that when local actors have the capacity to gather revenue, they are also more likely to be able to implement effective forest conservation policy.

Figure 9.1 displays the magnitude of the effect that de facto decentralization has on forest cover in our sample. Where the importance of local sources of income is "none," the predicted percentage of forest cover is about 15 percent. Forest coverage is much greater—about 55 percent on average—where local sources of income are "much more important than other sources." These predicted values assume that all other variables are held constant at their mean.

Discussion

The results of the regression analysis raise an important question that merits further discussion. The question is: Why would strong local capacity to raise taxes produce better forest outcomes? The problem that we confront is that even if the local capacity for taxation reflects the local capacity for implementing locally defined policies, that does not necessarily mean that protecting forests will be the priority goal of the local governance actors. If a municipality has the goal of expanding agriculture or cattle raising activities, for example, these land uses compete with forestry, and it would be difficult to achieve a simultaneous expansion of all these competing land uses. Indeed, if a municipal government's goals lie more strongly with agriculture or livestock, we would expect that greater capacity would actually *reduce* forest cover.

To test this idea, we assemble a third regression model (see table 9.2), in which we test the effect of de facto decentralization on forest cover

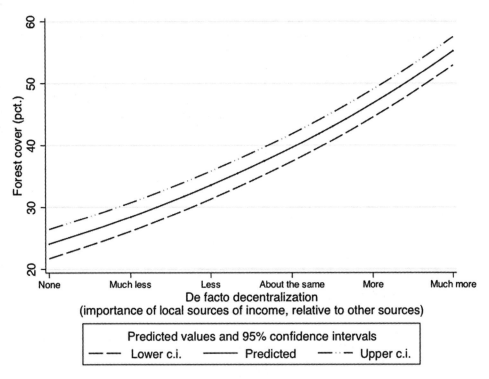

Figure 9.1

The effect of de facto decentralization on forest cover

across a range of values of motivation, as measured by the variable "relative importance of forestry," which is used as the dependent variable in our second model. Here, the independent variables of interest include an interaction term between "relative importance of forestry" and "de facto decentralization."

Before we discuss the results of this model, it may be important to note one problem with the interpretation of interaction terms such as this one. Social scientists have noted that the interpretation of the significance and direction of interaction terms is mathematically intensive and often cannot be inferred from the significance of the interaction term variable in a regression table. This is because most analysts, most of the time, are uninterested in the significance of a given interaction term; instead, they are interested in whether the relationship between some independent variable and some dependent variable differs significantly across values of some other independent variable (Brambor, Clark, and Golder 2006).

Table 9.2

The interactive relationship between motivation, decentralization, and forest cover

Dependent Variable: Forest Cover (pct.)	
Estimation technique	Extradispersed poisson regression regression (square root of dispersion adjustment) with unstructured correlation assumption
De jure decentralization	1.203
	(0.000)**
De facto decentralization	0.137
	(0.003)**
Relative importance of forestry as a policy priority	0.183
	(0.145)
Interaction: de facto decentralization × importance of forestry	0.003
	(0.950)
Community pressure	−0.001
	(0.982)
Frequency of central government supervision	0.035
	(0.280)
NGO pressure	0.066
	(0.075)+
Bolivia	−0.835
	(0.000)**
Elite education	0.009
	(0.461)
Municipal budget (thousands of US $)	−0.000
	(0.516)
Forest cover (lagged, pct.)	0.003
	(0.067)+
Road density (logged)	−0.188
	(0.000)**
Slope above 12% (pct.)	−0.003
	(0.083)+
Constant	1.870
	(0.000)**
Observations	371

p values in parentheses

+ significant at 10%; * significant at 5%; ** significant at 1%

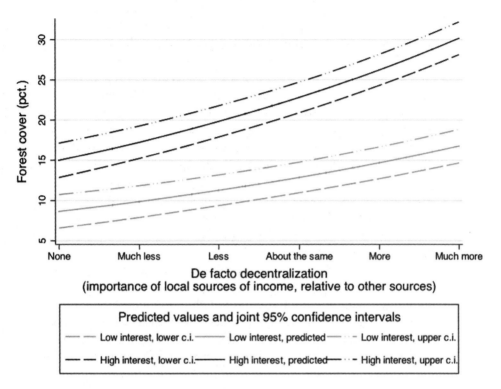

Figure 9.2

The differing effects of de facto decentralization in municipalities with high and low motivation to pursue forest-related activities

Such is the case in the model that we present here. Although the interaction term itself is not significant in the regression table, the effect of "de facto decentralization" varies significantly across the levels of motivation of municipalities to pursue forestry.

For a much clearer understanding of these results, see figure 9.2. Not only do municipalities with high levels of motivation to pursue forestry activities have significantly higher levels of forest cover, but those levels of forest cover increase significantly more quickly as de facto decentralization increases. That is, not only do more motivated municipalities have higher levels of forest cover, but the effect of de facto decentralization on forest outcomes is greater in these municipalities.

It may seem counterintuitive that de facto decentralization has an effect on forest cover even in the most weakly motivated municipalities (in terms of their motivation to provide forestry-related services). However, it may be important to note that, although these municipalities reported relatively less motivation to provide forestry services than other types of services, the indicator used here measures only *relative* motivation to carry out forestry-related activity. Although many municipalities reported relatively less motivation to pursue forestry than other activities, this does not mean that the *absolute* motivation to prevent deforestation (for example) is zero. The observed relationship between de facto decentralization and forest outcomes is also consistent with the assertion in chapter 10 that a "strong state," capable of enforcing horizontal agreements between users, may be more likely to prevent a tragedy of the commons–like situation. Where governments have the capacity and motivation to enforce harvest limits, more users will comply with those regulations of their own volition because they will not fear being the "sucker" who complies while everyone else does not (chapter 10).

Another puzzle is the reason for the positive and significant relationship between de facto decentralization and the priority placed on forestry. Although we can only speculate, it may be that mayors are more likely to emphasize a given policy area when they have the capacity to address it. Therefore, where local government capacity is strong and mayors feel they have the ability to address forestry problems effectively, they will emphasize the importance of forestry more, relative to other policy areas, compared to municipalities with weaker capacity, where effective forest conservation policy is not feasible.

In addition, the fact that our de facto decentralization variable positively affects the priority of forestry activities would suggest that municipalities that are better at raising their own revenues are also more interested in investing in forestry, *but only when controlling for the propensity to convert forests to other land uses.* Our findings were especially sensitive to the exclusion of these biophysical controls. This provides a second possible explanation to the positive effects of the de facto decentralization variable on forest cover stability (model 1).

In sum, it must be stressed that the positive effect of de facto decentralization is contingent on this particular model specification—especially the presence of controls such as road density and steep slope, which measure the propensity for forest-to-agriculture conversions.

The challenge is whether the analysis of de facto decentralization will account for pressures on the forest resource from alternative land uses. We address this concern by introducing control variables for slope and road density in the models. These variables are good proxies for forest conversion to agricultural and pasture lands, which means that the effect of the de facto decentralization variable takes into account the feasibility of land cover conversion from forest to competing land uses. In fact, when we drop these control variables, the effect of the de facto decentralization variable loses some of its statistical explanatory power in both models.

The second thing that our analysis does to discern the effect of de facto decentralization on forest conditions is to examine the influence of this variable on a more intermediate outcome variable, which would provide a plausible causal path between local taxation capacity and forest cover. The intermediate outcome variable is the dependent variable in model 2: importance of forestry as a policy priority. The effect of de facto decentralization on this outcome variable is positive and statistically significant ($p < 0.01$). This result points to a plausible link between stronger local capacity to raise taxes and more stable forest cover because stronger taxing capacity is associated with greater emphasis on forestry policy.

Conclusions

This study provides empirical tests of the extent to which decentralization has any discernable effects on local forest outcomes. Because of the comparative research design and comparable data on local institutions and forest conditions across all three countries, as well as rigorous analytical methods, we are in a position to assess the effects of decentralization policies, its local governance responses, and how these things affect the environmental outcomes on the ground. We see no systematic effect of de jure decentralization in these three countries, but we do find some support for the hypothesis that de facto decentralization can make a difference under some circumstances. While de facto decentralization is associated with greater forest cover and greater importance of forestry as a policy priority, de jure decentralization actually seems to decrease the importance of forestry as a municipal policy priority.

One of the main contributions of the study is the introduction of a nuanced measure of local government capacity to the study of decentralization—the local ability to raise their own public revenue. The use of this measure in the analysis is more informative of the role played by local governments in shaping public policy outcomes, such as forest governance investments and forest cover stability.

This research is of potential practical value to donors and governments around the world that grapple with the challenge to create policies that effectively govern forest resources and other common pool resources. This is particularly true in the context of the international initiative called Reduced Emissions from Deforestation and Forest Degradation (or REDD), which promises millions of dollars in incentives to developing countries that manage to reduce their current deforestation rates. What this analysis shows is that local governments have an important role to play in efforts to protect forests and that it would be a mistake to centralize decision making about forest conservation efforts. It also shows that constraining local governments' ability to raise their own taxes and service fees may be counterproductive to local empowerment and responsible local governance. In general, our findings, along with those of the other chapters in this part of the book, agree that state building—in particular, the construction of strong governments with the capacity to enforce stringent regulations—also will help users self-regulate, leading to improved outcomes for both government and users.

Finally, our analytical results showing the positive effects of de facto decentralization on forest outcomes raises new questions for future studies in this area. Our interpretation of the causal process that links state capacity with better resource management outcomes is consistent with the accounts given in chapters 10 and 11. Even so, our interpretations are largely speculative in nature. Future research is needed to test these plausible causal paths more rigorously. In particular, we find that municipalities with greater capacity see forestry as more important. Is this because these governments have settled into an enforce-comply equilibrium, as chapter 10 suggests, or are other motivational factors at work? Future research would benefit from studying the local-level factors that contribute to municipal support for forest conservation.

Acknowledgments

We are grateful to several individuals for offering constructive criticism on earlier drafts: Andreas Duit, Katarina Eckerberg, and Martin Sjöstedt, along with the other participants at the workshop called "Mapping the Politics of Ecology: Comparative Perspectives on Environmental Politics and Policy," held in Stockholm on June 28, 2010. Thanks also to an anonymous reviewer and Alan Zarychta for their helpful comments and suggestions.

Financial support from the National Science Foundation (grant SES-648447) is gratefully acknowledged as well.

Notes

1. We define "decentralization" as the devolution by a central (i.e., national) government of specific functions, with all the administrative, political, and economic attributes that these entail, to local (i.e., municipal) governments that are independent of the center and sovereign within a legally delimited geographic and functional domain.

References

Acemoglu, Daron, Davide Ticchi, and Andrea Vindigni. 2006. Emergence and Persistence of Inefficient States. Working paper, National Bureau of Economic Research, Cambridge, MA.

Agrawal, Arun. 2001. Common Property Institutions and Sustainable Governance of Resources. *World Development* 29(10): 1649–1672.

Agrawal, Arun, and Elinor Ostrom. 2001. Collective Action, Property Rights, and Decentralization in Resource Use in India and Nepal. *Politics & Society* 29(4): 485–514; doi:10.1177/0032329201029004002.

Agrawal, Arun, and Jesse Ribot. 1999. Accountability in Decentralization: A Framework with South Asian and West African Cases. *Journal of Developing Areas* 33:473–502.

Andersson, Krister. 2003. What Motivates Municipal Governments? Uncovering the Institutional Incentives for Municipal Governance of Forest Resources in Bolivia. *Journal of Environment & Development* 12(1): 5–27; doi:10.1177/1070496502250435.

Andersson, Krister. 2004. Who Talks with Whom? The Role of Repeated Interactions in Decentralized Forest Governance. *World Development* 32(2): 233–249.

Andersson, Krister, and Clark C. Gibson. 2007. Decentralized Governance and Environmental Change. *Journal of Policy Analysis and Management* 26(1): 99–123; doi:10.1002/pam.20229.

Andersson, Krister, Clark C. Gibson, and Fabrice Lehoucq. 2004. The Politics of Decentralized Natural Resource Governance. *PS: Political Science & Politics* 37(03): 421–426.

Andersson, Krister, Clark C. Gibson, and Fabrice Lehoucq. 2006. Municipal Politics and Forest Governance: Comparative Analysis of Decentralization in Bolivia and Guatemala. *World Development* 34(3): 576–595. doi:10.1016/j.worlddev.2005.08.009.

Andersson, Krister, and Elinor Ostrom. 2008. Analyzing Decentralized Natural Resource Governance from a Polycentric Perspective. Policy Sciences, 41(1): 1–23.

Andersson, Krister, and Frank Van Laerhoven. 2007. From Local Strongman to Facilitator: Institutional Incentives for Participatory Municipal Governance in Latin America. *Comparative Political Studies* 40(9): 1085–1111.

Arnold, J. E. Michael. 1992. *Community Forestry: Ten Years in Review*. Rome: UNFAO.

Bahl, Roy. 1999. Fiscal Decentralization as Development Policy. *Public Budgeting & Finance* 19(2): 59–75. doi:10.1046/j.0275-1100.1999.01163.x.

Bates, Robert. 1981. *Markets and States in Tropical Africa*. Berkeley: University of California Press.

Besley, Timothy, and Torsten Persson. 2009. The Origins of State Capacity: Property Rights, Taxation, and Politics. *American Economic Review* 99(4): 1218–1244; doi:10.1257/aer.99.4.1218.

Besley, Timothy, and Torsten Persson. 2010. State Capacity, Conflict, and Development. *Econometrica* 78(1): 1–34.

Besley, Timothy, Torsten Persson, and Daniel M. Sturm. 2010. Political Competition, Policy, and Growth: Theory and Evidence from the US. *Review of Economic Studies* 77(4): 1329–1352.

Bish, Robert L, and Vincent Ostrom. 1973. *Understanding Urban Government*. American Enterprise Institute Press.

Blair, Harry. 2000. Participation and Accountability at the Periphery: Democratic Local Governance in Six Countries. *World Development* 28(1): 21–39; doi:10.1016/S0305-750X(99)00109-6.

Boone, Catherine. 2003. Decentralization as Political Strategy in West Africa. *Comparative Political Studies* 36(4): 355–380.

Brambor, Thomas, William Roberts Clark, and Matt Golder. 2006. Understanding Interaction Models: Improving Empirical Analyses. *Political Analysis* 14(1): 63–82; doi:10.1093/pan/mpi014.

Chang, Ha-joon. 1993. The Political Economy of Industrial Policy in Korea. *Cambridge Journal of Economics* 17:131–157.

Chubb, John E. 1985. The Political Economy of Federalism. *American Political Science Review* 79(4): 994–1015.

Contreras-Hermosilla, Arnoldo, and Maria Teresa Vargas Ríos. 2002. *Social, Environmental, and Economic Dimensions of Forest Policy Reforms in Bolivia*. Bogor, Indonesia: Center for International Forestry Research.

Crook, Richard C, and James Manor. 1998. *Democracy and Decentralisation in South Asia and West Africa*. New York: Cambridge University Press.

de Mello, Luiz R, Jr. 2000. Fiscal Decentralization and Intergovernmental Fiscal Relations: a Cross-Country Analysis. *World Development* 28(2): 365–380; doi:10.1016/S0305-750X(99)00123-0.

Faguet, Jean-Paul. 2004. Does Decentralization Increase Government Responsiveness to Local Needs? Evidence from Bolivia. *Journal of Public Economics* 88(3–4): 867–893; doi:10.1016/S0047-2727(02)00185-8.

Faguet, Jean-Paul, and Fabio Sanchez. 2008. Decentralization's Effects on Educational Outcomes in Bolivia and Colombia. *World Development* 36(7): 1294–1316; doi:10.1016/j.worlddev.2007.06.021.

Ferejohn, John A., and Barry R. Weingast, eds. 1997. *The New Federalism*. Stanford, CA: Hoover Institution Press.

Fiszbein, Ariel. 1997. The Emergence of Local Capacity: Lessons from Colombia. *World Development* 25(7): 1029–1043.

Frees, Edward W. 2004. *Longitudinal and Panel Data: Analysis and Applications in the Social Sciences*. New York: Cambridge University Press.

Gibson, Clark C. 1999. *Politicians and Poachers*. Cambridge: Cambridge University Press.

Gibson, Clark C, and Fabrice Lehoucq. 2003. The Local Politics of Decentralized Environmental Policy in Guatemala. *Journal of Environment & Development* 12(1): 28–49.

Government of Peru. 1975. *Law of Forestry*.

Government of Peru. 2001. *Law of Forestry and Wildlife*.

Hoffman, John P. 2004. *Generalized Linear Models: An Applied Approach*. Boston: Pearson Education, Inc.

Hooghe, Liesbet, and Gary Marks. 2003. Unraveling the Central State, but How? Types of Multi-Level Governance. *American Journal of Political Science* 97(2): 233–243.

Horn, Murray J. 1995. *The Political Economy of Public Administration*. Cambridge: Cambridge University Press.

Inter-American Development Bank. 1994. Fiscal Decentralization: The Search for Equity and Efficiency. In *Economic and Social Progress in Latin America*, 330. Washington DC: Inter-American Development Bank.

International Timber Trade Organization. 2004. *Achieving the ITTO Objective 2000 and Sustainable Forest Management in Peru*. Yokohama, Japan: International Timber Trade Organization.

Knight, Jack. 1992. *Institutions and Social Conflict*. Cambridge, MA: Cambridge University Press.

Larson, Anne M. 2002. Natural Resources and Decentralization in Nicaragua: Are Local Governments Up to the Job? *World Development* 30(1): 17-31

North, Douglass C. 1990. *Institutions, Institutional Change, and Economic Performance*. New York: Cambridge University Press.

OECD. 1997. *Final Report of the DAC Ad Hoc Working Group on Participatory Development and Good Governance*. Paris, France: OECD.

Ostrom, Elinor. 1990. *Governing the Commons*. New York: Cambridge University Press.

Ostrom, Elinor. 2005. *Understanding Institutional Diversity*. Princeton, NJ: Princeton University Press.

Pacheco, Pablo. 2000. *Avances y Desafíos en la Descentralización de la Gestión de los Recursos Forestales en Bolivia*. La Paz: CIFOR: The Center for International Forestry Research.

Platteau, Jean-Philippe. 2004. Monitoring Elite Capture in Community-Driven Development. *Development and Change* 35(2): 223–246; doi:10.1111/j.1467-7660.2004.00350.x.

Putnam, Robert D. 1995. Bowling Alone: America's Declining Social Capital. *Journal of Democracy* 6(1): 65–78.

Rabe-Hesketh, Sophia, and Anders Skrondal. 2008. *Multilevel and Longitudinal Modeling Using Stata*. College Station, TX: Stata Press.

Ribot, Jesse. 2008. *Building Local Democracy through Natural Resource Interventions: An Environmentalist's Responsibility*. Washington, DC: World Resources Institute.

Ribot, Jesse C. 2002. *Democratic Decentralization of Natural Resources*. Washington, DC: World Resources Institute.

Richards, John F, and Richard P. Tucker. 1988. *World Deforestation in the Twentieth Century*. Durham, NC: Duke University Press.

Seabright, Paul. 1993. Managing Local Commons: Theoretical Issues in Incentive Design. *Journal of Economic Perspectives* 74(4): 113–134.

Singh, Ashbindu. 2001. *An Assessment of the Status of the World's Remaining Closed Forests*. Nairobi, Kenya: United Nations Environment Programme.

Singleton, Sara G. 1998. *Constructing Cooperation*. Ann Arbor: University of Michigan Press.

Smoke, Paul. 2003. Decentralisation in Africa: Goals, Dimensions, Myths, and Challenges. *Public Administration and Development* 23(1): 7–16; doi:10.1002/pad.255.

Soria, Carlos. 2003. *Marco Legal Para Aprovechar los Bosques Secos de la Costa Peruana*. Lima, Peru: Foro Ecológico.

Thies, Cameron G. 2009. National Design and State Building in Sub-Saharan Africa. *World Politics* 61(04): 623–669; doi:10.1017/S0043887109990086.

Thies, Cameron G, and David Sobek. 2010. War, Economic Development, and Political Development in the Contemporary International System. *International Studies Quarterly* 54(1): 267–287.

Treisman, Daniel. 2007. *The Architecture of Government: Rethinking Political Decentralization*. Cambridge: Cambridge University Press.

United Nations Food and Agriculture Organization. 1999. *State of the World's Forests*. Rome: Food & Agriculture Organization of the United Nations.

United Nations Food and Agriculture Organization. 2001. *The State of the World's Forests 2001*. Rome: Food & Agriculture Organization of the United Nations.

World Bank. 1988. *World Development Report 1988*. New York: Oxford University Press.

World Bank. 1997. *World Development Report 1997: the State in a Changing World*. New York: Oxford University Press.

World Bank. 2003. *World Development Report 2004*.. Washington, DC: Oxford University Press and the World Bank.

World Resources Institute. 2003. *World Resources 2002–2004*. Washington, DC: World Resources Institute.

Wunsch, James S, and Dele Olowu, eds. 1990. *The Failure of the Centralized State*. Boulder, CO: Westview Press.

Ziblatt, Daniel. 2008. Why Some Cities Provide More Public Goods than Others: A Subnational Comparison of the Provision of Public Goods in German Cities in 1912. *Studies in Comparative International Development* 43(3–4): 273–289.

10

Enforcement and Compliance in African Fisheries: The Dynamic Interaction between Ruler and Ruled

Martin Sjöstedt

Millions of African small-scale fishers and their families depend on fish and fishing for income and nutrition, and fish exports and licensing agreements with foreign fleets constitute an important source of foreign exchange and revenue for the cash-strapped African states. Yet many fisheries suffer from overexploitation and ecological stress, endangering the livelihoods of some of the poorest segments of society, putting pressure on public finances, and also severely threatening the overall health of the oceans (FAO 2006; Devine, Baker, and Haedrich 2006; World Bank 2004; Myers and Worm 2003).

Overexploitation certainly comes as no surprise to scholars of institutional theory and natural resource management (see Hardin 1968; Demsetz 1967). Yet, more than dismal predictions, institutional theory also provides remedies, and various forms of institutional arrangements such as marine protected areas and quota systems have been put forward as potential silver bullets in managing marine resources and fisheries (Scott 1988; Hilborn, Parrish, and Litle 2005; Andersen 1995). But the academic debate over which of these arrangements that might work better than the others—i.e., largely a debate concerning institutional design—by and large overlooks the fundamental issue of enforcement in the form of monitoring, control, and surveillance (MCS) and in particular lacks an understanding of the severe theoretical and practical challenges involved in establishing such enforcement.

The theoretical starting point in this chapter is that while fisheries management, as well as wildlife management in general, highlight two fundamental relationships in society—a horizontal relationship between citizens in a society and a vertical relationship between the government and its citizens (Gibson 1999; Sjöstedt 2012)—most theoretical and empirical effort has been focused on the horizontal one. As such, there has been a lot of research and policy effort deployed into studying and trying to solve classical collective action dilemmas among citizens. The vast institutions-as-remedy literature holds that institutions enable, guide, and motivate individual behavior—and by changing expectations of other people's behavior, high-quality institutions can produce and sustain long-term sustainable use of natural resources (Greif 2006).

The Significance of the Government-Citizen Relationship

The role of the government in this situation is basically to provide the institutions and to act as a third-party enforcer, making sure that the horizontal agreement between citizens is fulfilled. Yet, as correct and important as such insights may be, far less effort has been devoted to studying the theoretical problems involved in establishing such institutions in the first place, and especially the dynamic vertical relationship between ruler and ruled that enforcement of the institutionalized rules entails. In short, what makes the vertical relationship between the government and its citizens qualitatively different—and theoretically more difficult compared to the horizontal relationship between citizens—is the absence of third-party enforcement. More specifically, in this relationship, the actor normally supposed to enforce agreements (i.e., the government) is obviously part of the agreement, which in turn implies serious commitment problems and a need to go beyond the traditional view of institutions as politically determined rules that are only imposed coercively in a top-down manner (see Sjöstedt 2008). Instead, there is a need to focus on institutions as self-enforcing agreements and as outcomes of a dynamic interaction between the parties involved, where the incentives facing both parties in the interaction is taken into account.[1]

The chapter takes those incentives as a starting point, then develops a theoretical model and explores the dynamic interaction between ruler and ruled in the fisheries sector in Angola, Namibia, and South Africa. As such, this chapter is firmly rooted in the call in chapter 1, "Introduction: The Comparative Study of Environmental Governance," for bringing the state back into the analysis of environmental politics and natural resource management. In addition, the explicit comparative approach undertaken here makes this distinct from the case-study approach that dominates the field (the chapters in this book being important exceptions to this rule). The chapter is also firmly rooted in theories from the field of rational-choice political science and political economy.

The chapter is organized as follows. The first section gives a brief account of the current state of African fisheries and the challenges facing fishers and the industry at large. Then there is a brief review of the dominant institutions-as-remedy literature and the various policy instruments pushed for by governments and donors. The next section develops the core theoretical argument of this chapter. In short, this argument contributes to institutional theory by developing an understanding of institutions not only as rules that are imposed by the government, but rather as an equilibrium outcome in a dynamic relationship between the government and its citizens. As such, the argument contributes an understanding of institutional dynamics to situations where there is no third party to act as an enforcer. The rational-choice model developed here is a simplified account of real-world events, of course, but it is hoped that what gets lost in detail is balanced by a gain in abstraction and theoretical understanding. A more classical policy analysis—focusing on detailed processes and within-government dynamics—would contribute to our understanding as well, but the firm belief here is that such an analysis also would benefit from this chapter's somewhat stylized conceptualization of the vertical relationship between ruler and ruled.

Turning to the empirical investigation, this text tests the validity and applicability of the theoretical model by exploring the dynamic interaction between ruler and ruled in case studies from Angola, Namibia, and South Africa. This is done by conducting a comparative legal assessment and a desktop review of each respective country's fisheries regulations and enforcement strategies. Finally, a concluding discussion and potential directions for future research are provided.

Fisheries in Sub-Saharan Africa

Sustainable fisheries bring a wide range of benefits for some of the world's poorest population groups. First, the fisheries sector is an important source of income and employment. Food and Agriculture Organization (FAO) estimates show that more than 30 million people are employed as full-time fishers globally—of whom 95 percent are to be found in developing countries. Moreover, because fish is one of the most heavily traded food commodities, fisheries provide export earnings and enhance the overall macroeconomic performance of many states. Second, in some of the poorest countries of the world, fish supplies more than 50 percent of the animal protein intake and thus has an important effect on the nutritional status of many poor people (World Bank 2004). Yet, all the potential benefits from fisheries must be derived from a seriously decimated resource base. In 2005, one-quarter of the stocks monitored by the FAO suffered from serious depletion, while half the stocks were fully exploited and produced catches at their maximum sustainable limits (FAO 2006). Thus, coastal fisheries biomass suffers from substantial decimation worldwide (World Bank 2004). Given the potential benefits from fisheries, such ecosystem depletion naturally has serious repercussions on poverty reduction and development at large.

In sub-Saharan Africa, fisheries suffer from a double whammy consisting of pressure from long-distance industrialized fleets combined with chaotic expansion of small-scale fishing fleets within national borders. The European Commission has negotiated fishing licenses for vessels from Europe, targeting high-value stocks that command a ready market in Europe. Yet, these agreements have had a significant effect on the availability of high-quality, nutritious fish in local markets—traditionally the supplies that constitute a substantial portion of the nutritional intake of poor people. But overfishing cannot be blamed only on foreign fleets. In many cases, small-scale national fishing fleets have become too numerous and operate unregulated under open access conditions (World Bank 2004). Estimates of illegal, unregulated, and unreported (IUU) fishing are shaky to say the least, but they indicate that lack of compliance with fishery regulations poses a serious threat to both ecological systems and poverty reduction.[2]

The persistent problems of overfishing and IUU have inspired a lot of research on the underlying institutional reasons and motivations for fishers to comply or fail to comply with fisheries regulations. As we will see, while this research largely has focused on horizontal collective-action dilemmas among resource users—and on the potential for institutions to remedy the prevailing problems—the vertical relationship between the government and its citizens has not been given sufficient attention.

Institutions as Remedy

A growing consensus among researchers and policymakers now suggests that the problem of overfishing is, by and large, an institutional problem (Alcock 2002; Bailey 1988; Kaitala and Munro 1993). Weak or absent institutional arrangements have enabled overfishing to continue to this point of near-collapse, and to get fisheries back on a sustainable track, institutional reform is widely called for (FAO 2006; World Bank 2004; Townsend, McColl, and Young 2006; Iudicello, Weber, and Wieland 1999; Grafton et al. 2006; Bromley 2005).

"The fish in the sea are valueless to the fisherman, because there is no assurance that they will be there for him tomorrow if they are left behind today" (Gordon 1954). Preceding Coase's social cost, Olson's logic of collective action, Hardin's tragedy of the commons, and Alchian and Demsetz's property rights paradigm (Coase 1960; Olson 1965; Alchian and Demsetz 1973; Hardin 1968), Gordon recognized early the resource characteristics that make the management of fisheries and other common pool resources particularly problematic—namely, the inability to exclude outsiders. Intuitively, it would be in each fisher's interest to leave some fish behind so that the prospects for future revenues are safeguarded. But the numerous deteriorating resource systems that exist are evidence of open-access regimes that make all resource users expect that others are overharvesting the resource; for that reason, they engage in overuse themselves. This is the well-known "tragedy of the commons," also conceptualized as a collective action dilemma, a social trap, or as the prisoner's dilemma (Bromley 1992; Rothstein 2005; Axelrod 1984). In all these conceptualizations, the expectation that others will embark on an uncooperative path and free-ride on conservation efforts makes every individual reluctant to participate in conserving the collective good or employing a cooperative strategy themselves, even if this would have been the rational thing to do.

The current state of the world's fisheries accounted for above tends to support the dismal predictions concerning common pool resources and the inevitability of ending up in social traps or so-called suboptimal equilibria. Yet institutional theory not only comes with dismal predictions, it also provides potential solutions and instruments to avoid collective action dilemmas and the resulting overuse of natural resources. By and large, these proposed solutions have something to do with property rights (Christy 1996; Anderson 1995; Eggertsson 1990; Edwards 2003; Arnason 2005; Fox et al. 2003; Chavez and Salgado 2005). The early scholars of common pool resources predominantly suggested the introduction of private property rights because such an institutional arrangement was said to align social and private costs and, as such, to internalize externalities (Scott 1988; Demsetz 1967; Matthiasson 2003). But even though the logic of the private property rights paradigm is compelling, the literature has not provided convincing empirical evidence of its validity. In fact, the opposite is true: Growing evidence of the pitfalls—in particular the high costs and negative consequences for the poor—has in many cases motivated a reexamination of natural resource management institutions.

Although Hardin's tragedy of the commons has guided natural resource management policy (in unfortunate directions) for decades; developments in recent years have pushed governments and policymakers to reconsider the role of communities and actors in the state and market dichotomy (Acheson, Stockwell, and Wilson 2000; Bromley 2005; Ostrom 1990). In fact, as also recounted in chapter 9, "Decentralization and Deforestation: Comparing Local Forest Governance Regimes in Latin America," and chapter 11, "Causes and Consequences of Stakeholder Participation in Natural Resource Management: Evidence from 143 Biosphere Reserves in Fifty-Five Countries," communal organizations have proved to be able to handle problems that neither the state nor the market have been capable of managing effectively. Examples of this include the production of local public utilities and the internalization of ecological externalities (Hilborn, Parish, and Litle 2005; Hannesson 2005; Burton 2003).

In the fisheries sector, the institutional opportunity set thus includes a variety of ways to overcome open-access situations: individual transferable quotas (ITQs), individual effort quotas (IEQs), territorial use rights (TURFs), marine protected areas (MPAs), and various comanagement regimes (Newell, Sanchirico, and Kerr 2005; Anderson 1991; Arnason 1993; Huppert 2005). Yet, despite all these institutional innovations, serious problems of overuse persist, and in light of the severe IUU problems described above, the general interest in enforcement and MCS has increased remarkably in recent years. As such, there is now a growing awareness of the fundamental role played by the surrounding and complementary institutional framework in which resource systems and the institutional arrangements that govern their use are embedded (Morrow and Hull 1996; Burton 2003; Anderson 1989; Sjöstedt and Sundström, forthcoming). But this growing consensus generally suffers from a lack of understanding of the underlying theoretical challenges—and the incentives facing governments and resource users when it comes to enforcement and compliance hence are poorly understood.

A starting point for an increased understanding of compliance and enforcement is given by Margaret Levi, who argues that compliance is a calculated decision based on the behavior of others and, in order to comply, citizens need confidence that other citizens also will keep their part of the bargain. This is clearly the kind of collective-action dilemma that the institutions-as-remedy literature cited above has focused on primarily. However, Levi also stresses the importance of a second condition: that citizens feel confident that rulers will keep their part of the bargain (Levi 1997). Yet, this part of the equation has not been given the same attention in the enforcement and compliance literature. To some extent, however, it is recognized that institutions are themselves subject to the problem that they are designed to solve. That is, establishing institutions is in itself a collective-action dilemma: all citizens would like to benefit from such institutions, but because they can reap the benefits of these institutions without contributing to their creation, no one will put any effort into establishing them. The solution to this problem is usually given in the literature, once again, as for the government to act as a third-party enforcer. However, such conceptualizations tend to build on an idea of institutions only as politically determined rules that are imposed from the top down, not as an equilibrium in a dynamic interaction between ruler and ruled. In addition, such conceptualizations imply a heroic assumption of the government as either a benevolent third-party enforcer, always willing to enforce institutional arrangements for everyone's benefit, or as a Leviathan with unlimited resources and capacities to devote on coercion and repression. On the contrary, this chapter sees lack of enforcement—such as the fact that many MPAs become so-called paper parks—as the result of a possibly deliberate decision from the government not to enforce the particular institutional arrangement. Noncompliance from resource users, in turn, can be seen as the rational anticipation of such lack of enforcement; for an individual resource user, there is simply not much point to complying if the rules are not enforced and no one else is complying. In fact, to understand IUU, it is reasonable to see the various enforcement and compliance outcomes in the fisheries sector as equilibria in a dynamic enforce-comply game pursued by the government and its citizens. In line with Levi's second condition for compliance, noncompliance is the result of expectations from resource users that rulers will *not* keep their part of the bargain. Similarly, a deliberate decision from the government not to enforce institutional arrangements is likely to build on the anticipation that resource users would not comply anyway. The dynamics of this vertical agreement is developed further in the section below. Similar to other game theoretical accounts, this model is simplified, of course. Yet, the firm belief in this chapter is that such simplifications still can contribute greatly to our understanding of complex, real-world processes and events.

Enforcement Dynamics

To fully understand the vertical dynamics in natural resource management, we need a basic understanding of credible commitments. In short, commitment problems arise when a decision to act in a certain way depends on the action of other parties and when actors move sequentially in an intertemporal setting.[3] To enter into agreements under such circumstances, each actor needs some assurance from the other party that the agreement is going to be honored. Normally, such an assurance is given by an enforceable contract. For example, an agent's actions are usually constrained by the surrounding institutional framework: if the agent fails to fulfill his or her part of the agreement, then the other party can rely on a third party to enforce the contract. The other party would simply file a complaint, and an outside agency would make the first party fulfill the bargain. Yet, when it comes to the relationship between the government and its citizens, similar contractual relationships are not very useful. Because there is no outside agency to enforce the contract, the citizens cannot be absolutely assured that the government will fulfill its part of the bargain. If it does not, then " the agent violating the contract is precisely the party supposed to enforce it" (Acemoglu and Robinson 2006, 135).

To understand how and why agreements between the government and its citizens evolve and are sustained despite the severe commitment problems involved, institutional theory needs to develop an understanding of enforcement dynamics when the third-party enforcer is also the first or second party in an agreement. As such, institutional theory ought to focus on institutions as incentives and take into account the incentives facing both parties in an interaction, as well as the incentives facing the supposed enforcer. Only then can we understand so-called private order (i.e., situations where order prevails even though a third-party enforcer is missing). In this study of incentives and self-enforcing institutions, we thus need to see institutions and institutional dynamics as endogenous, in the sense that they are self-enforcing.[4]

The Enforce-Comply Game

The self-enforcing equilibrium described above can be seen as an outcome of a game pursued by the government and its citizens that I choose to call the "enforce-comply game."[5] In this game, the citizens can choose to comply or not comply, and the government can choose to enforce or fail to enforce the institutional arrangements. Of course, the assumption that both the government and the citizens are unitary actors that can enter easily into collectively binding agreements with other parties is a simplified version of real-world events. Yet, the ambition here is that such simplifications will help reveal the underlying logic of more complex interactions. The payoffs are given in table 10.1.

Table 10.1

The enforce-comply game

		Government	
		Enforce	**Not enforce**
Resource users	**Comply**	4 4	3 1
	Not comply	1 3	2 2

What this table shows is that unlike the classic prisoner's dilemma, the enforce-comply game has two Nash equilibria.[6] First, both resource users and the government benefit the most from an enforce-comply situation. Under such circumstances, the resource users benefit from the fact that the government has implemented sound fisheries regulations and that it makes sure that other fishers follow these regulations and that the common pool resources are sustainably managed for everyone's long-term benefit. For its part, the government benefits from citizens' compliance by not being forced to employ excessive resources to chase noncompliers. In the end, the government also gains the possibility of taking a part of citizens' output in the form of tax revenue or export earnings. Yet, compliance pays off for the citizens only if the government keeps its part of the bargain. Nonenforcement by the government affects the horizontal game played among the resource users—that is, it affects resource users' expectations of the actions of other resource users—and produces mutual expectations of noncompliance in the resource user collective. From the payoff matrix above, it is clear that the most costly outcome for an individual resource user is to comply while the rules are not enforced— i.e., this would make this particular resource user "the sucker." For the government, the most costly outcome is to put a great deal of effort into enforcement while noncompliance is widespread. Considering that both parties are risk averse, both anticipate the worst outcome and choose not to comply and not to enforce, respectively. That is, if both participants are risk averse, the dominant equilibrium is the one in which government does not enforce and citizens do not comply. Of course, the exchange is not as simple as portrayed here. Yet, while there are often multiple levels of government and multiple actors involved in fisheries management, the simple and general model developed here still provides an underlying logic of enforcement dynamics.

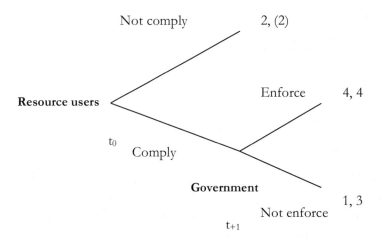

Figure 10.1

The enforce-comply game when resource users move first

The logic behind the above outcome is perhaps best shown by look-ing at the game as sequential. If resource users move first, it is clear that they possibly can benefit the most from choosing compliance. This higher payoff, however, depends on the government's choice of action, and the comply action carries with it the risk of ending up with the most costly outcome. Rational anticipation, however, would predict that resource users expect the government to enforce the regulations because the gov-ernment can gain the most from doing so. Yet, being risk averse, and lacking third-party enforcement, citizens choose not to comply in the first place. They are simply not willing to take the risk of nonenforcement from the government at t_{+1} and therefore choose noncompliance at t_0.

If government moves first, it encounters the same trust problem. Be-cause it risks ending up with its most costly outcome if it chooses enforce-ment at t_0, it chooses nonenforcement. It simply does not trust that the resource users will choose compliance at t_{+1} (even if this is the rational thing to do), and because the government is risk averse, the less beneficial equilibrium is once again the result.

Of course, there is still the possibility of ending up in the positive en-force-comply equilibrium. But whether a strategy leaning toward coer-cion or cooperation is more likely to produce this outcome is an empirical question that will be explored later in this chapter.

In conclusion, the model developed here illustrates that without third-party enforcement, the government and resource users risk ending up in

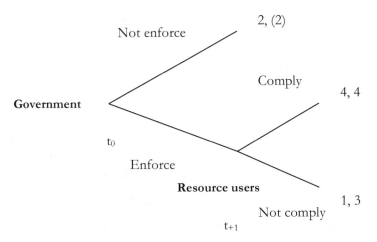

Figure 10.2

The enforce-comply game when the government moves first

a suboptimal equilibrium where the government does not enforce institutional arrangements and where resource users do not comply. Yet, what the game also shows is that—unlike in the prisoner's dilemma—this outcome is not the only Nash equilibrium, so there are rational reasons for both the government and the resource users to choose an enforce-comply outcome. But for this to happen, mechanisms need to be in place that reduce the uncertainty of the agents concerning each other's actions and gives each actor some assurance that the other party is going to keep its part of the bargain. As we will see below, the commitment literature holds that such mechanisms fall into the two closely interconnected categories of reputational and institutional mechanisms (Bardhan 2005).

Fostering Compliance

Given the growing awareness of the importance of enforcement—and the potential commitment problems that accrue—scholars in the field of political economy have increasingly started to pay attention to the mechanisms by which credible assurances can be established (Acemoglu and Robinson 2006; Frye 2004; Barzel 2002, 2000; Dixit and Nalebuff 1993). In terms of game theory, the mechanisms at hand—reputational and formal constraints—work as follows. When modeling reputation, the core argument is that "the long arm of the future provides incentives to honor the loan agreement today so as to retain the opportunity for funds tomorrow" (North and Weingast 1989, 807). Quite intuitively, and in line with the folk theorem, the relationship between citizens and rulers typically consists of more than one-shot interactions, and these interactions in turn could produce incentives for choosing a cooperative line of action. Moreover, even if they find themselves in the more negative equilibrium, the citizens or the government can move away from this situation by starting a trigger strategy. Such a strategy implies that both actors can trigger cooperative behavior by a credible threat of a less beneficial outcome (i.e., nonenforcement or noncompliance) forever. Recall that the less beneficial outcome is a Nash equilibrium, and therefore, the threat of withdrawing to this strategy forever is credible. And in light of this threat of a "grim punishment," a cooperative line of action can be self-enforcing.

In theory, however, reputational solutions have their limits, the most serious of which clearly is the anticipation that the game will come to an end and that actors deviate from the cooperative path for that reason (see Sjöstedt 2008; Greif 2006). Therefore, such solutions might need to be complemented by formal constraints in some form. If, for example, the resource users are about to move first in the abovementioned sequential game, the ruler needs a device that can enable him or her to commit *ex ante* (before the resource users make their initial choice) to be honest *ex post* (after the resource users have chosen to comply and the ruler is confronted with the choice to enforce or not enforce). Formal institutions of power sharing are excellent in this regard because they are durable and affect the future allocation of power: they act as a coordination device for the citizens and the government's promise in time t_0 is simply made credible by putting constraints on the ruler's ability to act opportunistically in time t_{+1} (Acemoglu and Robinson 2006). Devolving decision-making power might seem counterintuitive: it is clearly a cutback in the ruler's privileged position. Yet the model developed here gives an overall rationale for this: that such a strategy raises the possibility of ending up with an even more lucrative equilibrium. Paradoxically, by lessening its coercive capacities, the government thus can gain its citizens' trust, foster their compliance, and take part in their enhanced production. While chapter 9 provides an excellent overview of important aspects of decentralization processes, the particular aspect being focused on here is that power-sharing mechanisms constitute a coordination device for citizens and also has the potential to create an arena for two-way communication where citizens also can gain the trust of the government.[7]

Taken together, enforcement frameworks can lean toward either coercion and the ambition to deter people from noncompliance—i.e., what

North calls an authoritarian political order—or an approach that tries to encourage quasi-voluntary compliance through a more cooperative approach emphasizing comanagement and participation—i.e., what North calls a consensual political order (North 2006; see also chapter 11). Coercion is the Hobbesian solution to compliance problems and rests on the state's repressive apparatus of policing and a legal system where noncompliers are punished heavily (Becker 1968; Hauck 2008).

In this way, coercion may have important reputational and signaling effects because it shows that the government takes noncompliance seriously. Yet, in line with the logic of the model above, the use of coercion as the only compliance strategy risks producing an outcome where the government employs a vast amount of resources to enforcement, but noncompliance is widespread nevertheless. To truly foster compliance, more explicit trust-enhancing measures also might need to be in place. In line with Levi's reasoning, as well as the part of the literature reviewed in chapter 11, recent literature on compliance thus points out that, more than relying on coercion, the government can potentially foster compliance through such options as participatory management and the involvement of resource users in management and policy development (Hauck 2008; Gezelius 2003; Raakjær Nielsen 2003). In such a consensual political order, conformity is usually attributed to the internalization of social norms, which makes individuals want to behave in ways conducive to the existing social order (North 2006). Chapter 11 further discusses such a consensual political order as it deals with the potentially important role played by stakeholder participation, and also investigates the conditions under which participation is likely to occur.

The next section explores how the dynamic interaction between the government and the resource users plays out in the field, as well as which avenues for fostering compliance (i.e., compliance strategies) have been employed in Angola, Namibia, and South Africa.

Enforcement and Compliance in the BCLME Region

The model developed above conceptualizes the vertical relationship between the government and its citizens as a trust game and predicts that a leader who would like to move from a suboptimal equilibrium, where the government does not enforce fisheries regulations and the resource users do not comply, needs to bridge a trust gap. As explained above, according to institutional theory and the avenues open for the government to bridge this trust gap and foster compliance, enforcement frameworks can either lean toward coercion and deterring people from noncompliance, or they can lean toward an approach that encourages quasi-voluntary compliance through greater cooperation. To investigate how these institutional dynamics and theoretical predictions play out in the field, the analysis focuses on three neighboring countries in southern Africa: South Africa, Namibia, and Angola. These countries are all located along the Benguela current and thus constitute the so-called Benguela Current Large Marine Ecosystem (BCLME) region. In addition, they all have experienced high levels of overfishing and IUU. Yet, the characteristics of the countries' fisheries today differ on many accounts; while Angola arguably has been able to curb some of its IUU fishing in recent years, Namibia and South Africa represents two opposite cases, where the former generally is held to be much more successful than the latter when it comes to fostering compliance with fisheries regulations. A comparative assessment of these three countries thus provides ample opportunities for assessing the extent to which the abovementioned theoretical model is applicable as a framework for understanding enforcement and compliance dynamics in the fisheries sector. More specifically, the model predicts that trust-enhancing measures—either in the form of establishing a reputation that makes enforcement the expected behavior or through putting in place more formal constraints on the government's power—are prerequisites for ending up in a positive equilibrium where the government enforces and the citizens comply. The empirical question to be analyzed below is thus the extent to which this model can explain the different outcomes in Angola, Namibia, and South Africa; that is, which of the compliance strategies—or which kind of blending—has been adopted, and with what effects, in the three countries. Although there is a significant number of well-executed studies that focus on fisheries regulation, surprisingly few of these apply the comparative approach that is followed in this chapter. As highlighted in chapter 1, the research field is by and large dominated by single-case case studies. But by taking those studies as a point of departure, the ambition here is to put the findings into a comparative framework. Thus, the material consists of a review of each respective country's fisheries legislation, as well as a close reading of existing case studies conducted in the region. The actual comparison, in turn, is made along qualitative indicators where an approach geared toward coercion is recognized empirically by its focus on ensuring compliance through means such as inspections, investigations, and court activities to enforce the law. Indicators for the opposite approach are efforts toward encouraging quasi-voluntary compliance through a primary focus on raising public awareness, information campaigns, education, and participatory management to promote local ownership. The next sections describe the situation for each country and relate the cases to each other. Even more explicit comparisons are then made in the conclusion.

South Africa

The fishery sector in South Africa consists of a mix of sophisticated industrial fisheries, small-scale commercial fisheries, and subsistence fisheries, and it directly and indirectly employs around 43,000 people. Its value is estimated at US$520 million annually.[8] In addition, it is clear that IUU is widespread in South African waters. The dynamics, of course, differ between industrial and subsistence fisheries, but it is nevertheless well established that noncompliance with fisheries regulation is found in many fisheries, ranging from large-scale organized crime to misreporting and technical offenses concerning gear and minimum sizes. Estimates indicate that illegal fishing in South Africa in fact could amount to US$815 million annually (see Moolla 2010 for an extensive review of IUU in South Africa). Crime syndicates operate the abalone fishery, and the illegal trade in abalone is linked to drug trade and other illicit activities undertaken by Chinese triads and gangs in the Cape Flats area. Illegal harvesting of line fish and rock lobsters is another serious problem, and many recreational fishers are known to illegally sell their catches. Moreover, estimates show that IUU fishing for Patagonian toothfish in South African waters in the late 1990s was worth up to 2.6 billion rand over a two-year period (Moolla 2010). Hake fishery—South Africa's most important and valuable fishery—is also under severe IUU pressure although it is generally considered to be among the most well managed.

When it comes to enforcement, Moolla (2010) argues that until the year 2000, enforcement efforts in the fisheries were quite low-key and symbolically represented by a local fishery control officer. Yet enforcement turned more dramatic in 2001 and 2002, when the Hout Bay Fishing Industry was found guilty of having poached lobsters and Patagonian toothfish to a value of millions of rand, and of having bribed officials at the Marine and Coastal Management.[9] In the following years, the fight against illegal fishing continued to target organized crime, and two specialized environmental courts were set up in Hermanus and Port Elizabeth, where abalone poaching was prevalent. In addition, a specialized monitoring, surveillance, and enforcement unit called the MARINES was created.[10] Penalties were raised substantially as well: for example, the fine for possessing abalone without permission increased from 40,000 rand to 800,000 rand.

Yet, since 2005, enforcement activities reportedly have come to a standstill. The Department of Environmental Affairs and Tourism halted all investigation into triads and other organized crime involvement in the illegal trade in abalone[11] and also put a halt to the MARINES unit, as well as the environmental courts. At the same time, noncompliance is said to have increased: Moolla (2010) argues that "South Africa's significant illegal fishing challenges have noticeably ballooned between 2005 and 2009. Unsurprisingly, this has been directly linked to the institutional collapse at Marine and Coastal Management—the institution charged with regulating access to and conservation of, inter alia, fish stocks." In fact, at present, no MCS strategy is in place to prevent illegal fishing in any of the twenty-two commercial fishing sectors. MCS projects such as the environmental courts, the MARINES anti-poaching program, and partnerships with the abalone industry and NGOs have been abandoned, and the Marine and Coastal Management unit is said to have become a passive spectator on the sidelines. In turn, lack of enforcement has led to lack of compliance and contributed to high levels of illegal fishing in many fisheries, such as the abalone, West Coast rock lobster, and shark fisheries.[12] These conclusions are supported by other sources as well. A large regional study of MCS concludes that although there have been some high-profile successes targeting large-scale illegal fishing, the MCS system generally is not considered to operate as efficiently. There is a lack of management capacity due to inadequate training and widespread corruption. South Africa's record when it comes to marine resources, thus, is in stark contrast to the record in terrestrial conservation where South Africa is recognized as a center of excellence and an example for other African countries to follow (Lux Development 2005).

Taken together, the coercive approach of the early 2000s seems to have been replaced by an approach of nonenforcement. When it comes to trust building and cooperative assurances, the ministry has a Consultative Advisory Forum for Marine Living Resources, which provides input to the ministry on matters such as management and development of the fishing industry and issues related to total allowable catch. However, it was not until 1998 that small-scale subsistence fishing was recognized as a formal fishing sector in South Africa. Prior to the Marine Living Resources Act (MLRA) 18 of 1998, these fishers were by and large dealt with by law enforcement aimed at preventing their activities, and there were no comanagement regimes in place. The MLRA, however, allows the establishment of areas and zones where subsistence fishers may fish. The process leading up to the recognition of small-scale fishing, as well as the rather well developed cooperation between national, provincial, and local levels, was a highly consultative and participatory process. Yet implementation has been slow and comanagement approaches are still in their infancy, although cooperative traits are said to become more prevalent in the coming small-scale fisheries policy.

Namibia

The value of Namibia's fishing industry is about US$390 million per year, and about 15,000 people are directly and indirectly employed. The fisheries sector contributes to 5–10 percent of gross domestic product (GDP), and because more than 90 percent of catches are exported, the sector constitutes the second-biggest export earner of foreign currency (FAO 2010; Rukoro 2009). The fishery is almost exclusively industrial, and no artisanal fishing exists, more or less.[13] Namibia is reported to have reduced IUU threats and incidents to a minimum. Prior to the country's independence in 1990, however, uncontrolled fishing on a massive scale greatly reduced the abundance of all the major fish stocks in the Namibian water. Some concerns still exist when it comes to underreporting, and there are arguably some weaknesses related to the inspection phase of the landing process. Yet on the whole, Namibia is regarded as a highly successful example of combating IUU (Stop Illegal Fishing 2008).

The enforcement efforts started shortly after independence in 1990 when the government embarked on a "Namibianization" of the fisheries sector (Feike 2007). The establishment of an EEZ and the Marine Resources Act was followed by a directive to all unauthorized fishing vessels to leave Namibian waters. Some vessels refused, but they were soon confiscated and became the property of Namibia. In an interview, the Minister for Fisheries argued, "This sent a clear and unambiguous message that the new Namibia is serious about protecting its resources. These high-profile and dramatic arrests effectively ended the plunder of Namibian fish resources [. . .] We have just no choice but to invest in equipment, invest in personnel, and put up a very visible, very clear regime" (Stop Illegal Fishing 2008). Since then, very few incidents of unreported fishing have been reported, and Namibia's MCS system is generally regarded to have developed into a very effective system comprising an integrated program of inspection and patrolling at sea, on land, and in the air to ensure compliance with the fisheries laws. All catches are landed at two ports—Luderitz and Walvis Bay—making management and controls easier. Initially, the MCS activities were supported by outside donors, but the Namibian government itself soon resumed ownership of financial, human, and material support. In addition to patrol boats and aircrafts, there is a vessel monitoring system (VMS) in place. Also, the Namibian government has an observer program consisting of around 200 observers, and all vessels leaving the biggest port have observers on board. The observer program thus has almost 100 percent coverage, and there are few reports of infringements. The overall management strategy originates from a white policy paper from 1992 and from the Fisheries Act passed the same year. The Marine Resources Act was adopted in 2000 and is generally argued to contain a number of innovative management practices (Rukoro 2009). The Ministry of Fisheries and Marine Resources is responsible for the overall management of the industry and governs the sector via a system comprising fishing rights, annually set total allowable catches, and the allocation of quotas to rightholders.

Taken together, the Namibian government seems eager to enforce its fisheries regulations—it is, for example, the only country that has fully implemented its National Plan of Action on Illegal, Unreported, and Unregulated fishing (NPOA-IUU). Compliance seems to be correspondingly high, as reports of IUU have been more or less absent. When it comes to comanagement, part IV of the Marine Resources Act (MRA 2000) deals with the establishment of a Fisheries Observer Agency, and part V is about the establishment of a Marine Resources Advisory Counsel. However, the law does not provide for the appointment of individuals from coastal communities to monitor compliance with fishery laws. Enforcement, thus, is mostly achieved through a rather coercive approach focused on patrols and observation—which might be explained by the fact that Namibia is in many respects unique: the country has no significant small-scale fishery sector. However, penalties are rather low, which makes it possible to question whether the enforcement strategy really is focused on deterrence. In the same interview as quoted before, the minister stated, "At the moment, the penalties are business as usual. It is like you pat someone on the back for fishing illegally. It is not a deterrent." (Stop Illegal Fishing 2008). Nevertheless, it is clear that the Namibian government has established a reputation of strict enforcement by undertaking a number of early and high-profile arrests, an approach that has been described as an "aggressive policy to promote the recovery of the resources to previous levels." (Boyer and Boyer 2005, 4).

Angola

Angola has a combination of industrial and artisanal fisheries. Artisanal fisheries catch groupers, snappers, sea bream, croakers, and spiny lobsters. Semi-industrial and industrial fishers mainly target pelagic species such as horse mackerel, sardinella, and tuna, as well as shrimp and deep-sea crab (Stop Illegal Fishing 2008). About 40,000 people are employed directly in the fishery sector, with another 85,000 people in fishing-related activities. In total, the value of the commercial fisheries is estimated to be about US$180 million annually, and more than 90 percent of the fish caught are sold in the domestic market, which is quite different from the other two countries described in this chapter. As such, fisheries have a more crucial importance to food security in Angola than in Namibia and South Africa.

IUU fishing occurs to some extent in most Angolan fisheries. The major IUU activities include fishing in closed areas, illegal fishing methods, using illegal mesh sizes, and fishing without a license. Moreover, reported IUU activities also include the encroachment by industrial vessels into artisanal areas and unlicensed foreign vessels operating in Angolan waters (Stop Illegal Fishing 2008). The Angolan legislation defines four types of fisheries: subsistence, artisanal, semi-industrial, and industrial fisheries. Subsistence fishers—i.e., fishers fishing for direct consumption for their households rather than for commercial purposes—are not fully required to be licensed if they fish up to 20 kilograms of fish per person per day. Artisanal fishers are those who fish for commercial purposes, but whose boats are 14 meters or less in length. Semi-industrial boats can be up to 20 meters in length, while all vessels exceeding 20 meters are classified as industrial fishing vessels (du Preez 2009). In addition, according to law, the first 4 nautical miles from the shore is exclusively reserved for artisanal and subsistence fishing. Moreover, the Aquatic Biological Resources Law, passed in 2004, outlines a comparatively strong enforcement regime with a substantially improved MCS system that includes reviews of fishery procedures and possibilities for sanctions, as well as specifications of a number of "serious" and "common" offenses (Amador 2004).

When it comes to enforcement, the Angolan government started to regulate its fisheries actively in the 1990s. Almost every aspect of the industry is now governed by laws. An important feature of the Angolan fishery industry is that small-scale fishers are being encouraged to form cooperatives, which are legally recognized and can access credit from the government. Local traditional leadership usually participates in the process, and actions are implemented only when consensus is reached. The Institute for the Development of Artisanal Fisheries and Aquaculture (IPA) was established in 1992 and has the mandate to promote artisanal fisheries through the development of cooperatives, training and community development, technical assistance projects, and other initiatives. For example, IPA has established a community-based catch-monitoring system, and it also supported the appointment of village-based catch monitors chosen by the villagers themselves. The monitors receive a monthly salary from IPA and facilitate the reporting of fish landings to local authorities (du Preez 2009).

Of the three countries surveyed here, Angola is clearly the one making the most explicit provisions for artisanal fisheries and community involvement in its regulatory framework. Hence, significant steps have been taken toward establishing an enforcement framework built on co-management and cooperation. Noncompliance still exists, but the system of community-based observers is said to have increased compliance remarkably (du Preez 2009; Stop Illegal Fishing 2008).

Conclusions

The theoretical starting point of this chapter is that while fisheries management and wildlife management in general highlight two fundamental relationships in society—a horizontal relationship between citizens in a society and a vertical relationship between the government and its citizens—most theoretical and empirical effort has been focused on studying the horizontal one. The vertical relationship highlights important theoretical and practical challenges, however, so there is an urgent need to bring the state back into the analysis of natural resource management in general, and fisheries in particular. More specifically, in this relationship, the actor that is normally supposed to enforce an agreement (i.e., the government) is obviously part of that agreement, which in turn implies serious commitment problems and a need to go beyond the traditional view of institutional mechanisms as politically determined rules that are only imposed coercively, in a top-down manner. Instead, there is a need to focus on these institutions as self-enforcing agreements and as outcomes of a dynamic interaction between the parties involved, where the incentives facing both parties in the interactions are to be taken into account. The framework developed in this chapter thus holds that institutional theory needs to develop an understanding of enforcement when the supposed third party and enforcer is also part of the agreement. More specifically, the model developed here illustrates this vertical relationship between the government and its citizens as a trust game, and it predicts that a leader who would like to move from a suboptimal equilibrium, where the government fails to enforce fisheries regulation and the resource users do not comply, needs to bridge a trust gap. The main mechanisms to do this fall under either reputational or institutional mechanisms. More specifically, according to institutional theory and the avenues open for the government to foster compliance, enforcement frameworks can lean toward either coercion and the ambition to deter people from noncompliance, or an approach that tries to encourage quasi-voluntary compliance through a more cooperative approach emphasizing comanagement and participation.

The cases analyzed in this chapter are all part of the BCLME but represent fishing industries with different characteristics. Angola has a large artisanal fishery that primarily supplies fish to the domestic market; South Africa has a mix of artisanal and industrial commercial and export-oriented fisheries; and Namibia has an almost exclusively industrial fishing industry. The analysis by and large supports the model's predictions.

In South Africa, the government and its citizens seem to find themselves in a nonenforcement-noncompliance equilibrium. After an approach focused on coercion and deterrence—but where noncompliance reportedly was widespread—the Marine and Coastal Management has run out of funds, virtually collapsed, and withdrawn from enforcement altogether. This has in turn spurred even more noncompliance. At the same time, cooperative measures have been undertaken, but they remain in their infancy.

In Namibia, the government went from preindependence nonenforcement to enforcement quite dramatically, through a number of high-profile arrests. These actions can be understood as having an important signaling effect, and thus the reputation established by the Namibian government was that enforcement became its expected behavior. This in turn seems to have been credible, and the industry has by and large responded by employing a compliance strategy. A reputational mechanism—but without formal constraints such as devolved decision-making powers—thus seems to be the mechanism through which Namibia went from nonenforcement noncompliance to an enforce-comply outcome.

In Angola, there is a clear move toward an enforce-comply outcome, and the way this has been done is clearly through trust-building measures, such as by devolving decision-making powers to cooperatives. Apparently, the community-based approach is reported to have increased compliance, and Angola can be said to be on its way toward an enforce-comply outcome. While the main mechanism in Namibia's case was the reputation established through some high-profile arrests, Angola's strategy builds heavily on building trust through devolving decision-making powers. The cases thus fit well with the theoretical model. Yet, a remaining question for future research is to investigate more closely the underlying reasons for *why* countries employ different compliance strategies.

In conclusion, the theoretical model developed in this chapter provides a rationale for understanding the underlying institutional and motivational drivers of successful and less successful outcomes in the fisheries sector, as well as in natural resource management in general. By explicitly focusing on the vertical relationship between ruler and ruled, it also sheds light on enforcement and compliance dynamics more generally, and opens the door to further research into institutions as endogenous and self-enforcing agreements. Future research also would benefit from more in-depth and detailed comparative accounts of within-government dynamics and the political processes by which various ministries and governmental departments fight over control and resources.

Notes

1. The problem of having the third party as part of the agreement is analogous to what Weingast (1995, 1997) has described as the fundamental political dilemma of any economic system: "A government strong enough to protect property rights and enforce contracts is also strong enough to confiscate the wealth of its citizens."

2. A study by the Marine Resources Advisory Group (MRAG 2005) estimates the value of IUU fishing in sub-Saharan Africa to about US$0.9 billion annually. A more recent study built on a different methodology, however, estimated that the value of IUU fishing in South Africa alone amounted to about US$815 million annually (Moolla 2010). In addition, there is some conceptual confusion about what the concept really entails. For example, it does not differentiate between commercial and artisanal fisheries, nor does it take differences between need and greed motivation into account.

3. Assume that two actors are engaged in a potentially mutually beneficial transaction. In this setting, Agent X has the opportunity to undertake an activity Ab at time t_0. But given his interdependence on Agent Z, he will undertake this activity only if he expects that Agent Z will perform a certain action Bb at time t_{+1}. Agent Z, in turn, would like to see Agent X undertake activity Ab in time t_0, so he promises to perform action Bb. Once Agent X has performed the activity in time t_0, however, Agent Z gains the utility in time t_{+1} without performing action Bb. Consequently, he has no incentive to perform action Bb. Yet, looking ahead and reasoning backward, Agent X rationally anticipates this breach of agreement and logically chooses not to undertake activity Ab in time t_0. The inability of Agent

Z to commit credibly to fulfilling his part of the agreement in time t_{+1} thus holds back mutually beneficial transactions and makes both actors worse off. The logic is simply that Agent X has to give before receiving; but at the time of giving, all he gets from Agent Z is a promise of future rewards. And if Agent Z's commitment to follow through on this promise lacks credibility (i.e., if there are no clear-cut incentives for Agent Z to adhere to the promise), there will be no giving at all from Agent X (Greif 2006).

4. Greif (2006) argues that it is the behavior and the expected behavior of other agents that make each agent act in a manner that contributes to motivating, guiding, and enabling others to behave in a manner that led to the institutional elements to begin with. Institutions and the behavior that they generate thus constitute an equilibrium: Institutions mirror the actions of the interacting agents, but they also constitute the structure that influences each agent's behavior.

5. This game is a variant of an assurance game, also called a trust game or stag hunt (see Skyrms 2004 for the extended logic and technical details of this game).

6. While the only Nash equilibrium in a prisoner's dilemma is the uncooperative one, the game depicted here also have a cooperative Nash equilibrium. That is, it has one equilibrium that is payoff dominant (meaning it is Pareto superior to other equilibria) and one that is risk dominant (actors are risk averse, and the more uncertainty there is about the actions of others, the more likely this outcome is).

7. Firmin-Sellers argues, "most importantly, rulers can use their positions to signal an intention to behave cooperatively, establishing a credible commitment to abstain from pursuing their immediate self-interest. To this end, the ruler may create formal channels for two-way communication, establishing the ruler's interest in listening to the subjects' preferences and suggestions" (Firmin-Sellers 1995, 873).

8. The main demersal species are hake, snoek, monk, squid, ribbonfish, abalone, and rock lobster, while the main pelagic species are anchovy, pilchard, round herring, and horse mackerel. In addition, deepwater and Antarctic species such as Patagonian toothfish and orange roughy, as well as tuna, are caught by the national fleet. Industrialized fleets are found mainly along the west coast, while artisanal and subsistence fishers are more prominent on the east coast (Stop Illegal Fishing 2008).

9. A single shipment in 2001 comprised more than 1,660 kilograms of lobster, which is more than the 2009 commercial lobster TAC (Moolla 2010).

10. The specialized courts were able to achieve an 80 percent conviction rate, compared to a 10 percent conviction rate in the general magistrates' courts, and the success of the MARINES program was explained partly by the close collaboration with the former Directorate for Special Operations, the South African Revenue Service (SARS), and the Asset Forfeiture Unit (Moolla 2010).

11. In fact, there are arguments holding that the Marine and Coastal Management (one of the four branches of the Department of Environmental Affairs and Tourism) became financially reliant on the sale of confiscated abalone and thus abandoned all intention to reduce poaching.

12. Because of the lack of enforcement and control in many cases, pillaging has become so common that the abbreviation MPA is sometimes used to mean marine pillaging areas rather than marine protected areas.

13. Namibia's waters hold around twenty different commercial species, primarily small pelagic species such as pilchard, anchovy, horse mackerel, and mackerel, but also large pelagic species such as adult mackerel, demersal hake, monkfish, sole, crab, and lobster.

References

Acemoglu, Daron, and James Robinson. 2006. *Economic Origins of Dictatorship and Democracy*. Cambridge: Cambridge University Press.

Acheson, James M., Terry Stockwell, and James A. Wilson. 2000. Evolution of the Maine Lobster Co-Management Law. *Marine Policy Review* 9(2): 52–62.

Alchian, Armen, and Harold Demsetz. 1973. The Property Rights Paradigm. *Journal of Economic History* 33(1): 16–27.

Alcock, Frank. 2002. Bargaining, Uncertainty, and Property Rights in Fisheries. *World Politics* 54(4): 437–461.

Amador, Teresa. 2004. Review and Audit of the Legal Provisions and Institutional Arrangements That Impact on the Artisanal Fisheries Sector in the BCLME Region. In *Final Report for IPA*. Angola: Artisanal Fishing Institute.

Anderson, Lee G. 1989. Enforcement Issues in Selecting Fisheries Management Policy. *Marine Resource Economics* 6(3): 261–277.

Anderson, Lee G. 1991. Efficient Policies to Maintain Total Allowable Catches in ITQ Fisheries with At-Sea Processing. *Land Economics* 67(2): 141–157.

Anderson, Lee G. 1995. Privatizing Open Access Fisheries: Individual Transferable Quotas. In *The Handbook of Environmental Economics*, ed. Daniel W. Bromley, 453–474. Oxford, UK: Blackwell.

Arnason, Ragnar. 1993. The Icelandic Individual Transferable Quota System: A Descriptive Account. *Marine Resource Economics* 8(3): 201–218.

Arnason, Ragnar. 2005. Property Rights in Fisheries: Iceland's Experience with ITQs. *Reviews in Fish Biology and Fisheries* 15(3): 243–264.

Axelrod, Robert. 1984. *The Evolution of Cooperation*. New York: Basic Books.

Bailey, Richard. 1988. Third-World Fisheries: Prospects and Problems. *World Development* 16(6): 751–757.

Bardhan, Pranab. 2005. *Scarcity, Conflicts, and Cooperation: Essays in the Political and Institutional Economics of Development*. Cambridge, MA: MIT Press.

Barzel, Yoram. 2000. Property Rights and the Evolution of the State. *Economics and Governance* 1(1): 25–51.

Barzel, Yoram. 2002. *A Theory of the State: Economic Rights, Legal Rights, and the Scope of the State*. Cambridge: Cambridge University Press.

Becker, Gary. 1968. Crime and Punishment: An Economic Approach. *Journal of Political Economy* 76:169–217.

Boyer, David C., and Helen J. Boyer. 2005. Sustainable Utilization of Fish Stocks—Is This Achievable? In *FAO Fisheries Report No. 782*. Rome: FAO.

Bromley, Daniel W. 1992. *Making the Commons Work. Theory, Practice, and Policy*. San Francisco: Institute for Contemporary Studies.

Bromley, Daniel W. 2005. Purging the Frontier from Our Mind: Crafting a New Fisheries Policy. *Reviews in Fish Biology and Fisheries* 15(3): 217–229.

Burton, Peter S. 2003. Community Enforcement of Fisheries Effort Restrictions. *Journal of Environmental Economics and Management* 45(2): 474–491.

Chavez, Carlos, and Hugo Salgado. 2005. Individual Transferable Quota Markets under Illegal Fishing. *Environmental and Resource Economics* 31(3): 303–324.

Christy, Francis T. 1996. The Death Rattle of Open Access and the Advent of Property Rights Regimes in Fisheries. *Marine Resource Economics* 11(4): 287–304.

Coase, Ronald. 1960. The Problem of Social Cost. *Journal of Law & Economics* 3(1): 1–44.

Demsetz, Harold. 1967. Towards a Theory of Property Rights. *American Economic Review* 57(2): 347–359.

Devine, Jennifer A., Krista D. Baker, and Richard L. Haedrich. 2006. Fisheries: Deep-Sea Fishes Qualify as Endangered. *Nature* 439(7072): 29.

Dixit, Avinash, and Barry Nalebuff. 1993. *Thinking Strategically: The Competitive Edge in Business, Politics, and Everyday Life*. New York: W. W. Norton & Company, Inc.

du Preez, Mari-Lise. 2009. Fishing for Sustainable Livelihoods in Angola: The Cooperative Approach. *SAIIA Occasional Paper*, No. 45, October 2009.

Edwards, Steven F. 2003. Property Rights to Multi-attribute Fishery Resources. *Ecological Economics* 44(2–3): 309–323.

Eggertsson, Thrainn. 1990. *Economic Behavior and Institutions*. Cambridge: Cambridge University Press.

Feike. 2007. *Report on the Legislative*. Policy and Governance Frameworks in the BCLME Region.

Firmin-Sellers, Kathryn. 1995. The Politics of Property Rights. *American Political Science Review* 89(4): 867–881.

Food and Agriculture Organization (FAO). 2006. *The State of the World Fisheries and Aquaculture*. Rome: FAO.

Food and Agriculture Organization (FAO). 2010. *Fishery and Aquaculture Country Profiles, Angola*. Rome: FAO.

Fox, Kevin J., Quentin Grafton, James Kirkley, and Dale Squires. 2003. Property Rights in a Fishery: Regulatory Change and Firm Performance. *Journal of Environmental Economics and Management* 46(1): 156–177.

Frye, Timothy. 2004. Credible Commitment and Property Rights: Evidence from Russia. *American Political Science Review* 98(3): 453–466.

Gezelius, Stig. 2003. *Regulation and Compliance in the Atlantic Fisheries: State/Society Relations in the Management of Natural Resources*. Dordrecht, the Netherlands: Kluwer Academic Publishers.

Gibson, Clark C. 1999. *Politicians and Poachers: The Political Economy of Wildlife Policy in Africa*. Cambridge: Cambridge University Press.

Gordon, Scott H. 1954. The Economic Theory of a Common Property Resource: The Fishery. *Journal of Political Economy* 62(2): 124–142.

Grafton, Quentin R., et al. 2006. Incentive-Based Approaches to Sustainable Fisheries. *Canadian Journal of Fisheries and Aquatic Sciences* 63(3): 699–710.

Greif, Avner. 2006. *Institutions and the Path to the Modern Economy: Lessons from Medieval Trade*. New York: Cambridge University Press.

Hannesson, Rögnvaldur. 2005. Rights-Based Fishing: Use Rights versus Property Rights to Fish. *Reviews in Fish Biology and Fisheries* 15(3): 231–241.

Hardin, Garrett. 1968. The Tragedy of the Commons. *Science* 162:1243–1248.

Hauck, Maria. 2008. Rethinking Small-Scale Fisheries Compliance. *Marine Policy* 32(4): 635–642.

Hilborn, Ray, Julia Parrish, and Kate Litle. 2005. Fishing Rights or Fishing Wrongs? *Reviews in Fish Biology and Fisheries* 15(3): 191–199.

Huppert, Daniel D. 2005. An Overview of Fishing Rights. *Reviews in Fish Biology and Fisheries* 15(3): 201–215.

Iudicello, Suzanne, Michael L. Weber, and Robert Wieland. 1999. *Fish, Markets, and Fishermen: The Economics of Overfishing*. Washington, DC: Island Press.

Kaitala, Veijo, and Gordon R. Munro. 1993. The Management of High Seas Fisheries. *Marine Resource Economics* 8(7): 313–329.

Levi, Margaret. 1997. *Consent, Dissent, and Patriotism*. Cambridge: Cambridge University Press.

Lux Development. 2005. SADC-EU MCS Programme. http://www.mcs-sadc.org.

Marine Resources Advisory Group (MRAG). 2005. Review of Impacts of Illegal, Unreported, and Unregulated Fishing on Developing Countries. DFID.

Matthiasson, Thórólfur. 2003. Closing the Open Sea: Development of Fishery Management in Four Icelandic Fisheries. *Natural Resources Forum* 27(1): 1–18.

Moolla, Shaheen. 2010. Illegal Fishing of Marine Resources in South African Waters: In Search of Solutions. Draft paper. Cape Town, South Africa, Institute for Security Studies.

Morrow, Christopher E., and Rebecca Watts Hull. 1996. Donor-Initiated Common Pool Resource Institutions: The Case of the Yanesha Forestry Cooperative. *World Development* 24(10): 1641–1657.

Myers, Ransom A., and Boris Worm. 2003. Rapid Worldwide Depletion of Predatory Fish Communities. *Nature* 423(6935): 280–283.

Newell, Richard G., James N. Sanchirico, and Suzi Kerr. 2005. Fishing Quota Markets. *Journal of Environmental Economics and Management* 49(3): 437–462.

Nielsen, Jesper Raakjær. 2003. An Analytical Framework for Studying: Compliance and Legitimacy in Fisheries. *Marine Policy* 27(5): 425–432.

North, Douglass. 2006. *Understanding the Process of Economic Change*. Princeton, NJ: Princeton University Press.

North, Douglass, and Barry Weingast. 1989. Constitutions and Credible Commitments: The Evolution of the Institutions of Public Choice in 17th- Century England. *Journal of Economic History* 49(4): 803–832.

Olson, Mancur. 1965. *The Logic of Collective Action: Public Goods and the Theory of Groups*. Cambridge: Harvard University Press.

Ostrom, Elinor. 1990. *Governing the Commons*. Cambridge: Cambridge University Press.

Rothstein, Bo. 2005. *Social Traps and the Problem of Trust*. Cambridge: Cambridge University Press.

Rudd, Murray A. 2004. An Institutional Framework for Designing and Monitoring Ecosystem-Based Fisheries Management Policy Experiments. *Ecological Economics* 48(1): 109–124.

Rukoro, Raywood Mavetja. 2009. Promotion and Management of Fisheries in Namibia. In *Towards Sustainable Fisheries Law: A Comparative Analysis*, ed. Gerd Winter. Gland, Switzerland: IUCN Environmental Law Centre. http://data.iucn.org/dbtw-wpd/edocs/EPLP-074.pdf.

Scott, Anthony. 1988. Development of Property in the Fishery. *Marine Resource Economics* 5(4): 289–311.

Sjöstedt, Martin. 2008. Thirsting for Credible Commitments: How Secure Land Tenure Affects Access to Drinking Water in Sub-Saharan Africa. Doctoral dissertation, University of Gothenburg, Göteborg Studies in Politics 110.

Sjöstedt, Martin. 2012. Horizontal and Vertical Resource Dilemmas in Natural Resource Management: The Case of African Fisheries. *Fish and Fisheries*. doi:10.1111/j.1467-2979.2012.00481.x.

Sjöstedt, Martin, and Aksel Sundström. 2013. Overfishing in Southern Africa: A Comparative Account of Regime Effectiveness and National Capacities. *Journal of Comparative Policy Analysis*. doi: 10.1080/13876988.2013.835525.

Skyrms, Brian. 2004. *The Stag Hunt and Evolution of Social Structure*. Cambridge: Cambridge University Press.

Stop Illegal Fishing. 2008. *Stop Illegal Fishing in Southern Africa*. Gaborone, Botswana. Stop Illegal Fishing. Gaborone, Botswana.

Townsend, Ralph E., James McColl, and Michael D. Young. 2006. Design Principles for Individual Transferable Quotas. *Marine Policy* 30(2): 131–141.

Weingast, Barry. 1995. The Economic Role of Political Institutions: Market Preserving Federalism and Economic Development. *Journal of Law Economics and Organization* 11(1): 1–31.

Weingast, Barry. 1997. The Political Foundations of Democracy and the Rule of Law. *American Political Science Review* 91(2): 245–263.

World Bank. 2004. *Saving Fish and Fishers: Toward Sustainable and Equitable Governance of the Global Fishing Sector.* Washington, DC: World Bank.

11

Causes and Consequences of Stakeholder Participation in Natural Resource Management: Evidence from 143 Biosphere Reserves in Fifty-Five Countries

Andreas Duit and Ola Hall

A central assumption in many contemporary environmental management paradigms, such as adaptive management and adaptive comanagement, is that stakeholder participation is a crucial component in successful conservation programs. As a consequence, policy prescriptions tend to focus on increasing the number of stakeholders involved in conservation efforts of various kinds. The assumption that stakeholder participation leads to better management of ecosystems mainly has been investigated using case study techniques, and there have been few studies aimed at reaching more generalizable conclusions regarding the pivotal role attributed to stakeholder participation. A first objective of this chapter is thus to conduct an extensive test of the claim that stakeholder participation in natural resource management programs leads to better management of ecosystems.

A second knowledge gap regards the question of how institutional contexts affect local-level stakeholder participation in natural resource management: How does the quality of democratic institutions and levels of corruption affect stakeholders' propensity for participating in natural resource programs? Case study designs are generally not effective in estimating the effect of such system-level factors, which in turn has contributed to a lack of knowledge about the role of institutional contexts in stimulating and repressing the participation of stakeholders.

In an effort to address both these research needs, this chapter will engage in two tasks. First, we investigate the role of institutional contexts for stakeholder participation. Specifically, we test the hypothesis that the institutional environment of stakeholder participation, in terms of political rights and levels of corruption, has an impact on the level of stakeholder involvement in natural resource management. The second hypothesis to be scrutinized has to do with the outcomes of stakeholder participation—does it contribute to better conservation performance? Here, we test the hypothesis that more frequent participation of various stakeholder groups is associated with better management outcomes in Biosphere Reserves (BRs). We employ remote sensing estimates of changes in biodiversity levels combined with survey data on stakeholder participation and management patterns, as well as national level data for political rights and corruption, collected from a total of 143 BR areas in 55 countries.

Stakeholder Participation in Natural Resource Management: Causes and Consequences

Often contrasted to a traditional view of natural resource management, which takes as its primary goal to let experts and scientists preserve natural habitats or resources by restricting human access to them, the notion of stakeholder participation in natural resource management and conservation has emerged as a strong alternative management approach in recent years (Kasperson 2006). Adaptive management (Lee 1999), and in particular adaptive comanagement (Olsson et al. 2004; Armitage et al. 2007), are at once theories of, and policy prescriptions for, management of natural resources and conservation areas in which stakeholder participation plays a crucial role. There are several components associated with these approaches, but a key idea is to involve rather than expel people and groups who are either living within, using the resources of, or are in other ways involved in the natural area that is to be preserved. Several positive consequences are argued to follow from this. First, stakeholder involvement is thought to increase the legitimacy of the preservation effort itself, which in the long run will create a more benign context for continued preservation efforts. Because conservation schemes often intervene in the daily lives of local inhabitants, securing local legitimacy (if not direct support) is frequently an indispensable prerequisite for successful conservation. The legitimacy of preservation programs is thought to increase if local stakeholders are allowed to do things such as influence decision-making processes regarding the protected area, have access to the area, be allowed to use some of the resources located in the area, participate in preservation activities, and receive information about events and ongoing processes in the area.

There is also a growing recognition in both the research and practice of natural resource management of the fact that there is rarely such a thing as a "pristine" natural area or resource, and that the people living in it or using it are often a most influential force in shaping the landscape that the conservation program is trying to preserve (Nabahan 1997). The continued activities of local stakeholders, therefore, might be a direct requirement for conserving a landscape formed by human activities, which again means that local actors must be allowed to enter into the preservation process. Involving local stakeholders also might mean that conservationists get access to local knowledge about the natural resource or area, both in terms of historical information about how the ecosystem has behaved in the past and ongoing monitoring performed by individuals who often are highly skilled at detecting changes and unusual patterns in the natural environment. In addition to these instrumental justifications of stakeholder participation, it can be argued from a normative or democratic standpoint that involving stakeholders is simply the right thing to do when outside actors decide for one reason or another that a certain area in which people have been living and working for generations should be conserved.

The scientific literature devoted to questions of stakeholder participation is heavily indebted to Elinor Ostrom's studies of institution-building processes in common pool resources (CPRs), in the sense that most studies of stakeholder participation are studies of small- and medium-scale natural resource management programs (Duit 2011). Much of this literature has been developed within a case study framework, which means that concerns and findings from this line of research are not by default transferrable into the large-N approach taken in this chapter. For instance, several decades of research into the dynamics of sustainable CPRs have revealed the crucial role played by seemingly small differences in institutional rules and actor incentive structures (Ostrom 2005; Nelson and Agrawal 2008). Such subtleties are not easily captured in cross-national data sets (for an exception, see chapter 9, "Decentralization and Deforestation: Comparing Local Forest Governance Regimes in Latin America"), which is one reason why this chapter is primarily concerned with the identification of aggregate- or system-level factors associated with stakeholder participation in natural resource management and conservation. Conversely, those are very often the sort of factors overlooked by case study designs.

What Explains Stakeholder Involvement?

A large part of the reason why conservation programs differ in their ability to engage stakeholders lies in idiosyncratic and largely place-dependent factors such as leadership styles and philosophies, characteristics of local communities of stakeholders, historical patterns of resource use, and biophysical traits of the area in which the resource is located. But it is also probable that stakeholder participation, much like many other forms of collective action, is in part determined by institutional and cultural factors in the political environment in which the participatory effort is embedded. In this chapter, two such contextual factors—protection of *political rights* and levels of *corruption*—will be assessed with regard to their relative impact on stakeholder involvement in conservation activities. There is a large number of possible contextual factors that might be influencing the propensity of stakeholders to get involved in preservation programs, but because there is a limit to how many explanatory facts can be handled at once (even in an exploratory study such as the present one), some sort of criterion must be applied when selecting independent variables.

The reason why the level of corruption and the extent to which political rights are protected have been selected as explanatory factors is that they reflect two central traits of any political system that permeate almost every aspect of the daily life of individuals, whether it is economic exchange for one's own gain or collective action with the purpose of achieving a common goal. The basic idea is that actors, to be able to accumulate wealth through investments and mutual exchange or organize collectively to express their demands, need a set of rules—institutions—that record, regulate, and protect individual rights and which sanctions violations of these rights in an unbiased, fair, predictable, and transparent manner. In the absence of such institutions, actors cannot know that their dealings with other actors are carried out in a nonbiased fashion, which raises transaction costs and lowers the overall rate of collaborative ventures, ultimately preventing individual actors, as well as society as a whole, from reaping the benefits from cooperation and exchange (North 1990).

In a sense, corruption and protection of political rights can be seen as two different aspects of *institutional quality* (Alexander 2002; Rothstein and Teorell 2008). Recent research into issues of comparative development has emphasized the crucial role played by institutional quality, or fair, uncorrupt, and transparent core institutions (e.g., the judicial system, the public administration, the police force, and political assemblies) for achieving collective action on all levels of a society (Rothstein and Teorell 2008). In short, institutional quality is theorized as a prerequisite for reduced transaction costs and sustained collective action on a wide range of societal arenas, and therefore, it is likely that the outcome of natural resource management schemes is greatly influenced by the level of institutional quality in the surrounding society.

Corruption

The main reason why corruption is a serious problem for any society is not so much the fact that embezzlement of public funds is often involved, but rather that corruption, over time, erodes the capacity for collective action needed to address common problems. There is little theoretical cause for thinking that stakeholder participation in natural resource management programs should be an exception in this respect (Robbins 2000). The first reason for this is the fact that many, if not most, natural resource management schemes contain some sort of collective action problem (Ostrom 1990; Agrawal 2003; Ostrom 2005).

Problems related to corruption (e.g., nepotism, rent-seeking behavior, graft, and embezzlement) are known from case studies to be highly difficult obstacles for achieving sustainable usage of natural resources (Robbins 2000; Kolstad and Søreide 2009; Miller 2011). The well-known scenario of bribes paving the way for clear-cutting of protected rain forests, or poaching of endangered species such as elephants and rhinos made possible by park wardens turning a blind eye, are illustrations of the belief that corruption is a key problem in many resource management undertakings. Whenever there is a value attached to a public good type of natural area or resource, there are potential problems with corruption that need to be addressed for the resource to attain long-term sustainability. Several authors have made this assertion on theoretical grounds (e.g., Robbins 2000), and there are also a few empirical studies trying to assess the evidence of such a linkage. The evidence from large-N studies of aggregate data is inconclusive: Duit et al. (2009) were unable to discover any significant correlations between biodiversity management and institutional quality in a sample of twenty European countries, but this conclusion is limited to a single-species indicator from a small number of observations. Smith et al. (2003) demonstrated a link between corruption and measures of biodiversity management (such as forest cover loss, species richness, and endangered species protection), suggesting that conservation programs are less effective in corrupt countries. Studies by Welsch (2004), Pellegrini and Gerlagh (2006), and Esty and Porter (2005) all found that corruption was negatively related to environmental policy stringency or emission reductions. On the other hand, Morse (2006) found that corruption had only a marginal effect on environmental performance if the effect of economic development was taken into account.

Evidence from case studies points more conclusively to substantial problems caused by corruption for management of local natural resources (see the reviews by Kolstad and Søreide (2009) and Robbins (2000), as well as chapter 9 and chapter 10, "Enforcement and Compliance in African Fisheries: The Dynamic Interaction between Ruler and Ruled"). Corruption is identified many times as one of the mayor obstacles to sustainable resource management, and it is a clear conclusion from many case studies that that corruption can play a very negative role in resource management, but the question remains if this is also a general effect found across an extensive set of observations.

There are several indicators of corruption available for cross-national comparisons (Svensson 2005). One of the most widely used is Transparency International's Corruption Perception Index (CPI), which uses a combination of public and expert surveys to arrive at a measure of corruption in the public sector. As the name indicates, it measures the perception of how widespread and common corruption is in a country, rather than the actual rate of instances of corruption. A consequence of the survey approach is that the CPI does not distinguish between different corruption in different levels of society (high-level policymakers versus street-level officials), nor does it make a distinction between different forms of corruption (e.g., political versus administrative). These are limitations, of course, but the fact that most other corruption indicators are also perception based, as well as the CPI's broad coverage (178 countries), nevertheless makes that index the best available cross-national estimate of corruption.[1]

The prediction derived from theory, therefore, would be that corruption affects stakeholder participation negatively. As more resourceful actors are able to buy influence in corrupt societies, stakeholders will be less inclined to act collectively within the BR framework because their collective efforts might be trumped at any time by local strongmen, companies, and other powerful interests. In addition, more corruption will have a negative effect on management effectiveness of BRs. A BR in a more corrupt setting will be less able to protect its resources from over-exploitation, as a consequence of both corrupt external legal and political institutions and corruption within its own ranks.

Political Rights

Stakeholder participation involves a large element of collective action in the sense that it requires stakeholders to organize themselves, both internally and in relation to the BR. Even if the costs of organizing collective action are assumed to a certain extent by the BR organization (cf. Olson 1965), stakeholders nevertheless need to enter into some sort of organized collective action with other groups of stakeholders. A fundamental prerequisite for achieving collective action in natural resource preservation, therefore, is that citizens' basic rights to organize and act collectively are recognized and respected by other, often more powerful actors in the surrounding political system. Many times, natural resource management and protection comprise a contested activity, which means that if stakeholders lack even basic formal rights to organize and act politically, participation is likely to be lower or even nonexistent. The theoretical prediction is thus that better protection for political rights has an impact on an increased likelihood of stakeholder participation in resource management programs, as well as an effect on environmental quality in the natural area in which participation is taking place.

There are several different types of political rights, but in the context of stakeholder participation, it is more relevant to focus on the right to form associations and organizations. The right to form political organizations and associations freely lies at the heart of any polity that aspires to be perceived as democratic. As many stakeholder groups have their roots in civil society, a very basic requirement for more extensive stakeholder participation is that groups in civil society are granted the right to organize themselves without having to fear reprisals and persecution. Other aspects of democracy (e.g., freedom of speech, representation, etc.) are also important for environmental issues, but the right to organize is likely to be especially important in the case of stakeholder participation in environmental conservation schemes. A few previous large-N studies on the effect of democracy and political rights on environmental management paint a scattered and sometimes contradictory picture. Torras and Boyce (1998) found that civic liberties had a strong effect on water and air quality in developing countries. Midlarsky (1998), on the other hand, found that democracy was negatively correlated to deforestation, CO_2-emissions, and soil erosion; not related to freshwater availability and soil erosion by chemicals; and positively linked to protected land areas. In a more recent study, Pellegrini and Gerlagh (2006) demonstrated that democracy did not affect environmental policy stringency significantly, but that corruption had a detrimental effect on it.

Turning to case study findings, most studies in this design format focus on the interplay between democracy, decentralization, and natural resource management (cf. Ribot 2003, 2006; Andersson and van Laerhoven 2007). The main message from this literature is that decentralization and local democracy must be combined with actual transfers of competencies and accountability to local political actors if the benefits of stakeholder participation are to be realized. In general, green theorists often have placed great expectations on stakeholder participation as a way of achieving environmental objectives through grassroots activism (Dryzek 2005). Less common are studies that seek to investigate empirically the role of democracy for stakeholder participation. Recent review articles of stakeholder participation in general (Irvin and Stansbury 2004), and of stakeholder participation in natural resource management (Reed 2008) and environmental management (Koontz and Thomas 2006) in particular, make no mention of studies that have tried explicitly to assess the role of political rights in bringing about stakeholder involvement.

Hypotheses, therefore, must be formulated without clear guidance from previous research. One theoretically derived prediction is nevertheless that democracy, in the form of protection of the right to organize, can be expected to have a positive effect on stakeholder participation. The easier it is for stakeholders to organize themselves, the more likely they are to get engaged in BRs. As in the case of corruption, there are also reasons to expect political rights to have a direct effect on management effectiveness. Political rights are a prerequisite for collective action to challenge unsustainable resource use practices outside as well as inside the BR domain, and therefore we should expect a positive correlation between political rights and management performance. Data for political rights are taken from Freedom House's "Freedom in the World" data set. Specifically, we use the indicator "Associational and Organizational Rights" for 2005, which varies from 0 = worst to 12 = best.[2] The Freedom House data set has been criticized for being opaque and unreliable (Hochstetler 2012), but alternative data sets (e.g., Polity IV) do not contain estimates of organizational rights.

Data and Methods

Unit of Analysis—the World Network of BRs

There are currently 564 "Man and the Biosphere" reserves allocated in 109 countries. The "Man and the Biosphere" program was initiated by the United Nations Educational, Scientific, and Cultural Organization (UNESCO) in 1976. In the first two decades, BR sites were selected primarily for their scientific and natural value, but a new Statutory Framework was adopted in 1995 that stated that BRs are expected to fulfill three main functions: (1) to conserve biodiversity, (2) to foster sustainable social and economic development, and (3) to support research, monitoring, and education. The redirection of the BR mission meant that increased stakeholder involvement became a key task for already existing BR areas, as well as a strong selection criterion when awarding status as a BR to new conservation areas. Although the legal status of BRs is determined by national legislation, all BRs consist of a core area, in which biodiversity is preserved, and a surrounding buffer area, in which activities aiming at enhanced social and economic development are conducted. In addition, all BRs are required to adhere to guidelines expressed in the Statutory Framework. Taken together, this common regulatory framework means that BRs are similar enough to permit meaningful comparisons, while at the same time exhibiting a considerable degree of variation on the dependent variable stakeholder participation.

Survey Data
A survey of a population of 564 BR areas was carried out in 2008 using an online questionnaire, which proved to be a highly effective way of distributing and collecting surveys in a population consisting to a large part of respondents situated in remote areas. To get comparable information from a large set of cases, a self-administered questionnaire was developed that targeted coordinators, directors, and managers of BRs. The questionnaire was tested, revised, and uploaded for online access in English, French, Spanish and Chinese. In addition, hard copies were distributed extensively at the 3rd World Congress of Biosphere Reserves held in Madrid in February 2008 to compensate for the fact that 124 of the 531 BRs could not be reached via e-mail.

The questionnaire was designed to gather information about three main aspects of the BRs The first aspect is patterns of stakeholder involvement, defined as the number of different stakeholder categories (e.g., local farmers, indigenous people, resource users) participating in various BR functions (e.g., decision making, monitoring, implementation). The second aspect is management regimes, and this section of the survey sought to estimate the extent to which the BR employed a style of management that could be classified as either adaptive management or traditional conservation, as well as whether the BR was pursuing local development objectives in parallel with conservation efforts. The third aspect measured by the survey was self-assessed effectiveness in reaching sustainable development and biodiversity conservation goals, as well at the BR's perceived readiness for dealing with surprises and external shocks. The survey worked well for assessing participation patterns and management regimes, but using self-assessments as the chief tool for measuring performance effectiveness in biodiversity conservation is likely to generate a certain amount of bias. Combining the survey approach with remote sensing data on changes in habitats and biomass is an attempt at overcoming this bias while retaining a large-N research design.

Remote Sensing of Biomass Changes

In the second step of the analysis, we seek to relate patterns of stakeholder involvement in BR activities to the BR's management performance. Shedding light on this relationship requires measurements of management performance of the BRs. We have sought to respond to this need by including two separate measures of conservation effectiveness: (1) "hard" estimates of biomass increases in the BRs during the period 1996–2004, and (2) "soft" estimates, based on BR managers' self-assessment of effectiveness in conserving biodiversity, collected through a global BR survey.

A measure of BR management performance is derived from satellite data estimating relative changes in biomass (expressed in terms of percentage) as given by the Normalized Difference Vegetation Index (NDVI) within BR areas between 1996 and 2004. The NDVI index uses the fact that living plants absorb sunlight in specific wavelengths in the infrared spectrum to develop an indicator of biomass changes in landscapes. To account for seasonal differences in biomass, NDVI values were calculated for each month of the year and averaged. Each month has two composites (day 1–15 and 16–30); accordingly, each year was comprised of twenty-four NDVI composites. For all BR-areas, minimum and maximum values of NDVI were calculated, as well as the mean and standard deviations. As the precise extent was not available for most areas, circular buffer zones were created and used for calculating statistics. Centroids of most MAB-areas could be identified, generally, based on two or four available coordinate pairs. The radius of buffer zones was set to 0.16 decimal degrees to avoid contamination from surrounding non-MAB pixels. A water mask was added to the data set to exclude coastal areas where the buffer zone fell into the sea. BRs that were too small for remote sensing analysis (predominantly consisting of water surfaces), or were located in multiple countries, were excluded for the NDVI analysis, which brought down the total number of BRs in this part of the analysis to 109 (see chapter 9 for a similar use of satellite imagery).

The NDVI varies between −1 and +1, and higher values indicate larger amounts of biomass in the ecosystem. In practice, the NVDI rarely goes below 0. Very low values on the index (< 0.01) are found in areas with little or no vegetation (e.g., sand and snow); grasslands and shrubs typically result in values around 0.2, and tropical rain forest areas have values between 0.6 and 0.8.

The lowest index value (0.07) in our sample was found in the Wadi Allaqi Biosphere Reserve in the Egyptian desert; and the highest score (0.83) was found in the Croajingolong Reserve in northern Australia. The mean index score for 2004 in our sample was 0.43, with a standard deviation of 0.15. Most BR areas showed only small upward changes in the NDVI; the mean rate of change was an increase of 0.8 percent. The range goes from a −20.9 percent to a 19.1 percent change in the NDVI, and the standard deviation is 7.6 percent. This fact indicates that, despite the overall trend of relative stable levels of biomass, there are BRs that have experienced rather drastic changes in their natural systems. Based on satellite data, changes in the NDVI can be seen, to a certain extent, as a hard measure of BR management performance NDVI over time.

There are, nevertheless, a number of caveats to this assertion. First of all, the NDVI measures quantitative changes in total biomass, but not qualitative changes in the composition of the landscape ecology. In most cases, and especially within natural reserve areas, an increase in biomass can be interpreted as a sign of a well-functioning ecosystem. Moreover, increasing biomass can, according to the same logic, be assumed to be beneficial for most creatures living in the BR ecosystem, other than plants and trees. But it is also possible that some processes leading to an increase in overall biomass are not beneficial, especially if the management objective is to maintain the ecosystem within the boundaries of its stability domain. A second caveat about using the NDVI as an indicator of BR management performance is that not all changes in biomass are the result of management practices; rather, some of them might be caused by random factors such as variations in local climate, precipitation patterns, and pest outbreaks, against which management practices are more or less powerless. To the extent that such events have occurred in our sample of BRs during the study period, they are very likely to be randomly distributed and therefore not a source of systematic bias.

Our second measure of BR management effectiveness is based on BR manager's self-assessments of how "efficient [the BR] is in fulfilling BR objectives." Specifically, the survey respondents were asked to rate the performance of their BR for seven different dimensions[3], one of which was "conserving biodiversity." Because there are reasons to assume that managers might give an inflated assessment of their own management performance, we sought to address this bias by subtracting the biodiversity conservation performance rating from the mean rating of the remaining six performance dimensions. The resulting conservation effectiveness score is thus a measure of how BR managers rate their ability to conserve biodiversity relative to other management objectives. Although still subject to the problem of inflated self-assessment, this procedure is assumed to produce a better estimate of *relative* conservation effectiveness.

Controls

Two sets of control variables were included in all models. The first set of variables consisted of controls for background differences in BR size (hectares, natural log), BR age (years since designation as a BR area, base year 2008), and a dummy variable indicating whether the BR has been described by the respondent as consisting of a predominantly pristine (1) or cultural (0) landscape. We also included controls for overall management objectives of the BR by including two factor regression score variables based on a set of seven survey items intended to measure different aspects of BR management priorities (N = 143).[4] The first factor (eigenvalue 2.78; explained variance 39 percent) reflects the extent to which the BR prioritizes research and educational goals, whereas the second factor (eigenvalue 1.987; explained variance 28 percent) relies strongly on items that are associated with the promotion of social and economic development as the BR's main objective. By controlling for differences in management priorities, we aim to hold constant any effects of management strategies that actively seek to include stakeholders. Finally, in the models estimating management effectiveness, we included the mean NDVI score for the period 1996–2004 as a control for differences in landscape and ecosystem types. This control was not deemed relevant for the stakeholder participation models, and therefore was not included in those. On the national level, controls for overall levels of development and wealth, such as the World Bank's Human Development Index (HDI) and population density (in persons per square kilometer, from 2005), were included to control for the possibility that economic prosperity and population pressure might influence stakeholder participation and resource management in the BRs.[5]

What Is a Stakeholder?

Despite the fact that "stakeholder" is a key actor category in many contemporary theories and management paradigms, surprisingly little attention has been paid to finding an adequate and exhaustive definition of what a stakeholder actually is and how to recognize one. In the case of adaptive management and adaptive comanagement, the term "stakeholder" is usually attributed to any recognized group of individuals who have some form of claim in the natural area or resource being managed. This is, however, a very vague definition that leaves a number of questions unanswered: Are only civil society actors to be counted as stakeholders, or can other categories, such as business representatives and administration officials, be included in the tally? Should we count subjective or objective claims (or both) when deciding which groups have legitimate claims in the area?

It is not uncommon in studies of natural resource management for stakeholders to be identified on an ad hoc basis (Reed et al. 2009), often by simply defining a stakeholder as an individual or group that is interacting visibly with the resource or the resource governance system. This approach to identifying stakeholders runs the obvious risk of biasing the analysis to place uneven emphasis on salient and resourceful groups that are highly visible in the interaction with the resource, while underestimating less-resourceful actor groups. Or, as Roger Kasperson puts it:

Stakeholders in current parlance are usually local actors who have a clearly defined role in the planning or decision process or who have material interest (or "stake") in the outcome. In other words, they are local elites. (Kasperson 2006).

Thus, the problem of defining and identifying stakeholders in an empirical setting is not easily solved a priori. In-depth participatory approaches often can make more valid inferences about the true population of stakeholders in a given empirical setting, but this is not a feasible approach for a large-N study. In designing our survey, we therefore opted for a strategy that focused on determining the presence of a larger number of predetermined categories of stakeholders in different functions within the BR. This approach does not eliminate the problem with identifying stakeholders mentioned above, but by using a relatively detailed (and reasonably exhaustive as a result) list of stakeholder categories, the problem is nevertheless brought under a degree of control. Neither does this approach allow us to detect stakeholder groups that are not involved in BR activities but nevertheless are interacting with the BR ecosystem. Such groups are more likely to be resource-poor and directly dependent on the ecosystems for the subsistence that stakeholder groups that are partaking in BR management activities, which again introduces a possible but largely unavoidable bias toward local elites (or at least organized interests) in our sample.

Estimating Stakeholder Participation

The survey questionnaire gathered data for participation among six stakeholder categories: (1) scientists; (2) nongovernmental organizations (NGOs) and volunteers; (3) local farmers, fishers, foresters, and hunters; (4) other people living or working in the BR; (5) local governmental administrators related to conservation; and (6) local governmental administrators not related to conservation. For the analysis, categories 3 and 4 were collapsed into a category labeled "Locals," and categories 5 and 6 were merged into a category called "Local administration."

The activity of stakeholder groups were recorded for a total of seven BR functions: (1) representation in the BR coordination team, (2) representation in the BR steering committee/advisory board, (3) involvement in goal setting for the BR, (4) involvement in designing BR projects, (5) involvement in implementing projects or management measures, (6) involvement in day-to-day management efforts, and (7) involvement with monitoring biodiversity changes or other changes in the BR's ecosystem. The first four functions were classified as involvement in decision-making activities within the BR, and the remaining three functions were called implementation activities. The reason for making a distinction between these two activities is that they reflect different levels of engagement and influence on the part of the stakeholder. Involving stakeholders in decision-making activities in the BR means that stakeholders are given a large degree of influence and power over the entire BR, whereas involving stakeholders in implementation activities normally implies a more limited degree of influence. This means that we expect stakeholder participation in decision making to be the hardest to achieve, but also to produce the strongest effects on management outcomes. The theoretical prediction is thus that stakeholder participation in decision making should be affected more strongly by institutional factors than participation in implementation, but also to have a greater effect on management outcomes.

A single measure reflecting the overall level of stakeholder participation in a BR was constructed by dividing the number of stakeholder categories listed as participating in BR functions (four) with the total number of functions (seven). Although not an estimate of the exact number of individual stakeholders involved in the BR, it nevertheless gives a sufficiently valid impression of the level of stakeholder involvement. The average score on the stakeholder participation index for all 144 BRs was 3.07, which indicated that at least one stakeholder category was involved in about three BR functions. A total of 13 BRs reported no stakeholder

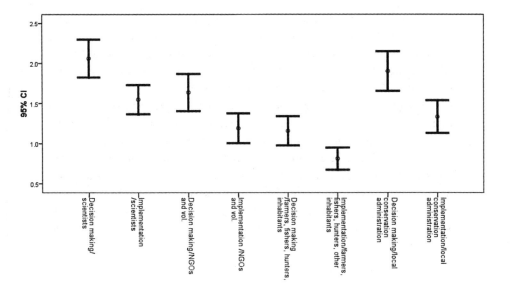

Figure 11.1

Stakeholder participation in BRs

Note: Shown here are mean participation scores and mean numbers of stakeholder categories participating in decision-making and implementation functions in BRs. N = 143.

involvement, and the highest recorded level of involvement in a BR was a stakeholder participation score of 7.71. Moving beyond the average level of stakeholder participation revealed some interesting differences in participation patterns between categories of stakeholders.

As shown by figure 11.1., there are marked differences between stakeholder categories in their participation patterns. The most common stakeholder category present in BR decision-making functions was scientists, closely followed by officials from local governmental administrations. On average, scientists were listed as participating in 2.06 out of 4 possible BR decision-making functions, whereas local administrators were present in 1.90 functions. NGOs and other volunteers participated to a somewhat lesser degree: this stakeholder category was present in 1.60 decision-making functions and in 1.36 implementation functions, on average. The lowest average level of participation was found among local resource users and inhabitants: this category was represented in only 1.15 decision-making functions and in 0.81 implementation functions.

Scientists and local administrators were thus the most common stake-holder categories to be found within BRs, whereas local inhabitants and resource users was the stakeholder category least likely to participate in BR activities. Given this difference in influence between the two catego-ries of BR functions, a reasonable assumption is that stakeholder partici-pation will tend to be more frequent in implementation than in decision making. Somewhat surprisingly, this prediction is not corroborated by the data. All stakeholder categories seem to be more involved in decision making and goal setting than in implementation and day-to-day manage-ment in the BRs. This might not reflect actual power relationships and levels of influence for stakeholders in BR decision-making processes, but it is nevertheless interesting to note that stakeholders more frequently seem to take the role of decision makers, rather than administrators and caretakers.

Results

The first part of the analysis investigates the role of institutional context for stakeholder participation. As the data for stakeholder participation is essentially count data (i.e., the number of stakeholder categories present in BR activities), a standard Poisson regression model was employed. BRs are grouped by countries, and to control for intra-class correlations stan-dard errors clustered by countries were used in all models.

Starting with the case of stakeholder participation in BR *decision-making* functions, the results displayed in table 11.1 show that scientists' participation in BR decision-making activities tends to be higher in BRs that prioritize research and educational activities over the promotion of sustainable development. The involvement of NGOs and local inhabit-ants and resource users in decision-making activities is positively influ-enced by a BR focus on sustainable development. Locals also tend to be more involved in younger BRs. The participation of local administration in BR decision making is more likely in BRs described as consisting of a cultural landscape. To sum up, the level of corruption or the protection of political rights in the country in which the BR is situated does not seem to affect the propensity of any stakeholder category to participate in BR decision making.

Participation in *implementation* activities displays a somewhat dif-ferent pattern. The participation of scientists and local administration is not affected by institutional contexts or by BR-specific controls (with the exception of local administrators, who participate significantly more

Table 11.1
Stakeholder participation

	Decision Making				Implementation			
	Scientists	NGOs	Locals	Local administration	Scientists	NGOs	Locals	Local administration
BR priority: Sustainable development	0.910	1.152*	1.250**	1.017	1.005	1.180*	1.297**	1.085
BR priority: Research and education	1.131*	0.921	1.019	0.940	1.109	0.934	0.928	0.965
BR age	0.992	0.994	0.980*	0.991	0.999	0.997	0.992	0.999
BR size (log)	0.988	1.057	0.970	1.000	0.973	1.003	0.997	0.946
Pristine (dummy)	0.889	0.842	0.980	0.733*	0.906	0.742	0.743	0.694*
Human development index	0.406	0.932	0.317*	0.792	1.490	0.734	1.922	0.338
Population density	0.999	1.000	0.998	0.999	1.000	0.999	0.999	0.999
Corruption	1.040	0.987	1.090	0.963	1.007	0.977	1.024	1.081
Political rights	0.970	1.054	0.976	0.958	.974	1.114*	.931**	0.966
Wald X^2	26.020**	17.640*	37.650***	44.550***	8.330	21.11**	28.04***	16.05

* $p < 0.05$; **, $p < 0.01$; ***$p < 0.001$. N = 143.
Table shows incident rate ratios. Poisson regression with clustered standard errors [clustered by countries (49)].

frequently in BRs situated in cultural landscapes), which indicates that the involvement of these stakeholder categories is largely idiosyncratic, or at least that they are determined by factors outside the model. However, NGOs participate more in implementation activities in BRs that are prioritizing sustainable development and are situated in countries that offer better protection for political rights. Local resource users are also more prone to participate in BRs that prioritize sustainable development, but in contrast to NGOs, they are negatively affected by a stronger institutional protection of the right to organize politically.

Judging from the results displayed in table 11.1, there is only, at best, a weak effect of institutional context on the propensity of stakeholders to get involved in conservation programs. However, in the course of testing the robustness of our model, we divided our sample of BRs into two groups based on the World Bank's classification of income levels of different nations.[6] Running the models in table 11.1 on two groups of countries—one consisting of sixty-seven high-income countries and another containing seventy-six low-, lower-middle-, and upper-middle-income countries—revealed some interesting findings. Estimates for scientists and local administration remained more or less unaffected as compared to the full sample, but participation of NGOs and locals exhibited some interesting differences between high- and low-income countries. In the full-sample model, participation of locals in implementation activities displayed a somewhat puzzling negative relationship to levels of political rights. When estimated in the group of medium- and low-income countries, this effect remained and was even strengthened (IRR 0.881; $p > 0.001$), but in the high-income group, there was instead a positive but insignificant relationship[7] between better protection for political rights and local stakeholders' involvement in implementation.

In addition, in the full model, there was a strong effect of a sustainable development management priority on the number of local stakeholders engaged in the BR. This effect disappeared in the low-income group but remained in the high-income sample, which indicates that the effect of management priorities (or perhaps the meaning of management priorities) was affected by levels of economic development as well. A similar but partially reversed pattern can be discerned for NGO participation in implementation. Political rights seemed to increase significantly NGO involvement in BRs situated in low-income countries (IRR 1.15; $p > 0.01$), but this effect disappeared in high-income countries. The positive effect of sustainable management priorities found in the full-sample model was absent in both low- and high-income models, but instead, a negative effect of research and education management priorities was revealed in the high-income sample.

At first glance, this could be interpreted as a crowding-out effect in which international developmental NGOs squeeze out local stakeholders in low-income BRs, but there was no negative correlation between participation rates of NGOs and local stakeholders to support this explanation. Another possible interpretation of this result is that local stakeholders are more prone to get involved in the absence of a well-functioning and democratic political system through which they can articulate their wishes and needs regarding the management of local natural resources. It might be the case that BRs can take on the function of "safe havens" for local stakeholders who lack other venues for expressing their concerns and needs regarding natural resources. Through their connection to international organizations such as UNESCO and the global network of other BRs, the BR may be able to offer some level of protection from local powers, as well as function as a forum for expressing views and wishes about how, from the perspective of the local stakeholder, vital natural resources should be managed. The function of a safe haven is likely to diminish as the quality of democratic institutions increases. This explains why the negative correlation between participation and political rights appears only in the group of low-income countries, as they are more likely to have less well-functioning democratic institutions.

Turning to the issue of the effects of stakeholder participation on conservation effectiveness, table 11.2 indicates a sharp difference between the objective estimates of biomass changes derived from satellite imagery and the subjective data based on BR managers' self-estimates of the BR's ability to conserve biodiversity. There is no correlation between the two dependent variables, which indicates that they are, at best, reflections of two separate dimensions of management performance. Political rights have a consistent effect on the NDVI in both the decision-making and implementation models, which indicates that better protection of political rights has an independent effect on management effectiveness of BRs. Political rights are also positively and significantly linked to higher scores on the conservation effectiveness index in the decision-making model. In the implementation model, the corresponding effect has roughly the same strength, but it fails by a small margin to achieve statistical significance ($p > 0.06$). Political rights thus have a significant effect on both indicators of management performance. In sum, it seems that protection of political rights is an important institutional factor influencing the effectiveness of ecosystem management.

Table 11.2
Effects of stakeholder participation

	Decision Making		Implementation	
	NDVI	Self-assessment	NDVI	Self-assessment
Scientists	−0.245	0.164	−1.331	0.024
NGOs	−0.189	−0.256	0.627	−0.250
Locals	−0.320	−0.073	1.938**	0.018
Local administration	0.422	0.187	0.618	0.110
BR priority: Sustainable development	−0.556	−0.357	−1.079	−0.435
BR priority: Research and education	−1.670	−0.006	−1.226	0.002
BR age	−0.055	0.024	−0.033	0.020
BR size (log)	−0.632	−0.042	−0.650	−0.059
Pristine (dummy)	−1.913	0.168	−1.359	-0.008
Mean NDVI score 1996–2004	−1.300	−2.175	−2.985	−2.309
Human development index	5.257	1.197	5.965	0.644
Population density	−0.015	0.001	−0.013	0.001
Corruption	−1.139*	−0.289	−1.340*	−0.297
Political rights	0.818*	0.233**	0.906*	0.219*
N	109	92	109	92
F	2.84**	1.79	3.40**	2.43**
Adj. R^2	0.13	0.25	0.30	0.24

*$p < 0.05$, **; $p < 0.01$; ***$p < 0.001$. N = 143.
Table shows unstandardized regression coefficients. OLS regression with clustered standard errors [clustered by countries (44–49)].

On the other hand, the models do not confirm theoretical predictions about the role of corruption. This variable shows an equally consistent negative effect on the NDVI, which indicates that lower levels of corruption bring down management performance (as reflected by changes in biomass in the BR). This is a puzzling finding, but it might be explained by the fact that higher levels of corruption also preclude a more rational and efficient exploitation of natural resources (cf. Barrett et al. 2006), which in turn places less pressure on natural resource systems, as well as on conservation areas. A weak but negative effect between the corruption index and the biomass index can be found even in a zero-order correlation ($r = -0.09$; $p = 0.37$), which indicates that the effect is not a model artifact.

The only stakeholder category exhibiting a significant effect on management performance is locals involved in implementation activities in the BR. Having a higher rate of local inhabitants and resource users participate in the implementation of BR policies is significantly associated with increases in the biomass index over time. The effect is relatively strong: involving a stakeholder in an additional BR implementation function is associated with an average increase in the NDVI by 1.9 percent. No other stakeholder category has any effect on either of the dependent variables, although scientists' participation in in implementation is negatively and almost significantly ($p < 0.06$) linked to the biomass index.

Conclusions

This chapter has analyzed the causes and consequences of stakeholder participation in 143 BRs in fifty-five countries. Starting with the question of what causes stakeholder participation in natural resource management programs, the findings from the present analysis point to a pivotal role for political rights, together with a less pronounced effect of corruption. Political rights, or more specifically the right to organize and form associations, do seem to have an relatively strong effect on stakeholder participation. However, this effect is confined to specific categories of stakeholders participating in certain BR functions and thus is not applicable across the entire spectrum of stakeholders. The effect appears only in the case of NGOs and local stakeholders participating in BR implementation activities, but here it is about as strong as the effect of specific management policies of the BR. In other words, institutional guarantees for political rights are about as important as a BR policy prioritizing combined social, economic, and ecological development in persuading NGOs and locals to get involved in BR implementation activities.

The overall beneficial role of political rights in facilitating stakeholder participation is in line with theoretical predictions, but the analysis also revealed an unexpected negative correlation between political rights and participation in low-income countries. This was interpreted as a reflection of a safe haven function of BRs in countries with weak democratic and legal institutions. In such a setting, a conservation organization, such as a BR with extensive contacts to a global community directed by an international organization, is able to provide an arena for voicing claims about the usage and management of important natural resources, and perhaps also offer a degree of protection against more powerful actors in the local context. This finding is relevant to the (mostly theoretical) discussion on the role of multilevel institutions in natural resource management, in which a core argument is that local conservation organizations often can derive a great deal of support and resources from becoming linked to institutions on higher administrative scales (Winter 2006; Pahl-Wostl 2009). The results from this study indicate that there might be something to this idea, but with the qualifier that the importance of multilevel institutions seems to be more pronounced in low-income countries with weak democratic institutions.

In contrast, the level of corruption does not appear to affect the involvement of any category of stakeholders. This is a somewhat surprising finding in light of the literature, from which a relatively consistent prediction would be that corruption has a detrimental effect on the willingness and ability of stakeholders to get involved in collaborative management efforts. This assumption does not find support in the BR data, and therefore the conclusion must be that stakeholders are just as prone to participate in BR natural resource programs in low- and high-corruption countries. It is, however, important to point out that this conclusion is limited to stakeholder participation in BR programs, and that other types of collaborative resource management programs might be more susceptible to the negative effects of corrupt institutions. For instance, as demonstrated in chapter 10, the impact of a corrupt state is quite different when it is the state that enacts rules and is responsible for monitoring compliance.

Another anomalous finding is the relatively strong and negative effect of corruption on changes in BR biomass. BRs situated in less corrupt countries tend to have lower rates of biomass growth over time. This finding goes against much of the literature, which predicts a positive effect of the absence of corruption on sustainable management of natural resources. One more obvious explanation for this finding—that biomass growth is more rapid in tropical and subtropical countries that also tend to be more corrupt—is not viable because the NVDI controls for differences in productivity between ecosystems. As argued above, one possible explanation is that corruption, in addition to aiding unlawful exploitation of natural resources, also precludes more rational, efficient, and large-scale resource utilization.

Turning to the issue of consequences of stakeholder participation in BR programs, the general pattern is again an absence of strong effects, but with one notable exception: involving local stakeholders in the implementation phase of BR programs is associated with significant increases in the biomass index. In other words, BRs that involve local stakeholders in implementation activities tend to experience higher levels of biomass growth over time. Equally noteworthy is the fact that the involvement of any other stakeholder category is not linked to management outcomes. One interpretation of this finding would be that the participation of any stakeholder groups other than local inhabitants and resource users is redundant in the sense that it does not improve conservation performance. This is, however, too drastic a conclusion—local stakeholders might be the only ones to have a direct effect on management outcomes, but that does not mean that the participation of other stakeholder categories does not contribute to things such as increased legitimacy and better information about local conditions (cf. Schultz et al. 2011).

The analysis presented in this chapter has two main conclusions. First, it was shown that institutional context, in terms of protection of democratic rights, does seem to stimulate the participation of those stakeholder categories that are usually considered the hardest to get involved: local resource users and inhabitants. The exact mechanisms underlying this result need to be investigated more thoroughly, but one possibility is that BRs are able to provide a safe haven for those stakeholder groups that are the most vulnerable to a low level of protection of political rights. By extension, this finding, along with corresponding results from chapters 9 and 10, also highlights how local-level natural resource management programs are never carried out in isolation from surrounding political and institutional contexts. Second, we found that the only stakeholder category whose participation can be linked to management effectiveness is local stakeholders involved in BR implementation activities. However, as no effect could be discerned for locals participating in decision-making activities, this finding can be seen as only a partial corroboration of theoretical predictions about the role of stakeholder participation.

Notes

1. Data for the CPI was collected from the Quality of Government (QoG) data set. Estimates for CPI are the average of the index scores during the period 2000–2007. The index ranges between 10 = "highly clean" and 0 = "highly corrupt." See Jan Teorell, Nicholas Charron, Marcus Samanni, Sören Holmberg, and Bo Rothstein, the Quality of Government data set (version 27), May 10, 2010, University of Gothenburg, the Quality of Government Institute, for data description and additional details.

2. The Freedom House data was collected from the QoG data set. See note 1.

3. The response alternatives were "conserving biodiversity," "fostering social development," "fostering economic development," "providing logistic support to research," "providing logistic support to monitoring," "providing logistic support to education," "facilitating dialogue, collaboration, and integration of different objectives." All dimensions were rated on a scale running from 1 = "very poor" to 10 = "very good."

4. A total of seventeen missing values in these seven variables were imputed by regressing each variable for the other six.

5. The HDI index was taken from the QoG data set, and population density estimates were collected from the UNdata website: United Nations, Department of Economic and Social Affairs, Population Division, World Population Prospects (2008 revision), New York, 2009.

6. World Bank. (2008). Country classifications. URL: http://data.worldbank.org/about/country-classifications/country-and-lending-groups.

7. This effect in the high-income subsample is in fact significant, but only due to one extreme outlier case (United Arab Emirates), which has a high income level but very low levels of political rights. Excluding this case from the model renders the positive effect of political rights on locals' participation in implementation insignificant.

References

Agrawal, Arun. 2003. Sustainable Governance of Common-Pool Resources: Context, Methods, and Politics. *Annual Review of Anthropology* 32:243–262.

Alexander, George. 2002. Institutionalized Uncertainty, the Rule of Law, and the Sources of Democratic Stability. *Comparative Political Studies* 35(10): 1145–1170.

Andersson, Krister, and Frank van Laerhoven. 2007. From Local Strongman to Facilitator: Institutional Incentives for Participatory Municipal Governance in Latin America. *Comparative Political Studies* 40(9): 1085–1111.

Armitage, Derek, Fikret Berkes, and Nancy Doubleday. 2007. *Adaptive Co-Management: Collaboration, Learning, and Multi-Level Governance*. Vancouver: University of British Columbia Press.

Barrett, Christopher B., Clark C. Gibson, Barak Hoffman, and Mathew D. Mc-Cubbins. 2006. The Complex Links between Governance and Biodiversity. *Conservation Biology* 20(5): 1358–1366.

Dryzek, John S. 2005. *The Politics of The Earth*. Oxford: Oxford University Press.

Duit, Andreas. 2011. Natural Resources. In *International Encyclopedia of Political Science*, ed. Bertrand Badie, Dirk Berg-Schlosser, and Leonardo Morlino. London: Sage.

Duit, Andreas, Ola Hall, Grzegorz Mikusinski, and Per Angelstam. 2009. Saving the Woodpeckers: Social Capital, Governance, and Policy Performance. *Journal of Environment & Development* 18(1): 42–61.

Esty, Daniel C., and Michael E. Porter. 2005. National Environmental Performance: An Empirical Analysis of Policy Results and Determinants. *Environment and Development Economics* 10(4): 391–434.

Hochstetler, Kathryn. 2012. Democracy and the Environment in Latin America and Eastern Europe. In *Comparative Environmental Politics: Theory, Practice, Prospects*, ed. Paul F. Steinberg and Stacy D. VanDeveer, 199–230. Cambridge, MA: MIT Press.

Irvin, Renee, and John Stansbury. 2004. Citizen Participation in Decision Making: Is It Worth the Effort? *Public Administration Review* 64(1): 55–65.

Kasperson, Roger E. 2006. Rerouting the Stakeholder Express. *Global Environmental Change* 16(4): 320–322.

Kolstad, Ivar, and Tina Søreide. 2009. Corruption in Natural Resource Management: Implications for Policy Makers. *Resources Policy* 34(4): 214–226.

Koontz, Tomas M., and Craig W. Thomas. 2006. What Do We Know and Need to Know about the Environmental Outcomes of Collaborative Management? *Public Administration Review* 66:111–121.

Lee, Kai N. 1999. Appraising Adaptive Management. *Conservation Ecology* 3(2): 3.

Midlarsky, Manus I. 1998. Democracy and the Environment: An Empirical Assessment. *Journal of Peace Research* 35(3): 341–361.

Miller, Michael J. 2011. Persistent Illegal Logging in Costa Rica: The Role of Corruption among Forestry Regulators. *Journal of Environment & Development* 20(1): 50–68.

Morse, Stephen. 2006. Is Corruption Bad for Environmental Sustainability? A Cross-National Analysis. *Ecology and Society* 11(1): 22. http://www.ecologyandsociety.org/vol11/iss1/art22.

Nabahan, Gary Paul. 1997. *Cultures of Habitat: On Nature, Culture, and Story*. Washington, DC: Counterpoint.

Nelson, Fred, and Arun Agrawal. 2008. Patronage or Participation? Community-Based Natural Resource Management Reform in Sub-Saharan Africa. *Development and Change* 39(4): 557–585.

North, Douglass C. 1990. A Transaction Cost Theory of Politics. *Journal of Theoretical Politics* 2:335–367.

Olson, Mancur. 1965. *The Logic of Collective Action: Public Goods and the Theory of Groups.* Cambridge, MA: Harvard University Press.

Olsson, Per, Carl Folke, and Tomas Hahn. 2004. Social-Ecological Transformation for Ecosystem Management: The Development of Adaptive Co-management of a Wetland Landscape in Southern Sweden. *Ecology and Society* 9(4): 2.

Ostrom, Elinor. 1990. *Governing the Commons: The Evolution of Institutions for Collective Action.* Cambridge: Cambridge University Press.

Ostrom, Elinor. 2005. *Understanding Institutional Diversity.* Princeton, NJ: Princeton University Press.

Pahl-Wostl, Claudia. 2009. A Conceptual Framework for Analysing Adaptive Capacity and Multi-Level Learning Processes in Resource Governance Regimes. *Global Environmental Change* 19(3): 354–365.

Pellegrini, Lorenzo, and Reyer Gerlagh. 2006. Corruption, Democracy, and Environmental Policy: An Empirical Contribution to the Debate. *Journal of Environment & Development* 15(3): 332–354.

Reed, Mark S. 2008. Stakeholder Participation for Environmental Management: A Literature Review. *Biological Conservation* 141(10): 2417–2431.

Reed, Mark S., Anil Graves, Norman Dandy, Helena Posthumus, Klaus Hubacek, Joe Morris, Christina Prell, Claire H. Quinn, and Lindsay C. Stringer. 2009. Who's In and Why? A Typology of Stakeholder Analysis Methods for Natural Resource Management. *Journal of Environmental Management* 90(5): 1933–1949.

Ribot, Jesse. 2003. Democratic Decentralisation of Natural Resources: Institutional Choice and Discretionary Power Transfers in Sub-Saharan Africa. *Public Administration and Development* 23(1): 53–65.

Ribot, Jesse. 2006. Choose Democracy: Environmentalists' Socio-political Responsibility. *Global Environmental Change* 16:115–119.

Robbins, Paul. 2000. The Rotten Institution: Corruption in Natural Resource Management. *Political Geography* 19(4): 423–443.

Rothstein, Bo, and Jan Teorell. 2008. What Is Quality of Government? A Theory of Impartial Government Institutions. *Governance: An International Journal of Policy and Administration* 21(2): 165–190.

Schultz, Lisen, Andreas Duit, and Carl Folke. 2011. Participation, Adaptive Co-management, and Management Performance in the World Network of Biosphere Reserves. *World Development* 39(4): 662–671.

Smith, R. J., R. D. J. Muir, M. J. Walpole, A. Balmford, and N. Leader-Williams. 2003. Governance and the Loss of Biodiversity. *Nature* 426(6962): 67–70.

Stringer, Lindsay C., Andrew J. Dougill, Evan Fraser, Klaus Hubacek, Christina Prell, and Mark S. Reed. 2006. Unpacking "Participation" in the Adaptive Management of Social-Ecological Systems: A Critical Review. *Ecology and Society,* 11(2): 39. http://www.ecologyandsociety.org/vol11/iss2/art39.

Svensson, Jakob. 2005. Eight Questions about Corruption. *Journal of Economic Perspectives* 19(3): 19–42.

Teorell, Jan, Nicholas Charron, Marcus Samanni, Sören Holmberg, and Bo Rothstein. 2010, May 27. *The Quality of Government Data Set.* University of Gothenburg: The Quality of Government Institute.

Torras, Mariano, and James K. Boyce. 1998. Income, Inequality, and Pollution: A Reassessment of the Environmental Kuznets Curve. *Ecological Economics* 25(2): 147–160.

United Nations, Department of Economic and Social Affairs, Population Division. 2009. *World Population Prospects: The 2008 Revision.* New York: United Nations.

Welsch, Heinz. 2004. Corruption, Growth, and the Environment: A Cross-Country Analysis. *Environment and Development Economics* 9:663–693.

Winter, Gerd, ed. 2006. *Multilevel Governance of Global Environmental Change—Perspectives from Science, Sociology, and the Law.* Cambridge: Cambridge University Press.

World Bank. 2008. *Country Classifications.* http://data.worldbank.org/about/country-classifications/country-and-lending-groups.

12

Conclusion: An Emerging Ecostate?

Andreas Duit

As argued in chapter 1, "Introduction: The Comparative Study of Environmental Governance," studies of environmental governance can be viewed as analyses of the large-scale transformation of society's relationship with nature that is currently unfolding. This transformation is visible in most areas of society, but the studies in this volume have primarily focused on understanding the role of the state in environmental governance from a comparative perspective. Using a variety of methods, study designs, and data sources, the work reported in all the chapters of this book have contributed new knowledge about how the state influences and is influenced by accelerating processes of environmental degradation. A clear message from these studies is that the state has been, and continues to be, an influential actor in and arena for environmental governance, which indicates a need to revisit some of the assumptions in previous research about the state's limited and diminishing capacity in environmental governance.

The State and Environmental Governance in a Larger Context

One question remains: How can these findings about the role of the state in environmental governance be placed within a larger conceptual framework? One way of theorizing the role of the state is through the notion of an emerging environmental state, or *ecostate* for short. The term itself is not new: ecological modernization theorists such as Mol and Buttel (2002) have used it to denote a state that provides environmental public goods. Dryzek et al. (2003) have investigated the rise of "green states" in a comparative perspective, and Barry and Eckersley (2005) have discussed the normative foundations of the "ecological state." More particularly, James Meadowcroft (2005, 2012) and chapter 4 of this book, "The Three Worlds of Environmental Politics," have suggested that the ecostate can be seen as analogous to the welfare state in some respects (see also Dryzek et al. 2003; Christoff 2005; Duit 2011). The welfare state and the ecostate are similar, they argue, in the sense that the state takes on the function of mitigating the effects of market externalities: social costs in the case of the welfare state, and ecological costs in the case of the ecostate. Thus, the massive buildup of environmental institutions, policies, and knowledge that has been observed in many countries since the 1960s can be understood in this light: it is basically an institutional and administrative response to a case of global-scale market failure, predominantly but not exclusively located in the nation-state. Alternatively, to paraphrase Esping-Andersen(1990), environmental governance is ultimately about the decommodification of nature through political interventions in consumption patterns and production processes. In this metatheoretical sense, then, all the contributions to this volume can be viewed as studies of how different societies are responding through policies, institutions, knowledge building, and value changes to an accelerating problem of externalizations of environmental costs.

How can the ecostate be defined, and what criteria should be used to determine if a state can qualify as an ecostate? Some suggestions can be found in the literature. Peter Christoff suggests that a distinction can be made between four basic types of ecostates. The most advanced of these is the hypothetical "green state," in which the state has developed a significant capacity to ensure that human activities are compatible with ecological limits. In contrast, the already existing "environmental welfare state" has a more constrained capacity and assumes a limited responsibility for mitigating environmental harm, but only for the sake of human well-being and sustained economic growth. Finally, Christoff envisions a neoliberal ecostate (which favors market solutions to environmental problems) and an ecofascist state (which prioritizes the environment over all other human concerns) (Christoff 2005, pp. 41–45). In chapter 4, Detlef Jahn takes his point of departure in green political theory and argues that a green state is characterized by decreasing energy consumption and dominance of renewable energy sources and rail transport, as these structural factors are reflecting the extent to which a state is based on green or productionist foundations. Neither author is thus able to provide a single benchmark criterion for what can be considered an ecostate, and both typologies include a hypothetical and (as yet) nonexistent ideal green state, which then is contrasted against existing ecostates with less encompassing environmental regimes.

Building on these two classifications, and with the aim of simplifying them somewhat, a threefold typology consisting of strong ecostates, partial ecostates, and weak ecostates can be discerned. A *strong ecostate* is a state in which the environment takes precedence over the economy in the sense that environmental sustainability is always promoted before economic growth by government regulation and redistribution. Environmental problems still might occur in such states (just as social inequality has not been eradicated in even the most encompassing welfare states); the key issue is whether the state uses its regulatory and redistributive powers to maximize environmental values over economic values. There are no examples of strong ecostates in the contemporary political landscape, as even countries with the most ambitious green agendas are not close to favoring the environment over economic growth consistently. The strong ecostate, therefore, should be thought of as an ideal type, which is mainly useful for providing a contrast to a class of real-world examples of partial ecostates. This is by far the most common type of ecostate in today's world, and there are numerous examples of states in which environmental concerns are sometimes, but not always, allowed to trump economic considerations.

The *partial ecostate* does internalize some of, but not all, the environmental costs externalized by the market and is usually characterized by a fairly well developed regulatory capacity in the environmental area. It is also important to note that this category—much like the welfare state category—probably contains a large amount of internal variation among different types of partial ecostates (see chapter 4 of this volume).

A final category, *weak ecostates,* consists of countries in which environmental concerns are rarely or never allowed to take precedence over economic gain, and in which the internalization of market externalizations of environmental costs is negligible. There might be contemporary examples of this type of state to be found (depending on how "negligible internalization of environmental costs" is defined), but in the remainder of this chapter, the term "ecostate" will be used in the sense of the median category "partial ecostates."

Once the analogy between the welfare state and the ecostate has been established, findings from the chapters in this volume also suggest that it is equally important to note the ways in which the analogy does *not* hold (cf. Meadowcroft 2005). Interestingly, the discrepancies between the welfare state and the ecostate also point to some very fruitful issues for future research in comparative environmental politics and policy. The following sections will review how the studies collected in this volume indicate some key differences between the welfare state and the ecostate, as well as elaborate on how these dissimilarities point to new research issues in comparative environmental politics and policy.

The Ecostate and Local Natural Resource Management

Whereas the welfare state is predominantly found in centralized administrations in Western industrialized countries, the ecostate is present in both developing and developed states, and often it has its most tangible effects on local administrative scales (cf. Steinberg 2010). Hence, a first set of differences between the welfare state and the ecostate relates to the geographical location and administrative scales of the ecostate. Being an institutional response to the market economy and industrialization, the welfare state emerged from a small group of wealthy industrialized countries, globally speaking. This is not necessarily the case for the ecostate. Undoubtedly, industrialization was a strong, driving force behind many environmental problems and therefore also behind a state's responses to these problems. But in addition to environmental problems rooted in industrial production, there is an additional large and growing category of environmental problems related to unsustainable resource utilization or management of natural resources (e.g., fisheries, forestry, biodiversity protection, water management, and land use). Not only are such problems more frequent in developing countries, they also are more pressing because people in developing societies tend to depend more directly on natural resources for their subsistence. An additional difference in scale has to do with the fact that while the welfare state relied on nationwide policy programs implemented by centralized and hierarchical administrations, the ecostate is charged with the task of managing ecosystems and natural resources situated on administrative and political scales ranging from the very local to the global. Therefore, we should not look for the ecostate exclusively in those areas of the world where the welfare state used to be, nor should we expect it to take on the same functions and organizational forms as the welfare state.

Chapter 9, "Decentralization and Deforestation: Comparing Local Forest Governance Regimes in Latin America," chapter 10, "Enforcement and Compliance in African Fisheries: The Dynamic Interaction between Ruler and Ruled," and chapter 11, "Causes and Consequences of Stakeholder Participation in Natural Resource Management: Evidence from 143 Biosphere Reserves in Fifty-Five Countries," provide examples of the new role and extended geographical scope of the ecostate in local-scale resource management. In their study of local forest governance regimes in three Latin American countries in chapter 9, Krister Andersson, Tom Evans, Clark C. Gibson, and Glenn Wright seek to understand how decentralization reforms play out in different institutional contexts. Their research is responding to an ongoing policy debate about the effects of decentralization on the environment. Specifically, Andersson et al. analyze why some local governance systems are more effective in maintaining a stable resource base than others, and they argue that the effects of policy change depend especially on the role played by local institutional arrangements. To date, studies of decentralization generally have explored how changes in the political power resources of local governments shape public policy outcomes. Most similar studies lack longitudinal data on microinstitutional variables as well as robust outcome measures for collective goods. Andersson et al. seek to overcome such shortcomings by conducting a comparative analysis of a unique longitudinal data set on environmental decision making and satellite images of forest cover from 300 local governments located in three countries with varying degrees of formal decentralization: Bolivia, Guatemala, and Peru. Their general hypothesis is that strong local governance arrangements—as evidenced by local capacity for generating local tax revenues—make it more likely that local governments will respond to positive incentives for protecting local collective goods, such as forest resources. The study finds evidence of an effect of de facto (but not de jure) decentralization on how sustainably forestry resources are managed. This in turn suggests that decentralization reforms have an impact on resource management only when they also entail a transfer of substantial decision-making competencies to the local level.

Relying on a combination of satellite image estimates of changes in biomass and a survey of Man and the Biosphere Area managers, the study by Andreas Duit and Ola Hall described in chapter 11 uses a similar methodological approach. Duit and Hall focused on investigating the role of stakeholder participation in conservation programs, specifically analyzing the effect of the institutional context on the level and type of stakeholder involvement, as well as the outcome of participation of different stakeholder groups. Stakeholder participation is often attributed a key role in theories of natural resource management, but the conditions under which stakeholder participation is stimulated, as well as the outcomes of stakeholder involvement, have received only scant attention in previous studies. Duit and Hall found that participation by stakeholder categories such as local resource users, nongovernmental organizations (NGOs), and volunteers were affected positively by institutions protecting political rights, whereas corruption seemed to play a lesser role. Furthermore, local stakeholders was the only stakeholder category that showed any connection to better environmental management within Biosphere Reserves (BRs), which indicates the existence of a link between institutional context and management outcomes in natural resource management. Institutional context, in the form of corruption and the trustworthiness of rulers, is also the topic of the study in chapter 10 of enforcement strategies and rule compliance in coastal fisheries in three African states (Angola, Namibia, and South Africa). Martin Sjöstedt argues that previous studies on resource management have tended to model the problem of sustainable resource use as a horizontal game between different resource users, in which the design of self-organized institutions is a crucial factor in determining whether the resource will be sustainably managed (cf. Ostrom 1990). This is in many ways a valuable and successful approach, but as Sjöstedt points out, it has neglected to incorporate the vertical relationship between government and citizens. Only rarely do we encounter common pool resources in which there is no government involvement: most natural resources are governed by state agencies to some extent. Sjöstedt then contends that the relationship between resource users and governments can be modeled as a trust game in which the ruling entity must be perceived as trustworthy by resource users if it is to be allowed to govern. This can be achieved through two strategies: by increased sanctioning and monitoring of rule infractions, or by a more cooperative tactic aiming at installing comanagement practices and participation. Users, in turn, can choose between compliance and noncompliance, which, in combination with the state's strategy, results in different equilibrium outcomes. The South African case seems to be stuck in a noncomply-nonenforce equilibrium, whereas the Namibian and Angolan cases, through a series of trust- or reputation-building measures, have been able to move to an enforce-comply situation.

In the case of natural resource management, these chapters all find that the behavior of the ecostate or the characteristics of its core institutions are crucial factors in explaining such various aspects as rates and consequences of stakeholder participation, rule compliance in fisheries management, and the outcome of decentralization reforms in forestry management. The state might not be directly present and actively intervening in many local processes, but it nevertheless plays a decisive role by establishing the rules of the game. As described in the chapters by Andersson et al. and Sjöstedt, the state primarily has an effect from a distance. In the study by Andersson et al., the state does this by deciding on and subsequently transferring varying degrees of decision-making competencies to municipalities; and in Sjöstedt's study, it does it by choosing different strategies for making local-level fishermen comply with regulations. In chapter 11, Duit and Hall do not take into account specific policy decisions made by the state in their study of stakeholder participation in conservation programs, but they do find a strong effect of the "character" of the state: states that protect political rights have a higher degree of participation among local stakeholders, which in turn leads to better management performance in conservation areas. In sum, the findings from the chapters in this part of the volume suggest that studies of local-level natural resource management would benefit from taking a closer look at how the character and actions of the state influences the outcome of local-level natural resource management programs. Research issues of particular interest includes the identification of the scope conditions of beneficial versus detrimental effects of state involvement, as well as the impact of institutional context on resource management at local scales.

Understanding and Measuring the Performance of Ecostates

As noted in chapter 1, a central topic in comparative environmental politics concerns the sources of cross-national variation in environmental performance: Why are some countries better than others in mitigating the effects of environmental problems, managing their natural resources, and limiting society's impact on natural systems? In fact, even if not addressed directly, most studies in comparative environmental politics ultimately try to address different aspects of this puzzle. The underlying assumption, of course, is that if we can figure out what makes a certain polity more responsive to the demands of nature, this knowledge subsequently can be used to improve environmental politics and policy in other, less successful countries.

As James Meadowcroft discusses in chapter 2, "Comparing Environmental Performance," one reason why there has been such limited progress on this central issue in environmental politics concerns the difficulties associated with measuring what societies are doing to address environmental degradation, as well as the outcome of these efforts in the natural environment. Meadowcroft systematically reviews the main contemporary approaches to measuring environmental performance of states and societies, and he finds that all approaches suffer from substantial difficulties in one respect or another. His overall conclusion is that, despite a wealth of data on the state of the environment, we presently lack even rough measures of the dependent variable (how countries are addressing environmental problems), which in turn means that the project of discovering the political correlates of good environmental governance is far from achieving its goal. This is as indicative of how environmental problems and their solutions have been perceived as it is problematic for scholars seeking to analyze ecostates. Meadowcroft concludes his chapter by outlining four dimensions in a new generation of estimates of environmental performance: (1) environmental governance, (2) quality of the lived environment, (3) ecosystems and natural resources, and (4) contributions to global environmental issues. By using multiple and disaggregated measures of environmental performance, it is possible to conduct more diversified analyses of environmental performance, in which a society can do well in certain areas and less well in others.

Meadowcroft's chapter also highlights another difference between the welfare state and the ecological state, which has to do with the role of measurement and the collection of data on various aspects of society. According to Ian Hacking, the development of social statistics in the nineteenth century was an important step toward the establishment of the welfare state: regularly collected statistical information about health, income, crime, births, and deaths made it possible for the state to accumulate detailed knowledge about the population's characteristics. This information subsequently could be employed in the design and evaluation of welfare programs, and social statistics soon became an integral part of the rational planning processes that lie at the heart of the welfare state (Hacking 1990). This historical process also contained instances of abusive data collection practices and prejudiced coding schemes, but as a whole, the welfare state developed ways of measuring itself and using collected information for planning, evaluation, and policy formulation. There has been no comparable development in the case of the ecostate. Beginning in the early 1970s, many countries set up environmental monitoring programs, initially for the purpose of keeping track of the spread of various industrial pollutants. As more and more types of environmental problems were discovered, environmental statistics came to record things such as waste recycling rates, endangered species, and rates of deforestation within the borders of nation-states. A key feature of all these environmental monitoring programs is that they are almost universally designed to measure the state of nature, but not the state of environmentally relevant activities in society. As a result, there has been little development in the measurement and monitoring of social aspects of sustainability; therefore, we have much cruder (or even nonexistent) measures of environmental performance in the social realm (e.g., environmental institutions and laws, policy instruments, the structure of environmental administration, and the environmental behavior of organizations and individuals). It is quite indicative that, as Meadowcroft points out, there are no systematic or comparable estimates of such basic aspects as how much different countries are spending on environmental protection, or the level of environmental taxes.

An additional challenge, as Meadowcroft notes, consists of establishing a theoretical model that can function as a guide for what can actually be considered a reflection of environmental performance. This is the true challenge here, for it requires a much more refined understanding of the society-nature relationship than we currently possess. Without such a model to guide data collection and hypothesis testing, research on environmental performance will find it difficult to move beyond the current state of non-directedness.

Responding to this need to base studies of environmental performance on a firmer theoretical foundation, Detlef Jahn sets out in chapter 4 to map the performance of different types of green states in addressing environmental concerns. Taking Esping-Andersen's studies of different types of welfare states as a model, Jahn analyzes the outcomes of environmental policies of twenty-one countries in the Organisation for Economic Cooperation and Development (OECD) in a similar fashion. The analytical focus centers on the link between basic environmentally relevant structural features and outcomes in terms of environmental performance. These basic structural features are deduced from theories of green political ideology and operationalized with quantitative data. Environmental performance is captured by a comprehensive index that includes environmental pollution and measures of improving environmental conditions.

The results of the analysis show that we may distinguish between three different worlds of green states. There are two worlds populated by green states that have relatively high levels of environmental performance. In the first, some countries more or less follow the ideas of green ideology. These countries combine a successful environmental performance with structural reforms adhering to green objectives. To a certain degree, such countries can be labeled "green states." In the second, there are countries that are environmentally successful but nevertheless are structured in the spirit of productionism. The third world contains a group of countries that combines productionist structures with low environmental performance.

Jahn's chapter raises two points of great relevance to the study of the ecostate. First, it demonstrates the utility of basing studies of environmental performance on a well-developed theoretical foundation. Second, it provides evidence for the same type of clustering effect of ecostates that Esping-Andersen found in the case of the welfare state.

In chapter 3, "Explaining Environmental Policy Adoption: A Comparative Analysis of Policy Developments in Twenty-Four OECD Countries," Christoph Knill, Susumu Shikano, and Jale Tosun focus on another key factor behind the performance of the ecostate: mechanisms underlying the diffusion of environmental policies among countries. By introducing the concept of "policy contagiousness" (the probability of a policy to spread), together with a novel way of measuring this policy trait and a sophisticated estimation technique borrowed from psychology, Knill et al. analyze longitudinal data on diffusion patterns of forty different environmental policies in twenty-four countries. The authors are able to identify a key role for institutional interdependence in the spread of environmental policies. Policies seem to spread primarily through international institutions and organizations; the more connections to such international bodies that a country has, the higher the likelihood of adopting a policy.

In the context of environmental performance, this finding is highly relevant to the debate on the transfer of policymaking power from national to international levels. It seems that states have not necessarily lost their power; rather, the impetus for new and more stringent policies is increasingly located outside any single state. Although not addressed directly in the study by Knill et al., this finding is highly relevant to the debate on what explains better environmental performance. Most previous suggestions for causal agents have focused on domestic structural factors such as corporatism and electoral systems, but the results from Knill et al. point to a hitherto less well studied effect of contacts between countries mediated through international institutions. It might be the case that a country's environmental performance is determined, at least in part, by how much contact it has with other countries through international institutions.

Taken together, the findings from chapter 3 and chapter 6, "Early Bird or Copycat, Leader or Laggard? A Comparison of Cross-National Patterns of Environmental Policy Change," are especially interesting in the light of claims about the detrimental effect of globalization on the capacity of the state to act as a forceful agent in environmental protection. Knill et al. show that globalization, in terms of increased contacts between countries through international organizations, seems to be beneficial for the spread of environmental policies between countries. In turn, Sommerer, in chapter 6, is able to demonstrate that both the number and the stringency of environmental policies are increasing over time in most countries. Putting these two findings together, it seems that globalization is actually *strengthening* the capacity of states to pursue increasingly stringent policies in response to various aspects of environmental degradation. Moreover, these results highlight another difference between the welfare state and the ecostate. Although emerging during the same time period, more or less, in different countries, the welfare state was very much a product of political processes taking place within the confinement of the nation-state. Judging from these researchers' findings, this does not seem to be the case for the ecostate, which in addition to domestic forces appears to be the result of increased internationalization.

Whereas the studies by Jahn and Knill et al. both use a large-N design, Roger Karapin's detailed longitudinal case study analysis of wind energy policies in Germany and the United States, found in chapter 5, "Wind-Power Development in Germany and the United States: Structural Factors, Multiple-Stream Convergence, and Turning Points," tackles an additional difficulty in studying environmental performance: the issue of separating and interrelating the effects of conscious efforts by political actors at addressing environmental problems from historical contingencies and good (or bad) fortunes. Germany is one of the leading industrialized countries in the development of wind power, with 7.4 percent of electricity generated from this source in 2012. By contrast, the United States takes a place among the laggards, with only 3.5 percent of electricity generated from wind. Structural differences between the countries help account for Germany's greater degree of wind-power development at present: the degree of dependence on fossil-fuel imports and manufacturing exports, electoral systems, and cooperativeness of business-government relations. However, these structural features cannot account for the earlier dominance of the United States in wind power (1978–1993) or for the rapid growth of wind power there after 2004. Wind power development in the two countries is marked by turning points that were due to a convergence of problem streams and political processes. The problem streams were driven by changes in fossil-fuel prices, focusing events concerning the environment, and technological developments. The political processes included election outcomes, political leadership, environmental movement activity, and the formation of advocacy coalitions. Karapin's comparative analyses suggest that although structures affect environmental outcomes, precisely which structures matter in a given period depends on this convergence of problem streams and political processes, which together create turning points in renewable-energy development and other environmental outcomes. Karapin concludes that structural and process factors should be related to each other to assess how much scope actors have and under what conditions they have it.

Chapters 2, 3, 4, and 5 point to a number of new directions for the study of environmental performance. A first lesson from these chapters is the value of basing explanations of environmental performance on a theoretical foundation. Many earlier studies in this genre were of a more exploratory nature, but to make further progress, it is necessary to start building and testing theoretical models for explaining environmental performance. A related second lesson concerns the need for better data on environmental performance, because without more valid, reliable, and comparable measurements of what states are doing (and not doing) in the environmental policy area theory, development is likely to remain hampered.

A third lesson to gain from chapters 2 and 6 is that new measures of environmental performance must allow a relatively large degree of intra-country and longitudinal variation. Not only do countries exhibit highly variable performance profiles across different policy areas and types of environmental problems, performance also seems to be highly variable over time. The notion of capturing this variation in a single measure such as the Ecological Footprint Index or the Environmental Performance Index (EPI), therefore, should be reconsidered. A final conclusion is that studies of environmental performance should seek, to a greater extent, to include not only nation-specific, but also supranational factors, as explanatory variables. It is evident from chapters 3 and 6 that environmental policy change is driven at least in part by processes that occur between, rather than within, countries, and it is important to account for such causal influences on environmental performance.

Policymaking and Citizenship in the Ecostate

A fundamental area of contrast between the welfare state and the ecostate with regards to policymaking relates to the fact that the welfare state precedes the ecostate in time, which means that the ecostate has emerged on top of—rather than parallel to—the welfare state. Given the strong structuring effect of the welfare state on policymaking and public administration, this means that we should expect the earlier type of welfare state to be highly influential in shaping later ecostates. Exactly how this structuring effect of the welfare state on the ecostate has played out is an interesting topic for further research, but evidence from some previous studies indicates that welfare state type has played an important role in shaping the ecostate in such diverse cases as Sweden and Australia (Lundqvist 2001, 2004; Christoff 2005). An example in this volume of the structuring effect of state institutions on environmental policy outcomes can be found in chapter 5, in which the institutional differences between neocorporativist Germany and the pluralist United States are identified as an important factor in explaining the proportion of wind energy in the respective countries.

A related difference between the welfare state and the ecostate concerns the organizational context of policymaking. The ecostate of today emerged in an organizational, administrative, and political context that is very different from that of the welfare state. The welfare state was built on a foundation of a large and fairly strong administrative apparatus disposing of substantial financial and organizational resources derived from redistributive taxation schemes (Pierson 1996). Policymaking at the apex of the welfare state was thus synonymous with strong governments within well-defined borders of nation-states, implementing large-scale and homogenous policy programs. This was also the organizational context of the first generation of environmental protection policies, which typically sought to target end-of-pipe emissions from industry and urban areas through the use of nationwide policy programs.

As the welfare state gradually lost some of its power from the 1970s and onward (Korpi 2003), the ecostate has continued to take on new tasks beyond point-source emission control such as biodiversity protection, climate change policies, and sustainable resource use. Partly as a response to the challenges posed by these new diffuse-source and lifestyle-related problems, and partly as a reflection of a more general transformative process in public administration, policymaking in the ecostate has moved from "government" toward "governance." This development is a well-researched topic in the literature (Jordan et al. 2003; Durant et al. 2004), but findings from chapter 6 and chapter 7, "The Role of the State in the Governance of Sustainable Development: Subnational Practices in European States," add two important aspects to this narrative: evidence for a continued key role for the state in environmental governance, and clear signs of a continuously growing policymaking capability.

Chapter 6 provides a most valuable mapping of the policy side of the evolving ecostate. In this chapter, Sommerer studies changes in seventeen environmental standards in twenty-four countries over a period of thirty-five years and finds that there has been a tremendous increase in the overall level of environmental regulation, and that an overwhelming majority (93 percent) of these changes have resulted in stricter standards. Therefore, worries that increased globalization and regulatory competition might lead to a race to the bottom among countries concerning relaxed environmental regulations are mostly unfounded. In addition, Sommerer is able to distinguish patterns of laggards (Bulgaria, Romania, and Mexico) and forerunners (Germany and France) among countries, as well as pioneers that have lost momentum over time (Japan) and latecomers catching up (Netherlands). On the whole, there has been a strong trend toward convergence between national policy portfolios; countries are becoming increasingly similar with regards to types and levels of environmental standards. Sommerer's study provides an excellent illustration of how preconceived notions of which countries are "greener" than others often do not hold up when confronted with empirical evidence, but it also points to an interesting policy dynamic unraveling over time. Finding explanations for why countries oscillate over time in terms of regulatory capacity is an interesting research task that might help to shed light on the drivers of policy expansion and contraction in the ecostate.

In chapter 7, Susan Baker and Katarina Eckerberg investigate the role of the state in the governance of sustainable development at the subnational, regional, and local levels in Europe. It presents a synthesis of empirical evidence from research undertaken in twenty-seven European states, alongside some critical commentary on the significance of these findings to achieve an understanding of the dynamics involved in contemporary governance processes. The empirical evidence shows that the state remains a key player in initiating and co-coordinating sustainable development planning processes, and that it contributes to capacity building through direct financing, institutional support, and the provision of expertise to subnational authorities. Moreover, the state initiates and coordinates policy networks and retains a great deal of power over the nature and functioning of network forms of governance. Consequently, the absence of state leadership is shown to have a detrimental effect on the implementation of sustainable development strategies and plans. Baker and Eckerberg deal directly with the key issue of the new role played by the state in environmental governance. A clear finding from their study is that although the state can no longer be called upon to remedy today's environmental problems in the same way as it cleaned up a first generation of industrial point-source emissions, the state is nevertheless an indispensable actor in emerging governance arrangements around second-generation environmental problems. The state is no longer the only actor on the environmental arena, but it remains the most important one.

Yet another difference between the welfare state and the ecostate involves the citizens who inhabit the two types of states. As T. H. Marshall pointed out, the emergence of social citizenship in the twentieth century was the product of a centuries-long struggle for the expansion of rights and obligations to broad strata of the population, but also an important precondition for the welfare state (Marshall 1950). In an innovative study of sustainable citizenship norms in a comparative perspective, Michele Micheletti, Dietlind Stolle and Daniel Berlin seek to define and operationalize a new type of citizenship: sustainable citizenship. In chapter 8, "Sustainable Citizenship: The Role of Citizens and Consumers as Agents of the Environmental State," they argue that this form of citizenship rests on three pillars: cosmopolitan responsibility beyond national borders, nonreciprocal responsibility toward other citizens disregarding one's own gain, and inclusion of the private sphere (family and market) as a relevant area of citizenship.

The emergence of norms of sustainable citizenship means that the ecostate operates in a partly new policymaking context. One key difference is that the sustainable citizen is more likely to accept far-reaching interventions into the private sphere for the sake of the environment. Contrary to expectations stemming from the dismantling of the welfare state, sustainable citizens might even demand that the state take on a more progressive role in dealing with environmental problems. At the very least, sustainable citizens expect the state to provide the informational (e.g., eco-labeling), institutional (e.g., environmental legislation) and infrastructural (e.g., public transport systems) tools needed to carry out their sustainable lifestyles. The move from social to sustainable citizenship also means that the social contract regulating interaction between rulers and ruled in the ecostate might come to look a whole lot different from the corresponding social contract of the welfare state. For instance, Micheletti et al. find that sustainable citizens seem less worried about securing the participation of others when choosing to participate in various forms of collective action in the environmental sphere. This indicates that one of the core concerns of environmental policymaking—the problem of collective action—might be losing some of its saliency in the environmental area.

Conclusion: The Ecostate and the Comparative Study of Environmental Goverance

By relating the findings and conclusions of the chapters in this volume to the welfare state, several defining traits of the ecostate have been highlighted. A first difference is one of size: for although the ecostate is expanding, it is nowhere near as dominating as the welfare state in directing state activities and resources. In 2006, public expenditure on environmental protection in the EU-25 ranged between 0.2 and 1.4 percent of GDP (Eurostat 2010, pp. 303–304). The corresponding figure for social protection in EU-25 countries ranged between 12 and 30 percent of GDP (Antuofermo and Di Meglio 2009, p. 4) These statistics do not reflect all of society's investments in and costs of environmental and social protection, but they nevertheless give a good illustration of the relative size of the contemporary ecostate. This might change as environmental problems become more salient, but for now, it is clear that the ecostate has not reshaped the state apparatus in the same way as the welfare state once did. Second, we can conclude that, unlike the welfare state, the emergence of the ecostate seems to be propelled by processes of policy transfer via international organizations, as well as political processes within the state. A third defining trait of the ecostate relates to its geographical distribution. While the welfare state from a global perspective can be described as a local phenomenon confined to a handful of Western states, the ecostate can be found in most parts of the world, albeit with vast differences in terms of problem pressures, performance, and policy instruments. This is not least true for developing countries, in which environmental politics, to a larger extent than in industrialized countries, revolves around the use and abuse of natural resources (Steinberg 2010), as well as in which actions and characteristics of the state is often pivotal to determining outcomes.

A fourth difference concerns the administrative and organizational context of the two states: Although the ecostate is built on top of the welfare state, it cannot rely on its centralized and hierarchical style of policymaking and implementation of policy programs. Instead, if it is to stand any chance of moving forward in environmental problem solving, the ecostate is forced to enter into cooperative ventures with civil society and market actors, as well as inventing and exploring new forms of policy instruments that meet the requirements of flexibility and openness required by such actors. However, this change in policymaking style does not necessarily mean that the ecostate lacks punch: many of the studies in this book point to the indispensable role of the state in environmental governance, as well as an increasing regulatory capacity over time.

A fifth key difference relates to changing norms of citizenship. New forms of citizenship such as sustainable citizenship tend to emphasize the nonreciprocity and nonterritorial responsibilities for the natural world to a higher extent than reciprocal, nation-specific, and anthropocentric norms that form the basis of social citizenship in the welfare state. With these changing norms comes a possible renegotiation of the social contract within the environmental realm, which might mitigate the ecostate's reduced policymaking ability to some extent. A sixth and final difference between the welfare state and ecostate is the simple fact that while we have a longstanding, well-established research program concerned with the former, we are only just starting to unravel the dynamics of the ecostate. The studies collected in this volume suggests that a starting point for such a comparative research effort can be found in two outstanding issues related to the ecostate: its origins and its role in combating environmental problems.

What has caused the emergence of the ecostate? If one accepts the notion that the ecostate is a useful framework for understanding the ongoing processes of transformation and adaptation of society's institutions and governance systems toward sustainability, one key question remains: what are the driving forces behind this development? As argued in chapter 1, the notion of a limited (if not nonexistent) ability of systems of representative democracy to provide protection for environmental resources has for decades functioned as a core assumption in many efforts to theorize and rethink the relationship between the modern society and nature (Barry 1999; Eckersley 2004). Representative democracy has frequently been criticized for being unable to reach the kind of long-term and often painful decisions needed to save the environment. Concerns for long-term consequences and the common good are systematically disadvantaged in collective decision making based on the principle of electoral competition between political parties (Underdal 2010).

As a general rule, voters in a Downsian-type electoral democracy are assumed to punish representatives that try to impose restrictions on present-day well-being in favor of long-term sustainability. This phenomenon is not confined to environmental matters; it is also present in fiscal policies and welfare schemes, for example. Consequently, there is not much electoral support to be built on the basis of environmental issues, which in turn means that we cannot assume that the representative democratic system will provide environmental protection beyond a very basic level

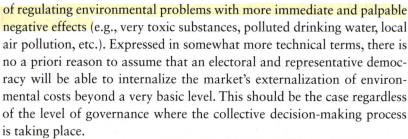

of regulating environmental problems with more immediate and palpable negative effects (e.g., very toxic substances, polluted drinking water, local air pollution, etc.). Expressed in somewhat more technical terms, there is no a priori reason to assume that an electoral and representative democracy will be able to internalize the market's externalization of environmental costs beyond a very basic level. This should be the case regardless of the level of governance where the collective decision-making process is taking place.

The empirical fact that environmental protection in many states and communities exceeds this basic level by an order of magnitude—as demonstrated most clearly in chapters 3, 4, and 6—therefore must be explained by other factors than the internal logic of representative democracies. While the present scope and level of environmental protection in the developed world is not nearly sufficient for solving the global environmental crisis, it is at the same time much more expansive than what would be predicted by the Downsian baseline model of democratic decision making by rational agents in a system of representative democracy. This mismatch between theory and empirical observations intensifies the question of what the driving forces are behind the evolution of the administrative and institutional capacity expansion that can be observed in almost all industrialized countries since the 1960s.

Exploring this question further will probably make it necessary to consider factors that go beyond the essentially functionalistic foundations of the ecostate model: exactly why do states respond to market externalizations of environmental costs? In addition to factors identified by studies in this volume (e.g., policy diffusion through international organizations and trade, emergence of sustainable citizenship norms, policy coalitions), further examples of such drivers might be the role of green parties and social movements, the internal dynamic of the market economy (e.g., the Kuznets curve and regulatory competition), technological developments, and increasing and more immediate negative effects of deteriorating natural resources and loss of ecosystem services (e.g., problem pressure).

Investigating the causes of the ecostate also implies a new direction in the research tradition of environmental performance studies: moving away from the search for correlations between political structures and environmental outcomes and toward an exploration of the drivers of regulatory growth that can be observed in most industrialized countries since the 1960s. This will require, as Meadowcroft and Micheletti et al. argue in their respective chapters, collection of new data on social aspects of environmental degradation, such as environmental policy outputs, institutions, and environmental behavior and norms of individual citizens. Also helpful in this endeavor would be historical and comparative case studies of the dynamics behind the emergence of different types of ecostates (see Schreurs 2002 for a good example of such a study).

The second remaining issue for further research has to do with the perennially contested role of the state in environmental protection. Although the appropriate role of the state is ultimately a normative issue, findings from the studies in this volume point to a continued and central position for the state in environmental management. At stake here, therefore, is not so much whether the state should be dismissed altogether or embraced wholeheartedly by scholars of environmental governance, but rather, how to gain a better understanding of under which circumstances the state is more and less able to address environmental problems effectively. A key research issue is thus to identify the scope conditions for state involvement in environmental governance. The rising influence of an international sphere of decision making and policymaking—a growing sense of awareness among citizens about the rights and responsibilities of individuals in global environmental change, and an increasing proportion of governance-type coordination and steering mechanisms—is indeed creating a new set of circumstances for state efforts at addressing environmental degradation. The fact that both policy processes and policy problems in the environmental sphere transcend national borders is undisputable, as is the fact that that the policy process has changed both in terms of actor constellations and policy instruments. The changing role of the state is also at the core of the discussion about emerging forms of governance, in which the argument is that public administration in most developed countries has undergone a shift from a Weberian administration toward more flexible and transitory forms of policymaking based on ad hoc networks consisting of a wide range of societal actors.

Environmental policy is often portrayed as an area in which governance-type solutions are particularly common and increasing (Jordan et al. 2003; Mol et al. 2009), but Susan Baker and Katarina Eckerberg argue in chapter 7 that many of these governance arrangements would not be able to persist and function without the state taking the role of initiator, financier, and coordinator. Governance arrangements depend on the state for carrying out key functions such as initiation of networks and collaborations; funding of larger projects, research, and policies; collective decision making, rule making, and lawmaking; and the sanctioning of transgressions. Roger Karapin's analysis of policy coalitions in wind power policies in Germany and the United States in chapter 5 is another case in point: policy outcomes were produced through multiple processes in which policy coalitions act within the boundaries set by state regulations and laws. Furthermore, sustainable citizenship can be realized to its fullest extent only in a state that is providing the necessary infrastructure. Local resource management schemes depend on the state for establishing and upholding the rules of the game, and increasingly stringent and numerous polices (much like viruses) need hosts in order to spread from country to country. As demonstrated by many of the chapters in this volume, a comparative approach often can be effective in identifying conditional (when the outcome is caused by a constellation of factors) as well as contextual effects (when the outcome is caused by surrounding political or institutional context), which indicates a continued role for comparative study designs when seeking to identify the scope conditions for the ecostate.

Perhaps the most important conclusion from the studies collected here is that if we were to stop including the state as an analytical category in studies of environmental politics and policy, we would probably be worse off in our understanding of how society and nature are intertwined. Conversely, more far-reaching studies of the state and its role would enrich our knowledge of the conditions for successful and unsuccessful environmental governance. The state should be brought back into comparative environmental studies—keeping in mind that the world and the state itself has changed considerably since the dawn of the welfare state.

References

Antuofermo, Mélina, and Emilio Di Meglio. 2009. Population and Social Conditions. In *Statistic in Focus*. Luxemburg: Eurostat.

Barry, John. 1999. *Rethinking Green Politics*. London: Sage.

Barry, John, and Robyn Eckersley. 2005. An Introduction to Reinstating the State. In *The State and the Global Ecological Crisis*, ed. John Barry and Robyn Eckersley, viiii–xxv. Cambridge, MA: MIT Press.

Christoff, Peter. 2005. Out of Chaos, a Shining Star? Towards a Typology of Green States. In *The State and the Global Ecological Crisis*, ed. John Barry and Robyn Eckersley, 25–52. Cambridge, MA: MIT Press.

Christoff, Peter. 2006. Ecological Modernization, Ecologial Modernities. In *Contemporary Environmental Politics. From Margins to Mainstream*, ed. Piers H. G. Stephens, John Barry, and Andrew Dobson, 179–200. New York: Routledge.

Dryzek, John S., David Downes, Christian Hunold, David Schlosberg, and Hans-Kristian Hernes. 2003. *Green States and Social Movements. Environmentalism in the United States, United Kingdom, Germany, & Norway*. Oxford: Oxford University Press.

Duit, Andreas. 2011. Adaptive Capacity and the Ecostate. In *Adapting Institutions: Governance, Complexity, and Social-Ecological Resilience*, ed. Emily Boyd and Carl Folke, 129–149. Cambridge: Cambridge University Press.

Durant, Robert F., Daniel J. Fiorino, and Rosemary O'Leary, eds. 2004. *Environmental Governance Reconsidered: Challenges, Choices, and Opportunities*. Cambridge, MA: MIT Press.

Eckersley, Robyn. 2004. *The Green State: Rethinking Democracy and Sovereignty*. Cambridge: Cambridge University Press.

Esping-Andersen, Gøsta. 1990. *The Three Worlds of Welfare Capitalism*. Cambridge: Polity Press.

Eurostat. 2010. *Environmental Statistics and Accounts in Europe*. Luxemburg: Publications Office of the European Union doi:10.2785/48676.

Hacking, Ian. 1990. *The Taming of Chance*. Cambridge: Cambridge University Press.

Jordan, Andrew, Ruediger K. W. Wurtzel, and Anthony R. Zito, eds. 2003. *New Instruments of Environmental Governance? National Experiences and Prospects*. London: Frank Cass Publishers.

Korpi, Walter. 2003. Welfare-State Regress in Western Europe: Politics, Institutions, Globalization, and Europeanization. *Annual Review of Sociology*, 29:589–609.

Lundqvist, Lennart J. 2001. A Green Fist in a Velvet Glove: The Ecological State and Sustainable Development. *Environmental Values*, 10(4): 455–472.

Lundqvist, Lennart J. 2004. *Sweden and Ecological Governance: Straddling the Fence*. Manchester, UK: Manchester University Press.

Marshall, Thomas H. 1950. *Citizenship and Social Class and Other Essays*. Cambridge: Cambridge University Press.

Meadowcroft, James. 2005. From Welfare State to Ecostate. In *The State and the Global Ecological Crisis*, ed. John Barry and Robyn Eckersley, 3–24. Cambridge, MA: MIT Press.

Meadowcroft, James. 2012. Greening the State. In *Comparative Environmental Politics. Theory, Practice, and Prospects*, ed. Paul F. Steinberg and Stacy D. VanDeveer, 63–87. Cambridge, MA: MIT Press.

Mol, Arthur P. J., and Frederick H. Buttel. 2002. The Ecostate under Pressure: An Introduction. In *The Ecostate under Pressure*, ed. Arthur P. J. Mol and Frederick H. Buttel, 1–11. Bingley, UK: Emerald Group Publishing Limited.

Mol, Arthur P. J., David A. Sonnenfeld, and Gert Spaargaren, eds. 2009. *The Ecological Modernisation Reader: Environmental Reform in Theory and Practice*. London: Routledge.

Ostrom, Elinor. 1990. *Governing the Commons: The Evolution of Institutions for Collective Action*. Cambridge: Cambridge University Press.

Pierson, Paul. 1996. The New Politics of the Welfare State. *World Politics,* 48(2): 143–179.

Schreurs, Miranda A. 2002. *Environmental Politics in Japan, Germany, and the United States*. Cambridge: Cambridge University Press.

Steinberg, Paul F. 2010. Comparative Environmental Politics: Beyond an Enclave Approach. *Review of Policy Research,* 27(1): 95–101.

Underdal, Arild. 2010. Complexity and Challenges of Long-term Environmental Governance. *Global Environmental Change,* 20(3): 386–393.

Appendix to Chapter 8

Descriptive Statistics, Question Wording, and Index Procedures for the Analysis of Swedish Survey Data

Variable	Range	Mean or Percent	Std.D	Cronbach's alpha	N
Sustainable citizenship	0–40	24	6.7	.73	1009
Concern	0–10	7.1	2.1	.73	1047
Interest	0–10	6.1	2.1	.80	1047
Attitudes	0–10	6.1	1.9	.74	1029
Behavior	0–10	4.4	2.9	.72	1026
Independent variables					
Solidarity citizenship	0–10	7.1	1.9	.81	1045
Duty citizenship	0–10	8.6	1.7	.74	1046
Information seeking citizenship	0–10	7.9	1.5	.66	1044
Trust representative institutions	0–10	5.3	2.4	.85	1023
Trust international institutions	0–10	4.9	2.0	.78	1038
Trust sustainability institutions	0–10	5.5	1.6	.78	1033
Efficacy people	0–10	5.3	2.2	.73	1009
Efficacy institutions	0–10	7.3	2.2	.91	996

Descriptive Statistics, Question Wording, and Index Procedures for the Analysis of Swedish Survey Data

Variable	Range	Mean or Percent	Std.D	Cronbach's alpha	N
Ideology (left=0)	0–10	5.1	2.4		1036
Age	18–78	50	16		1053
Economic situation	0–10	5.7	2.5		1042
Gender (woman=1)	0–1	53%			1053
Living area (large city=1)	0–1	31%			1029
Education (university=1)	0–1	38%			1035
Children (have children=1)	0–1	73%			1039

Concern

Question: "When you consider the situation today, what do you feel is the most worrying aspect for the future?" (4 categories: "very," "fairly," "not particularly," and "not worrying at all.") The items include worries about environmental pollution and violation of human rights.

Interest

Question: "In general, how interested are you in …? (a) environmental issues, (b) human rights, (c) developing countries/the third world (0–10 scale). *Index details*: The index used for interest is the mean value for each respondent that answered at least one of the items.

Attitudes

Question: "Below are a number of proposals which have been made in the course of political debates. What is your opinion of each of them? Focusing more on promoting: (a) a more environmentally friendly society even if this means low or no economic growth; (b) economic growth even if this means placing lower priority on environmental issues; (c) fair trade even if this means that many products in Swedish shops would be more expensive; (d) more environmentally friendly society even if this means a lower material standard of living; (e) providing for material needs in Sweden even if this means placing lower priority on aid to developing countries (5 categories from "very good to very bad proposal"). *Index details*: Items were reversed so that high numbers refer to positive attitudes towards sustainability reforms. The index used here is the mean value for each respondent who answered at least two of the five items.

Behavior

Question: "There are often many reasons for choosing to buy one product rather than another. In the last 6 months, how often have you chosen food, clothes or toys for the following reasons?" Here we use the alternatives (a) "the product was environmentally friendly"; (b) "product was manufactured under good working conditions"; and (c) "to support animal husbandry" (3 categories: "never/rarely," "occasionally" and "often/very often"). *Index details*: The behavioral index is the mean value for each respondent that rated the importance of at least one of the three reasons for buying food.

Citizenship Expectations

Question: "There are different views on what it takes to be a good citizen. In your personal opinion, how important is it to ..." (0–10 scale). *Index details*: Duty citizenship = "never try to evade paying tax"; "always obey laws and regulations"; and "never commit benefit fraud." Solidarity citizenship = "show solidarity with people in Sweden who are worse off than yourself"; "show solidarity with people in the rest of the world who are worse off than yourself"; "put other's interests before your own"; "do not treat immigrants worse than native Swedes." Information seeking citizenship = "develop your own opinions independently from other people's"; "stay well-informed about what is happening in society"; "don't expect the state to solve problems; instead act on your own initiative"; "actively try to influence societal issues." The index used for duty citizenship is the mean value for each respondent who answered at least one of the three items; for solidarity citizenship at least two of the four items and for information seeking citizenship two of the four items. The citizenship indexes were originally generated in a factor analysis (PCA) with varimax rotation conducted on the 11 survey items. The factors together explain 52% of the variance. Eigenvalues (initial) for the factors are: Duty citizenship: 1.7; Solidarity citizenship: 4.9; Info-seeking citizenship: 1.2.

Trust

Question: "How much trust do you have in the way in which the following institutions and groups manage their work?" (5 categories: from "a lot" to "very little trust"). *Index details*: The international organization index = "European Union," "United Nations," and "World Trade Organization." National representative institutions = "government" and "national parliament" and sustainability institutions = "public authorities responsible for environmental issues," "the consumer agency," "environmental organizations," and "consumer organizations." The dimensions were generated by factor analysis (varimax rotation). The index used for international institutions is the mean value for each respondent who answered at least one out of three items; and for sustainability institutions it is the mean value for those who answered at least two out of four items.

Efficacy

Question: "In your opinion, what opportunity do the following groups or people have to influence the development of society?" (1–7 scale. There was a "no opinion" alternative, which was included in the analysis and coded as a mid-scale answer.) *Index details*: The political institutions dimension = "politicians," "Swedish authorities," "European Union," "United Nations," "World Trade Organization." The people dimension = "yourself," "people in general/consumers," and "corporations." The dimensions were generated by a factor analysis with varimax rotation. The index used for institutional efficacy is the mean value for each respondent who answered at least two of the five items, and for people efficacy the mean value for those who answered at least one of three items.

Ideology

Question: "Political viewpoints are sometimes defined on a scale of left to right. Where about would you put yourself on a left to right scale?" (0–10 scale)

Economic Situation

Question: "On the whole, how would you describe the financial situation of your household?" (5 categories from "much worse" to "much better than the average Swede.") Variable rescaled 0-10.

Living Area

Question: "What type of area do you live in?" (6 categories. "Center of a large city" and "outskirts/suburbs of a large city" are coded as "1").

Education

Question: "What is your level of education? If you haven't finished your education, mark the level you are currently at?" (8 categories. "College/university studies," "college/university degree," and "post graduate degree" are coded as "1")

List of Contributors

Krister Andersson
is an associate professor of political science and codirects the Center for the Governance of Natural Resources at the University of Colorado at Boulder.

Susan Baker
is a professor of environmental policy at the Cardiff School of Social Science and lead academic at the Sustainable Places Research Institute, Cardiff University, Wales, UK.

Daniel Berlin
works at the University of Gothenburg, Sweden, and holds a Ph.D. in political science.

Andreas Duit
is an associate professor in the Department of Political Science and a research fellow at the Stockholm Resilience Center, both at Stockholm University.

Katarina Eckerberg
is a professor of public administration at the Department of Political Science, Umeå University, Sweden.

Tom Evans
is a professor in the Department of Geography at Indiana University. He serves as codirector of the Ostrom Workshop in Political Theory and Policy Analysis and director of the Center for the Study of Institutions, Population, and Environmental Change (CIPEC) at Indiana University.

Clark C. Gibson
is a professor of political science, director of the International Studies Program, and director of the Center of African Political Economy at the University of California, San Diego.

Ola Hall
is a senior lecturer in human geography at Lund University who specializes in geographic information systems (GISs) and remote sensing for the social sciences.

Detlef Jahn
is a professor of comparative politics at the University of Greifswald, Germany.

Roger Karapin
is a professor of political science at Hunter College and the Graduate Center, City University of New York.

Christoph Knill
is a professor of political science and public administration in the Institute of Political Science at the University of Munich, Germany.

James Meadowcroft
holds a Canada Research Chair in Governance for Sustainable Development and is a professor at the School of Public Policy and Administration, and in the Department of Political Science, at Carleton University in Ottowa.

Michele Micheletti
is Lars Hierta Professor of Political Science at Stockholm University and co-coordinator of the European Consortium of Political Science's Standing Group on Participation and Mobilization.

Susumu Shikano
is a professor at the Department of Politics and Public Administration of the University of Konstanz, Germany.

Martin Sjöstedt
is a research fellow at the Department of Political Science and the Quality of Government (QoG) Institute, University of Gothenburg, Sweden.

Thomas Sommerer
is a postdoctoral research fellow at the Department of Political Science, Stockholm University.

Dietlind Stolle
is an associate professor of political science at McGill University and the director of the Inter-University Centre for the Study of Democratic Citizenship.

Jale Tosun
is an assistant professor of international comparative political economy at the Institute of Political Science at the University of Heidelberg, Germany.

Glenn Wright
is an assistant professor of government at the University of Alaska Southeast in Juneau.

Index

American and Comparative Environmental Policy
Sheldon Kamieniecki and Michael E. Kraft, series editors

Michael E. Kraft and Sheldon Kamieniecki, editors, *Business and Environmental Policy: Corporate Interests in the American Political System*

Joseph F. C. DiMento and Pamela Doughman, editors, *Climate Change: What It Means for Us, Our Children, and Our Grandchildren*

Christopher McGrory Klyza and David J. Sousa, *American Environmental Policy, 1990–2006: Beyond Gridlock*

John M. Whiteley, Helen Ingram, and Richard Perry, editors, *Water, Place, and Equity*

Judith A. Layzer, *Natural Experiments: Ecosystem-Based Management and the Environment*

Daniel A. Mazmanian and Michael E. Kraft, editors, *Toward Sustainable Communities: Transition and Transformations in Environmental Policy*, second edition

Henrik Selin and Stacy D. VanDeveer, editors, *Changing Climates in North American Politics: Institutions, Policymaking, and Multilevel Governance*

Megan Mullin, *Governing the Tap: Special District Governance and the New Local Politics of Water*

David M. Driesen, editor, *Economic Thought and U.S. Climate Change Policy*

Kathryn Harrison and Lisa McIntosh Sundstrom, editors, *Global Commons, Domestic Decisions: The Comparative Politics of Climate Change*

William Ascher, Toddi Steelman, and Robert Healy, *Knowledge in the Environmental Policy Process: Re-Imagining the Boundaries of Science and Politics*

Michael E. Kraft, Mark Stephan, and Troy D. Abel, *Coming Clean: Information Disclosure and Environmental Performance*

Paul F. Steinberg and Stacy D. VanDeveer, editors, *Comparative Environmental Politics: Theory, Practice, and Prospects*

Judith A. Layzer, *Open for Business: Conservatives' Opposition to Environmental Regulation*

Kent Portney, *Taking Sustainable Cities Seriously: Economic Development, the Environment, and Quality of Life in American Cities*, second edition

Raul Lejano, Mrill Ingram, and Helen Ingram, *The Power of Narrative in Environmental Networks*

Christopher McGrory Klyza and David J. Sousa, *American Environmental Policy: Beyond Gridlock*, updated and expanded edition

Andreas Duit, editor, *State and Environment: The Comparative Study of Environmental Governance*